T0202820

LORENTZIAN WORMHOLES

From Einstein to Hawking

LORENTZIAN WORMHOLES

From Einstein to Hawking

Matt Visser

Washington University
St. Louis, Missouri

Springer

Library of Congress Cataloging-in-Publication Data
Visser, Matt
  Lorentzian wormholes : from Einstein to Hawking / Matt Visser.
    p. cm.
  Includes index.
  ISBN 978-1-56396-653-8
  1. General relativity (Physics). 2. Quantum field theory.
3. Quantum gravity. I. Title.
QC173.6.V57 1995   95-17876
530.1´1—dc20   CIP

Printed on acid-free paper.

9 8 7 6 5 4

ISBN 978-1-56396-653-8   SPIN 10875897

Springer-Verlag New York Berlin Heidelberg
A member of BertelsmannSpringer Science+Business Media GmbH

Dedication

— to my parents —
Herman and Cecilia,

— to my siblings —
Andy, Jenny, and Chris,

— to Elena —

Contents

List of Figures

Series Preface

The current rapid progress of research in such extremely important fields as dynamical systems, chaos and complexity, and nonlinear waves, combined with the steady development of state-of-the-art computer and graphical systems has opened new vistas for computational and mathematical physics. These vistas appear to be well beyond what was thought possible even a decade ago. In order to continue to stimulate and encourage these impressive developments, it is vital for applied mathematicians and physicists to communicate their achievements, results and ideas to scientists and engineers from a broad spectrum of backgrounds. The AIP Series in Computational and Applied Mathematical Physics (CAMP) is intended to serve this purpose.

The CAMP Series is comprised of graduate texts, monographs, and reference materials directed at researchers, teachers, and students in a range of pure and applied disciplines in order to achieve the above-stated goals. Books in the series address fields in computational physics such as numerical methods, novel computer hardware, nonlinear dynamics, and visualization. Topics in mathematical physics include statistical mechanics, quantum field theory, general relativity, or such topics in related mathematical disciplines as complexity theory, differential geometry, and group theory. Other interdisciplinary topics like artificial intelligence and neural networks are also included. Some volumes focus on narrow but engaging topics, while others provide a broad introduction or lay out the fundamentals of a field.

Preface

Wormholes, and their related offspring, spacewarps, timewarps, and time machines, have by now entered and thoroughly permeated the popular culture. These concepts have fired the imaginations of both authors and screenwriters. Perhaps the most visible recent manifestation of this is the television series "Star Trek: Deep Space Nine". Amid this welter of images from the various entertainment media it is easy to lose track of the fact that a relatively small but active band of general relativists and quantum field theorists has developed a considerable body of serious, and sober (though admittedly speculative), mathematical and physical analyses of the wormhole system.

This monograph deals with the physics and mathematics of Lorentzian wormholes. The general physical framework used for the discussion encompasses the intersection of classical Einstein gravity (general relativity) with the notions of quantum field theory. Individually these are extremely well tested theories. The intersection region between these theories, the field of "quantum gravity", is an area where considerable confusion, disagreement, and mutually incompatible opinions hold sway. Quantum gravity is also one of the major frontiers of modern theoretical physics. It is commonly expected (but certainly not proved) that a truly successful theory of quantum gravity would involve a revolution in modern physics similar in magnitude to the upheavals attendant to the individual and separate discoveries, early this century, of quantum mechanics and general relativity.

To start with, the very existence of Lorentzian wormholes is a highly speculative issue. There are plausible physical arguments that suggest that Lorentzian wormholes should exist at least at a microscopic scale (sizes of order the Planck length, about 10^{-35} metres). There are further plausible physical arguments that suggest that the existence of macroscopic Lorentzian wormholes (say, sizes of order a few metres) is not inconsistent with the rest of known physics. Be that as it may, there is certainly no positive experimental evidence, as of the time of this writing (1994) that might conclusively prove or disprove the existence of Lorentzian wormholes of any type.

There are several possible reactions to this state of affairs:

1. One might blandly assert the existence of wormholes, merely because one might wish such objects to exist to make the universe more interesting. Subject to this bold assertion one might profitably inquire as to what limits and constraints on the existence and behavior of such putative Lorentzian wormholes might be deducible from the general framework of Einstein gravity (general relativity) together with the notions of quantum field theory.

2. One might profitably view the whole business as a mathematical exercise, entertaining in its own right, but ultimately useful only insofar as it acts as a "test bed" for developing mathematical techniques that might be applied in other more conservative areas of investigation. Viewed in this light, Lorentzian wormholes might best be thought of as a teaching device.

3. One might, of course, merely throw up one's hands in horror at the ill-defined nature of the whole business—but then I would not be writing this monograph.

This monograph is intended for a broad audience: The minimum requirements are that the reader should have taken a first course in general relativity and an advanced quantum course. With such a background, and suitable reference texts at hand, the beginning student should be able to master a good three-quarters of the material without excessive difficulty. On the other hand, active researchers in the field may wish to focus their attention on the more technical chapters. There are many exercises scattered throughout the text. Most of these exercises are technically trivial, but should serve to ensure that the reader is kept awake. Several more substantial exercises are also provided. The energetically inclined may wish to attempt some of the suggested research problems.

Organization of this monograph

Part I—Background material: The monograph starts with an introductory sketch of the wormhole system, followed by a lightning survey of some key aspects of general relativity and quantum field theory. The survey will be painted with a very broad brush, and a few useful technical results will be collected in this section. (Beginning students should read this segment carefully; active researchers may wish to skim this segment.)

Part II—History: The monograph proper starts with a brief historical survey of the wormhole concept. In this historical survey we shall encounter the Einstein–Rosen bridge, Wheeler wormholes, Wheeler's spacetime foam, issues of topology change, and the cosmological constant problem. I will seek to carefully delineate the differences among traversable Lorentzian wormholes (macroscopic), quantum Lorentzian wormholes (microscopic), and their relatives, the Euclidean wormholes (gravitational instantons). I shall have comparatively little to say regarding Euclidean wormholes.

Part III—Renaissance: The renaissance of wormhole physics can be traced to the seminal paper by Morris and Thorne: "Traversable wormholes and interstellar travel: A tool for teaching general relativity" [190]. Considerable detail will be given concerning the interesting geometrical and gravitational properties of these classical and traversable Lorentzian wormholes. Some useful calculational techniques will be developed. For instance, the (extended) thin-shell formalism is very useful for describing traversable wormholes in the short-throat approximation. The techniques of this section are all classical and should be easily accessible to beginning students.

Part IV—Time travel: If, somehow, one manages to lay one's hands on a traversable wormhole, it *appears* to be very easy to turn such a system into a time machine. Objects that at first blush *appear* to be perpetual motion machines are also lurking in the background. This has serious, potentially paradoxical, implications for the state of physics as a whole. I shall explore some of the possibilities for dealing with this potential disaster. Various conjectures discussed are

1. the radical rewrite conjecture;

2. Novikov's consistency conjecture;

3. Hawking's chronology protection conjecture; and

4. the boring physics conjecture.

Part V—Quantum effects: The next section of the monograph isolates several highly technical discussions regarding quantum aspects of the Lorentzian wormhole system. (Beginning students may initially wish to avoid this segment of the monograph.) General ideas of semiclassical quantum gravity (and its limitations) are presented. The effect of the wormhole geometry on the renormalized vacuum expectation value of the quantum stress-energy tensor is discussed. In particular, the infinities that occur at the onset of time machine formation are of great interest. A second theme discussed in this part of the monograph is the canonical quantization of the wormhole system via minisuperspace techniques. This calculation has implications for the question of topology change.

Part VI—Reprise: Finally, I shall attempt to put everything in perspective. Many of the enigmas infesting quantum gravity will have been encountered in this monograph. I hope that I have carefully distinguished what is believed to be true from what is known to be true. In many of the more important and fundamental issues brought up by attempts to quantize gravity there is a frustrating inability to perform reliable calculations. Many nostrums promulgated on the basis of folklore have distressingly little in the way of reliable underpinnings.

To reiterate, the basic physics tools employed in this monograph are Einstein gravity (general relativity) together with quantum mechanics in its incarnation as quantum field theory. I shall not attempt to teach these subjects from scratch. As needed, the reader should refer to basic texts and papers. The single most important paper to read—one that focuses explicitly on Lorentzian wormholes—is that of Morris and Thorne, "Traversable wormholes and interstellar travel: A tool for teaching general relativity" [190]. More generally, the interface of general relativity with quantum field theory is described in several excellent review articles in Hawking and Israel, *General Relativity: An Einstein Centenary Survey* [134], and in Hawking and Israel, *300 Years of Gravitation* [135]. Textbooks addressing this interface region include Birrell and Davies, *Quantum Fields in Curved Space* [17], Fulling, *Aspects of Quantum Field Theory in Curved Space-Time* [99], and the very recent contribution by Wald, *Quantum Field Theory in Curved Spacetime and Black Hole Thermodynamics* [276]. Two recent collections of reprints of original technical articles are also of note: Hawking and Gibbons, *Euclidean Quantum Gravity* [133]; and Hawking, *Hawking on the Big Bang and Black Holes* [131].

As to more general background material: For textbooks on Einstein gravity consult some combination of Wald, *General Relativity* [275], Misner, Thorne, and Wheeler, *Gravitation* [186], Hawking and Ellis, *The Large Scale Structure of Space-time* [132], and Weinberg, *Gravitation and Cosmology* [279]. For textbooks on quantum field theory, consult Itzykson and Zuber, *Quantum Field Theory* [154], Ramond, *Field Theory: A Modern Primer* [219], or Ryder, *Quantum Field Theory* [232]. Interested readers from the mathematics community would be well-advised to consult Sachs and Wu, *General Relativity for Mathematicians* [233], and Glimm and Jaffe, *Quantum Physics: A Functional Integral Point of View* [115].

It would be helpful if the reader were to understand everything in all of these books. Since this is an unreasonable burden to place on even the best physicist, some attempt will be made (as indicated previously) to make the present monograph self-contained. (Naturally, in trying to please everyone, I run the risk of pleasing no one. If a section is either too elementary or too turgid, skip forward a few pages.)

I have tried to be forthright with my opinions and prejudices. When my opinions deviate from what I perceive to be the mainstream they are clearly labeled as such.

The references, while extensive, still only skim the surface of the literature on quantum gravity. On the more focused topic of Lorentzian wormholes the references should be reasonably complete—I hope that I have not made too many omissions. Nevertheless the reader should realize that (in the interests of manageability) I have been forced to pick and choose among many relevant topics. The discussion is accordingly confined to a broad, but nevertheless finite, number of issues.

Acknowledgments

In writing this monograph I have been greatly aided by the use of the physical and intellectual resources available to me at Washington University. I am indebted to Carl Bender for encouragement and advice. Both the particle physics and relativity groups contributed to a stimulating and active environment. The interest shown by Cliff Will, Wai-Mo Suen, Ely Shrauner, Mike Ogilvie, Leonid Grischuk, Maarten Golterman, and Claude Bernard is greatly appreciated.

I particularly wish to thank Elena Rivas for a careful reading of the manuscript—her independent perspective proved very useful in tidying up many obscurities. I also wish to thank Eric Poisson for his insightful comments. Michael Morris and Tom Roman also deserve special thanks for their careful reading of portions of the manuscript. Useful comments were also provided by Maarten Golterman and Ian Redmount.

I wish to thank Professor J.A. Wheeler and the Physical Review for permission to reproduce figure 6.1.

Many of the figures in this monograph were created with the aid of the Mathematica software system.

Some of the technical portions of the monograph are based on modifications, expansions, and rewrites of LaTeX fragments that have been extracted from my own technical publications. Appropriate thanks are due to the Physical Review and Nuclear Physics.

I also wish to thank the Aspen Center for Physics for providing me with the opportunity to attend the 1994 summer program, which provided an excellent environment for some of the final stages of manuscript preparation. Support for my research has been provided by the U.S. Department of Energy.

Part I

Background

Chapter 1

Introduction

Two of the greatest success stories of twentieth century physics are Einstein's theory of gravitation [general relativity (GR)] and the quantum theory of fields (QFT) as utilized in describing the physics of elementary particles. Individually the successes of these two approaches to modeling reality are truly impressive—both of these theories have withstood the test of time and experimental verification. Within their respective realms of validity, general relativity and quantum field theory (as embodied in the standard model of particle physics) are excellent descriptions of empirical physical reality.[1]

Unfortunately, when one attempts to merge Einstein gravity with quantum field theory one quickly encounters a rat's nest—a veritable trackless wasteland of messy inconsistencies, imponderables, incomprehensibilities, and incalculables collected under the rubric of "quantum gravity". This monograph will not attempt to settle any of the deep issues of quantum gravity—rather I shall be content to explore the edges of the swamp, carefully testing the firmness of the ground at each step.

The central topic of interest in this monograph is wormholes of the Lorentzian variety. Reduced to its most basic elements, a Lorentzian wormhole is a short-cut through space and time. The concept of a Lorentzian wormhole is essentially synonymous with that of a spacewarp—a warping, bending, or folding of space and possibly time. It is a historical accident that physicists typically talk of wormholes while the science fiction community typically speaks of spacewarps. While we do not have any direct experimental evidence for the existence of such objects, it is commonly believed that such objects *might* be formed in regions of intense gravitational fields, where the highly curved nature of the spacetime manifold *might* allow

---

[1]It would be possible at this point to insert a long, turgid, and ultimately pointless philosophical debate on the nature of reality. Such games do not interest me, and I hope they do not interest the reader.

3

for the existence of nontrivial topology.

Ideas along these lines have been floating around in the physics literature for the past sixty years. The earliest significant contribution I am aware of is the introduction, in 1935, of the object now referred to as an Einstein–Rosen bridge [77]. The field then lay fallow for twenty years until the period 1955–1957 when Wheeler coined the term "wormhole" and introduced his idea of "spacetime foam" [284, 187, 285]. A thirty year interregnum followed, punctuated by isolated contributions, until the major revival of interest following the 1988 paper by Morris and Thorne [190]. The last six years have seen a considerable amount of activity, and the field is now sufficiently mature to warrant an overall summary being presented.

Lorentzian wormholes/spacewarp come in at least two varieties:

1. Inter-universe wormholes (wormholes that connect "our" universe with "another" universe).

2. Intra-universe wormholes (wormholes that connect two distant regions of our universe with each other).

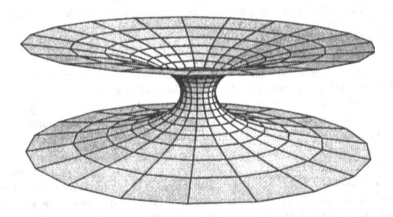

Figure 1.1: Inter-universe wormhole: Schematic representation of a wormhole/spacewarp/bridge connecting two universes in the multiverse.

A certain abuse of the English language has ensued—a conglomeration of many universes is now often referred to as a "multiverse". (Be thankful for small mercies—the competing term that does not seem to have caught on is "googolplexus" [82].) A "universe" is now rather crudely defined as any reasonably large, reasonably flat region of spacetime. Inter-universe wormholes permit one to wander around the various universes in the multiverse. On the other hand, intra-universe wormholes connect two distant regions of one universe with each other. Such a "self-connecting" Lorentzian wormhole is a method of getting from "here" to "there" quickly, indeed so quickly that one might effectively circumvent the speed of light barrier.

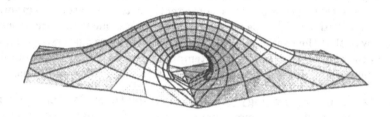

Figure 1.2: Intra-universe wormhole: Schematic representation of wormhole/spacewarp connecting distant regions of one universe.

The difference between these two classes of wormhole arises only at the level of global geometry and global topology. Local physics, near the "throat" of the wormhole, is insensitive to issues of intra-universal or inter-universal travel. An observer confined to making local measurements in the vicinity of the wormhole would not be able to tell whether he/she/it was traveling to another universe or to a distant part of our own universe. For calculational purposes this is actually an advantage; one can concentrate on the local behavior near the throat of the wormhole and not worry about global issues until later on in the game. The chief result of such calculations (as championed by Morris and Thorne) is that wormholes are described by "plausible" physics. Wormholes are certainly strange and peculiar objects but, locally at least, they do not seem to violate basic physical principles.

Global issues will be of critical importance in later parts of this monograph. Because general relativity is a theory of spacetime, any wormhole/spacewarp could equally well be thought of as a space-time-warp. In general relativity "space" and "time" are treated in an essentially symmetrical manner. If spacewarps/wormholes are "good" physics, then what about timewarps/time machines?

Adding time travel to the weird and wonderful effects infesting modern speculative physics opens up a whole new level of complexities—Pandora would be most gratified. Time travel and its attendant logical paradoxes cut to the very foundations of all of physics, both classical and quantum. There are at least four responses to the problem of time travel currently and actively under consideration.

1. **The radical rewrite conjecture:** One might make one's peace with the notion of time travel and proceed to rewrite all of physics (and logic) from the ground up. This is a very painful procedure not to be undertaken lightly.

2. **Novikov's consistency conjecture:** This is a slightly more modest way of making one's peace with the notion of time travel. One simply *asserts* that the universe is consistent, so that whatever temporal transpositions and trips one undertakes, events must conspire in such a way that the overall result is consistent. (A more aggressive version of the consistency conjecture attempts to *derive* this "principle of self-consistency" from some appropriate micro-physical assumptions.)

3. **Hawking's chronology protection conjecture:** Hawking has conjectured that the cosmos works in such a way that time travel is completely and utterly forbidden. Loosely speaking, "thou shalt not travel in time", or "suffer not a time machine to exist". The chronology protection conjecture permits spacewarps/wormholes but forbids timewarps/time machines.

4. **The boring physics conjecture:** This conjecture states, roughly: "A pox upon all nonstandard speculative physics. There are no wormholes and/or spacewarps. There are no time machines/timewarps. Stop speculating. Get back to something we know something about".

I shall explore these and other issues in some detail. Several points are worth making:

- Even though wormhole physics is speculative, the fundamental underlying physical theories, those of general relativity and quantum field theory, are both well tested and generally accepted. If we succeed in painting ourselves into a corner surrounded by disastrous inconsistencies and imponderables, the hope is that the type of disaster encountered will be interesting and informative—hopefully providing us with some clues as to how a consistent theory of quantum gravity might be formulated.

- Ultimately the questions raised are to be answered by experiment. Theoretical considerations will at best delineate potentially profitable avenues of experimental investigation. Unfortunately our chance of being able to test these ideas in the near future is slim to none. Nevertheless, if someone were to succeed in building a time machine the physics community would grit its collective teeth and proceed to settle the issue of the radical rewrite conjecture versus the consistency conjecture by experiment. Similarly, if someone were to succeed in building a wormhole/spacewarp, the status of Hawking's chronology protection conjecture would be settled by the ultimate arbitrator—experiment.

- Many of the intriguing questions surrounding the behavior of wormholes and time machines can be asked within the context of classical general relativity, without invoking quantum effects. Though the questions are classical, the answers (when there are answers) will often involve invoking the quantum realm to some greater or lesser extent. Some questions can be answered by simply considering "semi-classical quantum gravity"—quantum field theory on a fixed background geometry. Semiclassical quantum gravity is relatively well understood and there is broad overall agreement as to the salient features of the field. On the other hand, other questions can only be answered by exploring the uncharted wastelands of full-fledged quantum gravity. Answers to these questions are much more speculative.

To begin to address these issues, a certain amount of mathematical and physical machinery will be required as general background. Essentially all of this technical background material is collected in the remaining chapters of this first part of the monograph. These chapters also serve as a repository for several useful results that will be needed later.

I shall start by providing a brief sketch of general relativity. This is followed by a brief sketch of some key ideas of (flat-space) quantum field theory. Finally, there is a brief discussion of the Planck scale, the electrogravitic scale, and various systems of units in common use.

Chapter 2

General relativity

2.1 Basic notions

General relativity is a purely classical theory—it is *the* classical field theory *par excellence*. The mathematical machinery underlying the physics is that of differential geometry, in olden days referred to as the "absolute differential calculus". Space and time are merged into one unified arena, spacetime, which is endowed with curvature. The bending of space and time near an object *is* the gravitational field of that object. The "potential" describing the gravitational field is the spacetime metric (also known as the metric groundform). In component-based notation one commonly writes

$$ds^2 = g_{\mu\nu}\, dx^\mu dx^\nu. \tag{2.1}$$

Greek indices run from 0 to 3. The index 0 is normally taken to denote the "time" direction (loosely $dx^0 = c\, dt$). Latin indices, when used, run from 1 to 3 and denote the three "space" directions. The more mathematically inclined may wish to rephrase this in abstract tensorial notation:

$$\mathbf{g} = g_{\mu\nu}\, dx^\mu \otimes dx^\nu. \tag{2.2}$$

A spacetime is simply a differentiable manifold that possesses a metric. More technically, the formal definition is as follows:

Definition 1 *A spacetime is a four-dimensional manifold equipped with a Lorentzian (pseudo-Riemannian) metric. The metric should have Lorentzian (pseudo-Riemannian) signature $(-,+,+,+)$. A spacetime is often referred to as having $(3+1)$ dimensions. Experts: The manifold should be Hausdorff and paracompact.*

Starting from the potentials, one constructs the connexion (also known as the connection, affine connection, or Christoffel symbol)

$$\Gamma^{\alpha}{}_{\beta\gamma} \equiv \frac{1}{2} g^{\alpha\sigma} \left(g_{\sigma\beta,\gamma} + g_{\sigma\gamma,\beta} - g_{\beta\gamma,\sigma} \right). \tag{2.3}$$

Notation: The inverse metric, $g^{\mu\nu}$, is the matrix inverse of the metric $g_{\mu\nu}$. Commas denote partial derivatives: $X_{\mu,\nu} \equiv \partial_{\nu} X_{\mu} \equiv \partial X_{\mu}/\partial x^{\nu}$.

Loosely speaking, the connexion governs the "acceleration" of a freely-falling particle in a gravitational field. A small test particle, free to move under the influence of gravity only, will follow a geodesic of the spacetime metric. Geodesics are natural extensions of Euclid's notion of a straight line. Generalizing the notion of the shortest path between two points, a geodesic is defined as a curve that extremizes the total proper distance, that is

$$\delta \left(\int_{\Gamma} ds \right) \equiv \delta \left(\int_{\Gamma} \sqrt{g_{\mu\nu} \, dx^{\mu} \, dx^{\nu}} \right) = 0. \tag{2.4}$$

If one parameterizes the curve in terms of proper distance and denotes the tangent to the curve by the four-vector $V^{\mu} \equiv dx^{\mu}/ds$, then the evolution of the geodesic is governed by the geodesic equation

$$\frac{d^2 x^{\mu}}{ds^2} + \Gamma^{\mu}{}_{\alpha\beta} \frac{dx^{\alpha}}{ds} \frac{dx^{\beta}}{ds} = 0. \tag{2.5}$$

For a massive test particle, following a timelike path, the curve is more naturally parameterized in terms of the proper time $(d\tau)^2 = -(ds)^2$.

Definitions: Timelike $\iff g_{\mu\nu} V^{\mu} V^{\nu} < 0$. Null (lightlike) $\iff g_{\mu\nu} V^{\mu} V^{\nu} = 0$. Spacelike $\iff g_{\mu\nu} V^{\mu} V^{\nu} > 0$.

For a timelike geodesic in a Lorentzian *spacetime* the total proper time,

$$\int_{\Gamma} d\tau \equiv \int_{\Gamma} \sqrt{-g_{\mu\nu} \, dx^{\mu} \, dx^{\nu}}, \tag{2.6}$$

is a local maximum. For a geodesic in a Riemannian *space* the proper distance is a local minimum.

The Riemann curvature tensor is defined to be

$$R^{\alpha}{}_{\beta\gamma\delta} \equiv \partial_{\gamma} \Gamma^{\alpha}{}_{\beta\delta} - \partial_{\delta} \Gamma^{\alpha}{}_{\beta\gamma} + \Gamma^{\alpha}{}_{\sigma\gamma} \Gamma^{\sigma}{}_{\beta\delta} - \Gamma^{\alpha}{}_{\sigma\delta} \Gamma^{\sigma}{}_{\beta\gamma}. \tag{2.7}$$

Exercise: Using the convention that indices are to be raised and lowered using the inverse metric tensor and metric tensor, respectively, show that this definition is equivalent to

$$\begin{aligned} R_{\alpha\beta\gamma\delta} \equiv \quad & - \frac{1}{2} \left(g_{\alpha\gamma,\beta\delta} + g_{\beta\delta,\alpha\gamma} - g_{\alpha\delta,\beta\gamma} - g_{\beta\gamma,\alpha\delta} \right) \\ & - g^{\sigma\rho} \left(\Gamma_{\alpha\gamma\sigma} \Gamma_{\beta\delta\rho} - \Gamma_{\alpha\delta\sigma} \Gamma_{\beta\gamma\rho} \right). \end{aligned} \tag{2.8}$$

(This is a nice exercise in index gymnastics.)

The Riemann tensor governs the difference in acceleration of two freely-falling particles that are near to each other. (If you think this is too abstract, consider this: Earth's Moon is a relatively large hunk of matter. Therefore it distorts spacetime around itself, and in doing so contributes to the Riemann tensor in the vicinity of the Earth. In particular the Moon is responsible for the dominant time dependent piece of the Riemann tensor at the Earth's surface. This provides the dominant driving force behind the ocean tides.)

The Ricci tensor, Ricci scalar, and Einstein tensor are obtained by contraction:

$$R_{\mu\nu} \equiv R_{\alpha\mu\beta\nu} \, g^{\alpha\beta}, \tag{2.9}$$

$$R \equiv R_{\alpha\beta} \, g^{\alpha\beta} \equiv R_{\alpha\mu\beta\nu} \, g^{\alpha\beta} \, g^{\mu\nu}, \tag{2.10}$$

$$G_{\mu\nu} \equiv R_{\mu\nu} - \frac{1}{2} R \, g_{\mu\nu}. \tag{2.11}$$

Finally the Einstein field equations relate the curvature of spacetime (as measured by the Einstein tensor $G_{\mu\nu}$) to the distribution of matter and energy (as measured by the stress-energy tensor $T_{\mu\nu}$). Explicitly,

$$G_{\mu\nu} = \frac{8\pi G}{c^2} \, T_{\mu\nu}. \tag{2.12}$$

Here G is Newton's constant of universal gravitation, experimentally measured (Cavendish experiment [201]) to be

$$G = 6.672\,57(85) \times 10^{-11} \text{m}^3 \text{ kg}^{-1} \text{ s}^{-2}. \tag{2.13}$$

In any approximately Cartesian coordinate system it is convenient to make the components of the metric dimensionless, while the coordinates themselves carry dimensions of length $[L]$. Then the components of Γ have dimension $[L^{-1}]$, and those of the Riemann and Ricci tensors have dimension $[L^{-2}]$. The components of the stress-energy tensor have dimensions [energy/volume]. Explicitly,[1]

$$T_{\mu\nu} \equiv \left[\begin{array}{c:c} \rho & S_j \\ \hdashline S_i & \pi_{ij} \end{array} \right]. \tag{2.14}$$

Here ρ is the energy density, S_i is the energy flux (essentially a generalization of the Poynting vector), and π_{ij} is the stress (essentially a generalization of the notion of pressure).

---

[1] Remember, Greek indices run from 0 to 3, Latin indices run from 1 to 3.

The Einstein field equations can be derived from an action principle:

$$S = -\frac{c^3}{16\pi G} \int_\Omega R\sqrt{g}\, d^4x - \frac{c^3}{8\pi G} \int_{\partial\Omega} K\sqrt{^3g}\, d^3x$$
$$+ \int_\Omega \mathcal{L}\sqrt{g}\, d^4x. \tag{2.15}$$

The total action consists of the following:

1. The original Einstein–Hilbert Lagrangian. The integration runs over the entire four-dimensional region Ω. [Notation: $\sqrt{g} \equiv \sqrt{\det(g_{\mu\nu})}$.]

2. The Gibbons–Hawking surface term [110]. This integration runs over the three-dimensional boundary $\partial\Omega$. This boundary term is required for technical reasons. It can often, but not always, be safely ignored. Here ^3g denotes the induced three-metric, while K denotes the trace of the second fundamental form. Loosely speaking, the second fundamental form is the normal derivative of the metric. I will be more explicit later.

3. The contribution coming from the Lagrangian $\mathcal{L}$, describing whatever matter is present. Varying this action, keeping the metric fixed on the boundary, reproduces the Einstein field equations. The stress-energy tensor is given by

$$T^{\mu\nu}(x) \equiv \frac{2}{c} \frac{\delta}{\delta g_{\mu\nu}(x)} \int_\Omega \mathcal{L}\sqrt{g}\, d^4x. \tag{2.16}$$

2.2   Weak fields

If the gravitational field is "weak", then general relativity can be reduced to a "linearized theory" that exhibits deep analogies with ordinary electromagnetism. If the gravitational field is both "weak" and time independent then, modulo some technical fiddles, general relativity reduces to Newton's theory of universal gravitation. Some details will be useful in the subsequent discussions.

In weak fields the metric is approximately flat. That is, spacetime is closely approximated by the flat Minkowski spacetime of special relativity. Define

$$\eta_{\mu\nu} \equiv \mathrm{diag}[-1,1,1,1] \equiv \begin{bmatrix} -1 & 0 & 0 & 0 \\ 0 & 1 & 0 & 0 \\ 0 & 0 & 1 & 0 \\ 0 & 0 & 0 & 1 \end{bmatrix}. \tag{2.17}$$

Then in approximately Cartesian coordinates one has

$$g_{\mu\nu} \equiv \eta_{\mu\nu} + h_{\mu\nu}; \qquad\qquad h_{\mu\nu} \ll 1. \tag{2.18}$$

And so

$$\Gamma^{\alpha}{}_{\beta\gamma} = \frac{1}{2}\eta^{\alpha\sigma}\left(h_{\sigma\beta,\gamma} + h_{\sigma\gamma,\beta} - h_{\beta\gamma,\sigma}\right) + O(h^2). \qquad (2.19)$$

The Riemann tensor is then

$$R_{\alpha\beta\gamma\delta} = -\frac{1}{2}\left(h_{\alpha\gamma,\beta\delta} + h_{\beta\delta,\alpha\gamma} - h_{\alpha\delta,\beta\gamma} - h_{\beta\gamma,\alpha\delta}\right) + O(h^2). \qquad (2.20)$$

Contracting indices,[2]

$$
\begin{aligned}
R_{\mu\nu} = &- \frac{1}{2}[h_{,\mu\nu} + \Box h_{\mu\nu} - \eta^{\sigma\rho}(h_{\mu\sigma,\rho\nu} + h_{\nu\sigma,\rho\mu})] \\
&+ O(h^2).
\end{aligned}
\qquad (2.21)
$$

It is very convenient to use the freedom to make coordinate changes to adopt the Hilbert gauge, the gravitational analog of the electromagnetic Lorentz gauge ($\partial_\mu A^\mu = 0$). The Hilbert gauge condition is[3]

$$\partial_\mu\left(h^{\mu\nu} - \frac{1}{2}h\,\eta^{\mu\nu}\right) = 0. \qquad (2.22)$$

See, for instance [186, pp. 433–441 and pp. 944–945], or [203, p. 173]. Equivalently

$$\partial_\mu h^{\mu\nu} = \frac{1}{2}\partial^\nu h. \qquad (2.23)$$

(Aficionados will recognize that the Hilbert gauge is the linearization of the so-called harmonic gauge. In more recent books, the Hilbert gauge is often simply called the Lorentz gauge, or the gravitational Lorentz gauge, or sometimes the Hilbert–Lorentz gauge.) In this gauge the linearized Ricci tensor simplifies to

$$R_{\mu\nu} = -\frac{1}{2}\Box h_{\mu\nu} + O(h^2). \qquad (2.24)$$

The linearized Einstein tensor is then

$$G_{\mu\nu} = -\frac{1}{2}\Box\left(h_{\mu\nu} - \frac{1}{2}h\eta_{\mu\nu}\right) + O(h^2). \qquad (2.25)$$

Finally, the Einstein field equations for weak fields reduce to

$$\Box\left(h_{\mu\nu} - \frac{1}{2}h\eta_{\mu\nu}\right) = -\frac{16\pi G}{c^2}\,T_{\mu\nu} + O(h^2). \qquad (2.26)$$

---

[2] Notation: $\Box \equiv \eta^{\mu\nu}\partial_\mu\partial_\nu = -\partial_0^2 + \nabla^2 = -c^{-2}\partial_t^2 + \nabla^2$. Indices are now raised and lowered using the flat space metric $\eta_{\mu\nu}$.

[3] Notation: $h \equiv \eta^{\mu\nu}h_{\mu\nu}$.

A minor rearrangement gives the alternative form[4]

$$\Box h_{\mu\nu} = -\frac{16\pi G}{c^2}\left(T_{\mu\nu} - \frac{1}{2}T\,\eta_{\mu\nu}\right) + O(h^2). \tag{2.27}$$

Compare this with the flat space Maxwell equations in Lorentz gauge

$$\Box A^\mu = \frac{1}{\epsilon_0}J^\mu. \tag{2.28}$$

This implies that essentially all of the technical machinery developed to do ordinary electromagnetism can be carried over to weak field gravity.

For example, consider an isolated point particle of rest mass m_0, and four-velocity $V^\mu(\tau)$, following a path $y^\mu(\tau)$. The parameter τ is taken to be the proper time along the particle's worldline. Then

$$T_{\mu\nu}(x) = m_0\int d\tau\, V_\mu(\tau)V_\nu(\tau)\,\delta^4(x - y(\tau)). \tag{2.29}$$

The linearized field equations become

$$\Box h_{\mu\nu} = -\frac{8\pi G}{c^2}\,m_0\int d\tau\,(2V_\mu V_\nu + \eta_{\mu\nu})\,\delta^4(x - y(\tau)) + O(h^2). \tag{2.30}$$

The solution to this set of differential equations is the gravitational analog of the familiar Liénard–Wiechert potentials of electromagnetism [156, pp. 654–656]

$$h_{\mu\nu}(x) = -\frac{2G\,m_0}{c^2}\frac{(2V_\mu V_\nu + \eta_{\mu\nu})}{V\cdot[x - y(\tau)]}\bigg|_{[x-y(\tau)]^2=0}. \tag{2.31}$$

The right hand side is to be evaluated at the point where the particle path $y(\tau)$ intersects the past light cone with apex at x. If, furthermore, the point mass is moving with constant four-velocity, this expression simplifies to

$$h_{\mu\nu}(x) = +\frac{2G\,m_0}{c^2\,r}\,(2V_\mu V_\nu + \eta_{\mu\nu}), \tag{2.32}$$

where r is now the distance from the point x to the particle generating the gravitational field, as measured in the rest frame of that particle. The connection between the linearized Einstein field equations and Newton's law of universal gravitation are now manifest. More generally, for any static distribution of matter (assuming internal stresses are small compared to the energy density) the field $h_{\mu\nu}$ can be calculated in terms of the Newtonian gravitational potential, $\phi(x)$. Going to the rest frame and adopting Cartesian coordinates gives [186, pp. 445–446]

$$h_{tt} = h_{xx} = h_{yy} = h_{zz} = -2\phi(x)/c^2. \tag{2.33}$$

Equivalently

$$h_{\mu\nu}(x) = -\frac{2\phi(x)}{c^2}\,(2V_\mu V_\nu + \eta_{\mu\nu}). \tag{2.34}$$

---

[4]Notation: $T \equiv \eta^{\mu\nu}T_{\mu\nu}$.

2.3 Strong fields

The full Einstein field equations are nonlinear. This is the principal difficulty in extracting exact solutions. Nevertheless, an extensive collection of exact solutions is available [170]. For our purposes it will be sufficient to be aware of several simple basic solutions: the Schwarzschild and Reissner–Nordström solutions [29]. It is also useful to get some qualitative feel for strong field physics under our collective hats.

2.3.1 The Arnowitt–Deser–Misner split

In any well-behaved coordinate patch one can use the "time" coordinate to decompose the $(3+1)$-dimensional Lorentzian metric via the Arnowitt–Deser–Misner (ADM) split [186, pp. 507–508]. (Technical point: I wish to exclude coordinate charts that involve null coordinates.) The ADM split yields:[5]

$$g_{\mu\nu}(t,\vec{x}) \equiv \left[\begin{array}{c:c} -(N^2 - g^{ij}\beta_i\beta_j) & \beta_j \\ \hdashline \beta_i & g_{ij} \end{array} \right]. \tag{2.35}$$

(Technical point: If one wishes to make this a global decomposition for the whole of the manifold, then one must require the spacetime to be foliated by three-dimensional hypersurfaces. The hypersurfaces should also be spacelike, at least near spatial infinity.) The function $N(t,\vec{x})$ is known as the lapse function, while $\vec{\beta}(t,\vec{x})$ is known as the shift function. The three-metric $g_{ij}(t,\vec{x})$ describes the geometry of "space" while the lapse and shift functions describe how the space slices are assembled to form spacetime. For the inverse metric one has

$$g^{\mu\nu}(t,\vec{x}) \equiv \left[\begin{array}{c:c} -(1/N^2) & (\beta^j/N^2) \\ \hdashline (\beta^i/N^2) & (g^{ij} - \beta^i\beta^j/N^2) \end{array} \right]. \tag{2.36}$$

The geometry of the space slices depends strongly on the choice of the time function. This decomposition is most useful if there is some natural choice of time lurking in the background. If this is the case, the ADM split can be used as the basis for a quick and dirty definition of the notions of black holes and event horizons.

2.3.2 Black holes and horizons

Horizons are theoretical constructs that qualitatively exhibit two salient characteristic features:

---

[5] Remember: Greek indices run from 0 to 3; Latin indices run from 1 to 3.

1. Horizons are one-way membranes, permitting the passage of light and matter in one direction only.

2. Time slows to a stop at a horizon.

Precise mathematical details and definitions are somewhat subtle. See for instance [132, 186, 275]. There are at least five interesting types of horizon that occur in general relativity. They are typically associated with strong gravitational fields, though it is certainly possible to have a strong gravitational field without forming a horizon. (To add to the confusion, it is also possible to obtain some particular types of horizon without any gravitational field being present.)

2.3.3   Event horizon (absolute horizon)

The notion of an event horizon, also known as an absolute horizon, can only be defined for a spacetime containing one or more asymptotically flat regions.

Definition 2 *For each asymptotically flat region the associated future/past event horizon is defined as the boundary of the region from which causal curves (that is, null or timelike curves) can reach asymptotic future/past null infinity.*

Finding the event horizon/absolute horizon (if one exists at all) requires knowledge of the spacetime geometry arbitrarily far into the future. Crudely speaking, anything behind an event horizon/absolute horizon is trapped for all eternity and can never get out. Further technical details may be found in any of the standard textbooks [132, 186, 275]. A black hole may be defined as whatever is inside an event horizon/absolute horizon.

2.3.4   Apparent horizon

In contrast, the notion of an apparent horizon can be defined locally in terms of trapped surfaces. Pick any closed, spacelike, two-dimensional surface (two-surface). At any point on the two-surface there are two null geodesics that are orthogonal to the surface. They can be used to define inward and outward propagating wave fronts. If the area of both inward and outward propagating wave fronts decrease as a function of time, then the original two-surface is a trapped surface and one is inside the apparent horizon. More precisely, if the expansion of both sets of orthogonal null geodesics is negative, then the two-surface is a trapped surface. Crudely speaking, anything behind an apparent horizon is forced (at least initially) to travel inwards. This does not, by itself, guarantee that things behind an apparent horizon remain trapped forever. Further technical details may be found in any of the standard textbooks [132, 186, 275]. With suitable additional technical assumptions it is possible to prove a confinement theorem [152, 151].

2.3.5 Cauchy horizon

Cauchy horizons are associated with the onset of unpredictability. Suppose one is given a spacelike hypersurface Σ, and some initial data specified on this surface. These initial data might consist of information such as particle positions and velocities, some field configuration and its time rate of change, or even the spatial geometry and its time rate of change. One can solve appropriate equations of motion to get unique predictions for some region $D(\Sigma)$ known as the domain of dependence of Σ. The boundary of the domain of dependence is the Cauchy horizon of the surface Σ.

Definition 3 *For a spacelike surface Σ the associated future Cauchy horizon is defined as the boundary of the region from which all past-directed causal curves (that is, null or timelike curves) intersect Σ.*

For my purposes, making this definition more precise would be counterproductive. Those readers interested in further technical details may consult any of the standard textbooks [132, 186, 275].

2.3.6 Particle horizon

A particle horizon is an observer-dependent concept. A particle horizon occurs whenever a particular observer never gets to see or be influenced by the whole spacetime.

Definition 4 *For a causal curve γ the associated future particle horizon is defined as the boundary of the region from which causal curves (that is, null or timelike curves) can reach some point on γ.*

Finding the particle horizon (if one exists at all) requires knowledge of the spacetime geometry arbitrarily far into the future. Particle horizons commonly occur in systems undergoing prolonged acceleration and also in cosmological situations. Further technical details may be found in any of the standard textbooks [132, 186, 275]. I will not be interested in this class of horizons.

2.3.7 Putative horizon

It is often convenient to adopt a quick and dirty definition of horizon and worry about tidying up the details later. If one has a natural notion of time then the ADM $(3 + 1)$ split does the job. In every asymptotically flat region $N \to 1$, $\beta \to 0$, and $g_{ij} \to \delta_{ij}$ asymptotically as one approaches spatial infinity. Associated with each asymptotically flat region one may define a putative horizon simply by the vanishing of the lapse function ($N = 0$). This quick and dirty definition of a horizon often tells you more about your choice of time function than it does about the causal structure of the spacetime. Crudely speaking: When $N = 0$ time has slowed to a stop.

2.3.8 Relationships

In any stationary spacetime the event horizon and outermost apparent horizon coincide. Loosely speaking, stationary $\iff$ time independent. More precisely, stationary $\iff$ the metric possesses a (globally defined) timelike Killing vector. Even more precisely, stationary $\iff$ the metric possesses a globally defined Killing vector which is timelike near spatial infinity.

Because the spacetime is stationary, it has a unique time function such that the components of the metric are time independent—this time function is a natural choice for constructing the putative horizon (or putative horizons). In this case the outermost putative horizon will be an apparent horizon and will coincide with the event horizon.

In any spherically symmetric, but possibly time-dependent spacetime, the metric can (through a suitable choice of the time coordinate) be cast into the canonical form [186, pp. 594–595]

$$ds^2 = -N^2(t,r)dt^2 + g_{rr}(t,r)dr^2 + r^2(d\theta^2 + \sin^2\theta\, d\varphi^2). \qquad (2.37)$$

With this choice of time function the putative horizon coincides with the apparent horizon. If the black hole is gaining mass through the accretion of matter, the event horizon (absolute horizon) will be somewhere outside the putative/apparent horizon. If the black hole is losing mass (that is, the Hawking evaporation process dominates over accretion) then the event horizon (absolute horizon) is somewhere inside the putative/apparent horizon.

For the purposes of this monograph I shall commonly adopt the quick and dirty definition—horizons are taken to be putative horizons simply defined by looking for a zero in the lapse function. This is only done as a first step, and more refined analyses should then be invoked to clarify the actual state of affairs.

2.3.9 Schwarzschild geometry

Perhaps the simplest non-trivial solution to the Einstein field equations is the Schwarzschild geometry. This geometry is the unique spherically symmetric vacuum solution. In standard Schwarzschild coordinates (t, r, θ, φ)

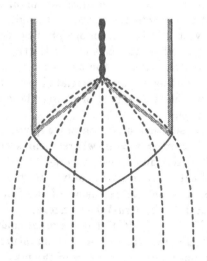

Figure 2.1: Schematic picture of black hole formation: The solid line represents the event horizon, while the grey line represents the apparent horizon. Worldlines of the infalling matter (pressureless dust) are represented by dashed lines. The infalling matter converges on the central singularity.

one has

$$ds^2 = -(1 - r_s/r)\, dt^2 + \frac{dr^2}{1 - r_s/r} + r^2(d\theta^2 + \sin^2\theta\, d\varphi^2). \qquad (2.38)$$

It is convenient to absorb a factor of c into the definition of the Schwarzschild time coordinate t. Then t has the dimensions of length: $[t] = [L]$. Also, by definition, $r > 0$. The Schwarzschild radius, r_s, is a constant with the dimensions of length that characterizes the overall scale of the spacetime geometry.

If one calculates the Ricci tensor $R_{\mu\nu}$ for this metric one finds that it vanishes. This implies that the Einstein tensor also vanishes. Via the Einstein field equations, this implies that the stress-energy tensor is also zero—the region away from $r = 0$ is empty vacuum. By going to large distances, where the field is weak, the Schwarzschild radius r_s can be related to the net mass M of the central gravitating body:

$$r_s = \frac{2GM}{c^2}. \qquad (2.39)$$

The metric *appears* to have a singularity at $r = r_s$. In pre-1960s books and papers this is often referred to as the "Schwarzschild singularity". This is a complete misnomer as there is no physical singularity at $r = r_s$; all that happens is that the coordinate system is breaking down there. The coordinate singularity at $r = r_s$ is no more physical than the singularity occurring at the north and south poles of the Earth when using latitude and longitude coordinates. Even if the surface $r = r_s$ is not physically singular, there is interesting physics associated with this surface: since the lapse function vanishes there, it is of course the event horizon of the Schwarzschild black hole.

The Schwarzschild solution does contain a physical singularity: the physical singularity occurs ar $r = 0$, where various curvature invariants constructed from the Riemann tensor diverge to infinity.[6] Indeed the Schwarzschild coordinate system covers only a part of the full spacetime manifold. The maximal analytic extension of the Schwarzschild solution was explored in great detail by Kruskal and Szekeres (see for example [186, pp. 833–835] and [275, pp 148–156]). The maximal analytic extension contains *two* asymptotically flat regions and *two* curvature singularities. The Schwarzschild coordinate patch covers *half* of the maximal analytic extension. Different coordinate patches cover different pieces of the manifold. For instance, it is possible to find coordinate patches that only cover the two asymptotically flat regions and avoid the regions containing the curvature singularities.

The isotropic coordinate system is an example of such behavior. Take

$$r = \rho \left(1 + \frac{r_s}{4\rho}\right)^2. \tag{2.40}$$

Note that ρ covers the range from $0 \to r_s/4 \to +\infty$, while r covers the range $+\infty \to r_s \to +\infty$. This new coordinate patch covers the region "outside" the event horizon twice, and discards the region "inside" the event horizon. For convenience define

$$d\Omega^2 \equiv d\theta^2 + \sin\theta^2 \, d\varphi^2. \tag{2.41}$$

Then

$$ds^2 = -\left(\frac{1 - r_s/4\rho}{1 + r_s/4\rho}\right)^2 dt^2 + \left(1 + \frac{r_s}{4\rho}\right)^2 [d\rho^2 + \rho^2 d\Omega^2]. \tag{2.42}$$

The horizon ($g_{tt} = 0$) has "moved" to $\rho_H = r_s/4$. This version of the metric appears to contain no curvature singularities. Indeed the metric is

---

[6]The only physical singularities encountered in this monograph will be curvature singularities. If a curvature singularity is surrounded by an event horizon, then the whole system is called a black hole. If a curvature singularity is not surrounded by an event horizon, it is said to be a "naked singularity".

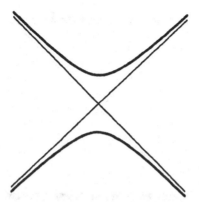

Figure 2.2: Kruskal diagram for the maximally extended Schwarzschild spacetime. The solid crossed lines denote the event horizons, while the dark hyperboloids represent the two curvature singularities of the maximally extended spacetime. There are two asymptotically flat regions.

invariant under the inversion

$$\rho \leftrightarrow \frac{r_s^2}{16\rho}. \tag{2.43}$$

Thus the region near $\rho = 0$ is actually a second asymptotically flat region. The point $\rho = 0$ is *not* a singularity. It is a second point at spacelike infinity. The use of the isotropic coordinate patch has not, of course, abolished the curvature singularities. The curvature singularities are still part of the maximally extended manifold; it's just that the present set of coordinates does not reach that region containing the singularities.

2.3.10   Reissner–Nordström geometry

The Reissner–Nordström geometry is the unique spherically symmetric solution to the coupled Einstein–Maxwell equations (gravity plus electromagnetism). The central gravitating body possesses mass, electric charge, and magnetic charge. The geometry includes radial electric and magnetic fields. In Schwarzschild coordinates the metric takes the form

$$ds^2 = -\left(1 - \frac{r_s}{r} + \frac{r_Q^2}{r^2}\right) dt^2 + \left(1 - \frac{r_s}{r} + \frac{r_Q^2}{r^2}\right)^{-1} dr^2 + r^2 d\Omega^2. \tag{2.44}$$

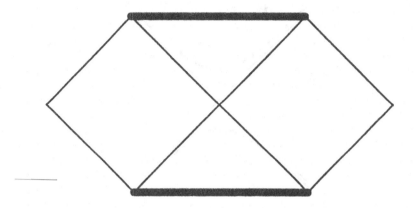

Figure 2.3: Penrose diagram for the maximally extended Schwarzschild spacetime. The solid crossed lines denote the event horizons, while the dark horizontal lines represent the two curvature singularities of the maximally extended spacetime. There are two asymptotically flat regions.

To interpret the parameters r_s and r_Q one goes to large distances where the weak field limit allows one to identify

$$r_s = \frac{2GM}{c^2}; \qquad r_Q^2 = \frac{G}{4\pi\epsilon_0 c^4}(Q_e^2 + Q_m^2). \qquad (2.45)$$

Here Q_e and Q_m are the electric and magnetic charges, respectively. The electromagnetic field strength tensor is given (in the natural local orthonormal frame) by

$$F^{\hat{t}\hat{r}} = E^{\hat{r}} = \frac{1}{4\pi\epsilon_0}\frac{Q_e}{r^2}. \qquad (2.46)$$

$$F^{\hat{\theta}\hat{\varphi}} = B^{\hat{r}} = \frac{1}{4\pi\epsilon_0}\frac{Q_m}{r^2}. \qquad (2.47)$$

Technical point: Setting up the orthonormal frames is a little tedious. Coordinate basis vectors are defined by considering the vector separation between two nearby points:

$$\Delta\mathbf{x} = \Delta t\,\mathbf{e}_t + \Delta r\,\mathbf{e}_r + \Delta\theta\,\mathbf{e}_\theta + \Delta\varphi\,\mathbf{e}_\varphi. \qquad (2.48)$$

Orthonormal basis vectors are then set up by defining

$$\mathbf{e}_{\hat{t}} \equiv \sqrt{-1/g_{tt}}\, \mathbf{e}_t; \quad \mathbf{e}_{\hat{r}} \equiv \sqrt{1/g_{rr}}\, \mathbf{e}_r; \quad \mathbf{e}_{\hat{\theta}} \equiv r^{-1}\mathbf{e}_\theta; \quad \mathbf{e}_{\hat{\varphi}} \equiv (r\sin\theta)^{-1}\mathbf{e}_\varphi. \tag{2.49}$$

These are the general relativistic equivalent of unit vectors pointing in the (t, r, θ, φ) directions, respectively. In fact

$$g_{\hat{\mu}\hat{\nu}} = \mathbf{e}_{\hat{\mu}} \cdot \mathbf{e}_{\hat{\nu}} = \eta_{\hat{\mu}\hat{\nu}} = \mathrm{diag}[-1, +1, +1, +1]. \tag{2.50}$$

Expressing tensors in terms of their components in the orthonormal basis cuts down on the clutter and makes it easier to compare results with their flat space special relativistic analogs.

The Reissner–Nordstöm geometry is a black hole and contains two horizons, an inner and outer horizon, located by setting $g_{tt} = 0$. The outer horizon is an event horizon, while the inner horizon is a Cauchy horizon. The horizons occur at

$$r_H = \frac{r_s}{2}\left\{1 \pm \sqrt{1 - 4(r_Q^2/r_s^2)}\right\}. \tag{2.51}$$

For $r_Q \ll r_s$ this approximates to

$$r_H^+ \approx r_s - \frac{r_Q^2}{r_s}; \qquad r_H^- \approx +\frac{r_Q^2}{r_s}. \tag{2.52}$$

It is possible to put the Reissner–Nordstöm solution into isotropic coordinates (see, for instance, [187, pp. 592–593]). Define a new radial coordinate ρ by

$$r = \rho\left[\left(1 + \frac{r_s}{4\rho}\right)^2 - \frac{r_Q^2}{4\rho^2}\right], \tag{2.53}$$

in which case

$$ds^2 = -\left[1 - \frac{r_s^2 - 4r_Q^2}{16\rho^2}\right]^2 \left[\left(1 + \frac{r_s}{4\rho}\right)^2 - \frac{r_Q^2}{4\rho^2}\right]^{-2} dt^2$$
$$+ \left[\left(1 + \frac{r_s}{4\rho}\right)^2 - \frac{r_Q^2}{4\rho^2}\right]^2 [d\rho^2 + \rho^2 d\Omega^2]. \tag{2.54}$$

This form of the metric has an inversion symmetry under

$$\rho \leftrightarrow \frac{r_s^2 - 4r_Q^2}{16\rho} \tag{2.55}$$

In this coordinate system there appears to be only *one* horizon, at

$$\rho_H = \tfrac{1}{4}\sqrt{r_s^2 - 4r_Q^2}. \tag{2.56}$$

The inner horizon appears to have vanished. This is an illusion. It is another example of the entertaining diseases that can afflict one if one is careless about switching coordinate systems without inquiring as to the region covered by the new and old coordinate patches. Whenever there is even one horizon in the game, great care must be taken to explore the full maximally extended spacetime. (The full maximally extended Reissner–Nordström spacetime is discussed, for instance, in [186, pp. 920–921].) Lack of care in this regard can easily make curvature singularities and even inner horizons seem to vanish into the mist.

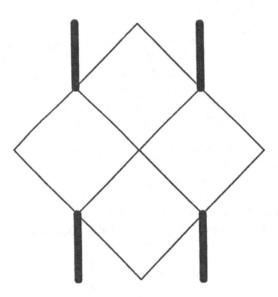

Figure 2.4: Part of the Penrose diagram for the maximally extended Reissner–Nordström spacetime $(M > Q)$. The solid crossed lines denote event horizons, while the dark vertical lines represent the curvature singularities of the maximally extended spacetime. This portion of the diagram should be repeated an infinite number of times in the vertical direction. There is an infinite stack of event horizons, singularities, and asymptotically flat regions.

Exercise: (Trivial) What happens to the Reissner–Nordström geometry once $r_s < 2r_Q$? Rephrase this condition in terms of the mass, electric and magnetic charges. Discuss the resulting geometry using both Schwarzschild and isotropic coordinates.

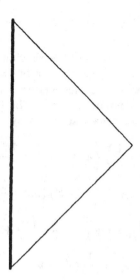

Figure 2.5: The Penrose diagram for the maximally extended Reissner–Nordström spacetime ($M < Q$). There are no event horizons. The dark vertical line represents the single curvature singularity, which is naked. There is only one asymptotically flat region.

2.3.11 Generic static spherically symmetric spacetime: With horizon

Let us, for the time being, not attempt to enforce the Einstein field equations. Any generic static spherically symmetric spacetime can without loss of generality be cast into the canonical form

$$ds^2 = -N^2(r)dt^2 + g_{rr}(r)dr^2 + r^2 d\Omega^2. \qquad (2.57)$$

Loosely speaking, static $\iff$ time independent and nonrotating. (Symmetric under time reversal: $g_{ti} = 0$.) Technically, static $\iff$ the metric possesses a globally defined Killing vector that is hypersurface orthogonal and is timelike near spatial infinity.

There is an art to further specifying the functional form of N and g_{rr} in such a way as to keep both computations and their interpretations simple. If the geometry contains a horizon, it is useful to take (see, for example, [267])

$$N^2 = e^{-2\phi(r)} \left\{ 1 - \frac{b(r)}{r} \right\}; \qquad g_{rr} = \left\{ 1 - \frac{b(r)}{r} \right\}^{-1}, \qquad (2.58)$$

that is,

$$ds^2 = -e^{-2\phi(r)}\left\{1 - \frac{b(r)}{r}\right\} dt^2 + \frac{dr^2}{1 - \frac{b(r)}{r}} + r^2 d\Omega^2. \qquad (2.59)$$

This metric has horizons at values of r satisfying $b(r_H) = r_H$, if any. If a horizon is present, the geometry describes a black hole.

It is useful to have various geometrical quantities associated with this metric at hand for future reference purposes. In an orthonormal frame the components of the Riemann tensor are given by

$$R^{\hat{t}}{}_{\hat{r}\hat{t}\hat{r}} = \left(1 - \frac{b}{r}\right)\left\{\phi'' - (\phi')^2 - \frac{3\phi'}{2r} - \frac{1}{r^2}\right\}$$

$$+ \frac{1}{2r}\left\{(1 - b')\left[3\phi' + \frac{2}{r}\right] + b''\right\}, \qquad (2.60)$$

$$R^{\hat{t}}{}_{\hat{\theta}\hat{t}\hat{\theta}} = R^{\hat{t}}{}_{\hat{\varphi}\hat{t}\hat{\varphi}} = +\left(1 - \frac{b}{r}\right)\left\{\frac{\phi'}{r} + \frac{1}{2r^2}\right\} - \frac{1 - b'}{2r^2}, \qquad (2.61)$$

$$R^{\hat{r}}{}_{\hat{\theta}\hat{r}\hat{\theta}} = R^{\hat{r}}{}_{\hat{\varphi}\hat{r}\hat{\varphi}} = \frac{1}{2r^3}(b'r - b), \qquad (2.62)$$

$$R^{\hat{\theta}}{}_{\hat{\varphi}\hat{\theta}\hat{\varphi}} = \frac{b}{r^3}. \qquad (2.63)$$

(Note: a prime denotes differentiation with respect to the coordinate r; that is, $' \equiv \partial_r$.) All other components of the Riemann tensor, apart from those related to the above by symmetry, vanish. Calculating these expressions is an exercise in brute force that every researcher should do *once*, and once only.

The Einstein tensor is easily computed by the usual standard but tedious manipulations. (See, for instance, [267]). In an orthonormal frame

$$G_{\hat{t}\hat{t}} = \frac{b'}{r^2} \qquad (2.64)$$

$$G_{\hat{r}\hat{r}} = -\frac{2}{r}\left\{1 - \frac{b}{r}\right\}\phi' - \frac{b'}{r^2} \qquad (2.65)$$

$$G_{\hat{\theta}\hat{\theta}} = G_{\hat{\varphi}\hat{\varphi}} = \left\{1 - \frac{b}{r}\right\}\left[-\phi'' + \phi'\left(\phi' - \frac{1}{r}\right)\right]$$

$$- \frac{3}{2}\phi'\left(\frac{b}{r^2} - \frac{b'}{r}\right) - \frac{1}{2}\frac{b''}{r}. \qquad (2.66)$$

All other components of the Einstein tensor are zero. Note: if we set $b(r) = r_s$, and $\phi(r) = 0$, then the metric reduces to the Schwarzschild solution. Inserting these special values of $b(r)$ and $\phi(r)$ into the above expression for the Einstein tensor shows that $G_{\mu\nu} = 0$, as was previously asserted.

Define the nonzero components of the stress-energy tensor to be

$$T_{\hat{t}\hat{t}} = \rho; \qquad T_{\hat{r}\hat{r}} = -\tau; \qquad T_{\hat{\theta}\hat{\theta}} = T_{\hat{\phi}\hat{\phi}} = p. \qquad (2.67)$$

Here ρ denotes the energy density, τ denotes the radial tension, and p denotes the transverse pressure. All other components of the stress-energy tensor vanish by the assumed spherical symmetry. The Einstein equations are simply rearranged to give

$$b' = \frac{8\pi G}{c^2} \rho \, r^2, \qquad (2.68)$$

$$\phi' = -\frac{8\pi G}{2c^2} \frac{(\rho - \tau)r}{(1 - b/r)}, \qquad (2.69)$$

$$\tau' = (\rho - \tau)\left[-\phi' + \frac{1}{2}\{\ln(1 - b/r)\}'\right] - 2(p + \tau)/r. \qquad (2.70)$$

If the black hole contains many horizons, it is typically only the outermost horizon that is of primary interest. For the outermost horizon one knows that $\forall r > r_H$ one has $b(r) < r$. Consequently $b'(r_H) \leq 1$. Assume for now that $b'(r_H) < 1$, and that $\phi(r_H)$ is finite. Then near the outermost horizon the metric takes the approximate form

$$ds^2 = -e^{-2\phi(r_H)}[1 - b'(r_H)]\frac{r - r_H}{r_H} dt^2$$

$$+ \frac{dr^2 \, r_H}{[1 - b'(r_H)](r - r_H)} + r^2 d\Omega^2 + O([r - r_H]^2). \qquad (2.71)$$

Exercise 1: (Tedious) Check all of these formulas.

Exercise 2: (Easy) Check that $\phi = 0$, $b(r) = r_s$, reproduces the Schwarzschild solution. Show that in this case $R_{\mu\nu} = 0$.

Exercise 3: Check that $\phi = 0$, $b(r) = r_s - (r_Q^2/2r)$, reproduces the Reissner–Nordström solution. Calculate $G_{\mu\nu}$.

Exercise 4: Develop an argument to show that the outermost horizon, if present, is an event horizon. Inner horizons, if present, are typically not event horizons.

2.3.12 Generic static spherically symmetric spacetime: Without horizon

Consider the generic static spherically symmetric spacetime

$$ds^2 = -N^2(r)dt^2 + g_{rr}(r)dr^2 + r^2 d\Omega^2. \qquad (2.72)$$

If the geometry does not contain an event horizon, it is advantageous to set

$$N^2 = e^{+2\phi(r)}; \qquad g_{rr} = \left\{1 - \frac{b(r)}{r}\right\}^{-1}, \qquad (2.73)$$

that is,

$$ds^2 = -e^{+2\phi(r)}dt^2 + \frac{dr^2}{1 - \frac{b(r)}{r}} + r^2 d\Omega^2. \qquad (2.74)$$

This is the form of the metric chosen by Morris and Thorne for their analysis of the spherically symmetric traversable wormhole system (subsequently we shall see a lot of this class of objects). This form of the metric is also very useful when investigating the internal structure of spherically symmetric stars or planets. (See, for example, [186, pp. 602–605].)

For future reference: In an orthonormal frame the components of the Riemann tensor are given by [190, p. 401]

$$R^{\hat{i}}{}_{\hat{t}\hat{i}\hat{t}} = \left(1 - \frac{b}{r}\right)\{-\phi'' - (\phi')^2\} + \frac{1}{2r^2}(b'r - b)\phi', \qquad (2.75)$$

$$R^{\hat{i}}{}_{\hat{\theta}\hat{i}\hat{\theta}} = R^{\hat{i}}{}_{\hat{\varphi}\hat{i}\hat{\varphi}} = -\left(1 - \frac{b}{r}\right)\frac{\phi'}{r}, \qquad (2.76)$$

$$R^{\hat{r}}{}_{\hat{\theta}\hat{r}\hat{\theta}} = R^{\hat{r}}{}_{\hat{\varphi}\hat{r}\hat{\varphi}} = \frac{1}{2r^3}(b'r - b), \qquad (2.77)$$

$$R^{\hat{\theta}}{}_{\hat{\varphi}\hat{\theta}\hat{\varphi}} = \frac{b}{r^3}. \qquad (2.78)$$

All other components of the Riemann tensor, apart from those related to the above by symmetry, vanish.

The nonzero components of the Einstein tensor are [190, p. 401]

$$G_{\hat{t}\hat{t}} = \frac{b'}{r^2}, \qquad (2.79)$$

$$G_{\hat{r}\hat{r}} = -\frac{b}{r^3} + 2\left\{1 - \frac{b}{r}\right\}\frac{\phi'}{r}, \qquad (2.80)$$

$$G_{\hat{\theta}\hat{\theta}} = G_{\hat{\varphi}\hat{\varphi}} = \left\{1 - \frac{b}{r}\right\}\left[\phi'' + \phi'\left(\phi' + \frac{1}{r}\right)\right]$$
$$\qquad\qquad - \frac{1}{2r^2}[b'r - b]\left(\phi' + \frac{1}{r}\right). \qquad (2.81)$$

The Einstein field equations are equivalent to [190, p. 401]

$$b' = \frac{8\pi G}{c^2}\rho\, r^2, \qquad (2.82)$$

$$\phi' = \frac{b - (8\pi G/c^2)\tau r^3}{2r^2(1 - b/r)}, \qquad (2.83)$$

$$\tau' = (\rho - \tau)\phi' - 2(p + \tau)/r. \qquad (2.84)$$

The results collected here do not depend on the details of the system under discussion. These results apply to any static spherically symmetric system, whether it be a star, a planet, a wormhole, or even a black hole. However, this particular form of the analysis is most useful for a system that does not contain an event horizon.

Exercise 1:   (Tedious) Check all of these formulas.

Exercise 2:   (Easier) Check that these formulas are compatible with those of the previous section by making the substitution

$$\phi_{\text{here}} = -\phi_{\text{previous}} + \frac{1}{2}\ln\left(1 - \frac{b(r)}{r}\right). \qquad (2.85)$$

Chapter 3

Quantum field theory

Quantum field theory (QFT) in flat Minkowski spacetime provides the mathematical underpinnings for the modern theory of elementary particles. As incarnated in the standard model of particle physics [SU(3) × SU(2) × U(1)], quantum field theory is an extremely well-tested model of physical reality. Unfortunately, the recent demise of the superconducting supercollider means that it is going to be difficult if not impossible to test the most important remaining issue of standard model particle physics— whether or not the generation of particle masses is indeed mediated by the Higgs mechanism. Quantum field theory also has a number of other uses, for example, in statistical mechanics and condensed matter physics, which will not concern us in the present monograph.

Technically, there are two complementary equivalent ways of approaching quantum field theory—either via canonical quantization or via Feynman's functional integral prescription. Both techniques have their advantages and their disadvantages.

3.1 Canonical quantization

In the canonical quantization approach to quantum field theory a quantum field is taken to be a collection of position-dependent operators, one for each point in space.[1] I adopt the Schrödinger picture. Operators are time independent, and the time evolution of the quantum system is encoded in the state vector. (Qualitatively similar comments can be made in the Heisenberg picture. In that picture operators are time dependent, while quantum states are time independent.) Each quantum field $\hat{\Phi}(\vec{x})$ has associated with it a conjugate momentum density $\hat{\pi}(\vec{x})$. The canonical

---

[1]Technical point: Actually, a quantum field is an operator valued tempered distribution, but purely technical details of this nature are not germane to the present discussion.

commutation relations between the field and its momentum are

$$[\hat{\pi}(\vec{x}), \hat{\Phi}(\vec{y})] = -i\hbar\, \delta^3(\vec{x} - \vec{y}). \tag{3.1}$$

This generalizes the canonical commutation relations of ordinary quantum mechanics:

$$[\hat{p}_i, \hat{q}^j] = -i\hbar\, \delta^i{}_j. \tag{3.2}$$

Some comments: (1) That this formalism, which lacks manifest Lorentz invariance, can nevertheless be so suitably massaged as to produce Lorentz invariant relativistic field theories is one of the technical triumphs of the canonical quantization program. (2) If one defines local average values for the field and its momenta by

$$\Phi_\Omega \equiv \frac{1}{\mathrm{Vol}(\Omega)} \int_\Omega \Phi(\vec{x}) d^3x, \tag{3.3}$$

$$\pi_\Omega \equiv \frac{1}{\mathrm{Vol}(\Omega)} \int_\Omega \pi(\vec{x}) d^3x, \tag{3.4}$$

then

$$[\hat{\pi}_\Omega, \hat{\Phi}_\Omega] = -i\frac{\hbar}{\mathrm{Vol}(\Omega)}. \tag{3.5}$$

This observation is central in establishing the notion of a classical limit.

In quantum mechanics one is used to "solving" the canonical commutation relations by going to the Schrödinger representation

$$\hat{q}^i \mapsto q^i; \qquad \hat{p}_i \mapsto -i\hbar\frac{\partial}{\partial q^i} = -i\hbar\partial_i. \tag{3.6}$$

A quantum mechanical "state" is then just some function $\psi(q^i)$ of the coordinates q^i.

The same thing can be done in quantum field theory;

$$\hat{\Phi}(\vec{x}) \mapsto \Phi(\vec{x}); \qquad \hat{\pi}(\vec{x}) \mapsto -i\hbar\frac{\delta}{\delta\Phi(\vec{x})}. \tag{3.7}$$

One must now deal with a functional derivative with respect to the field $\Phi(\vec{x})$. Similarly, a quantum field theoretic "state", $\Psi[t, \Phi(\vec{x})]$, is now a *function* of time t but a *functional* of the field configuration $\Phi(\vec{x})$.

For definiteness, consider the simple example of a free massless scalar field. Though this is a tremendously simplified toy model, it already exhibits some of the key features that will subsequently be of interest to us. The classical Lagrangian is

$$L = \frac{1}{2} \int d^3x \left[c^{-2}(\dot{\Phi})^2 - (\nabla\Phi)^2 \right] = -\frac{1}{2} \int d^3x\, [\Phi\Box\Phi]. \tag{3.8}$$

(Notation: an overdot denotes a time derivative, d/dt.) The conjugate momentum is

$$\pi(t,\vec{x}) = \frac{\delta L}{\delta \dot{\Phi}(t,\vec{x})} = c^{-2}\dot{\Phi}(t,\vec{x}). \tag{3.9}$$

So the classical Hamiltonian is simply

$$H = \frac{1}{2}\int d^3x \left[c^{-2}(\dot{\Phi})^2 + (\nabla\Phi)^2 \right]. \tag{3.10}$$

Quantizing, the functional Schrödinger equation is

$$-i\hbar\frac{\partial}{\partial t}\Psi[t,\Phi(\vec{x})] = \hat{H}\,\Psi[t,\Phi(\vec{x})]. \tag{3.11}$$

Explicitly

$$\hat{H} = \frac{1}{2}\int d^3x \left\{ -\hbar^2 c^2 \left(\frac{\delta}{\delta\Phi(\vec{x})} \right)^2 + (\nabla\Phi)^2 \right\}. \tag{3.12}$$

By analogy with the ordinary quantum mechanical simple harmonic oscillator one can now, by inspection, write down the ground state wave functional for this quantum field theory:

$$\Psi_0[t,\Phi(\vec{x})] \propto \exp\left\{ -\frac{1}{2\hbar c}\int d^3x\, \Phi(\vec{x})\,\sqrt{-\nabla^2}\,\Phi(\vec{x}) \right\}\, \exp\{-iE_0 t/\hbar\}. \tag{3.13}$$

Technical aside: Here $\sqrt{-\nabla^2}$, the formal square root of the operator $-\nabla^2$, is to be interpreted as a pseudodifferential operator. Its meaning is best understood by doing a Fourier transform, noting that $\nabla^2 \mapsto \mathcal{F}[\nabla^2] = -k^2$, and then defining $\mathcal{F}[\sqrt{-\nabla^2}] \equiv \|\vec{k}\|$.

The ground-state energy is formally

$$E_0 = \frac{1}{2}\hbar c\,\mathrm{tr}(\sqrt{-\nabla^2}). \tag{3.14}$$

This ground-state energy is infinite; this will not be a problem at this stage. (A more precise discussion of the meaning of this formal expression will be deferred until I get around to discussing the cosmological constant.)

Furthermore, the ground-state wave functional is not normalizable in the usual sense—this also is no great problem. The key point is that the ground-state wave functional is peaked at $\nabla\Phi = 0$.

I wish to use this formalism to develop some sort of estimate for the typical size of a quantum fluctuation when the quantum field $\Phi(\vec{x})$ is probed at a distance scale L. To this end *define* the correlation function

$$\langle\Phi(\vec{x})\Phi(\vec{y})\rangle \equiv \langle\Psi_0|\hat{\Phi}(\vec{x})\hat{\Phi}(\vec{y})|\Psi_0\rangle. \tag{3.15}$$

In terms of this correlation function, *define*

$$\Delta\Phi_L \equiv \sqrt{\langle\Phi(\vec{x})\Phi(\vec{y})\rangle}\Big|_{|\vec{x}-\vec{y}|=L}. \tag{3.16}$$

To evaluate this object, it is more efficient to Fourier transform to momentum space:

$$\Phi(\vec{x}) = \int \frac{d^3k}{(2\pi)^3} \, \Phi(\vec{k}) \, e^{i\vec{k}\cdot\vec{x}}. \tag{3.17}$$

Then

$$\Psi_0[\Phi(\vec{k})] \propto \exp\left\{-\frac{1}{2\hbar}\int \frac{d^3k}{(2\pi)^2} \, k \, \Phi(\vec{k})^2\right\}. \tag{3.18}$$

The momentum space two-point correlator is

$$\langle\Phi(\vec{k}_1)\Phi(\vec{k}_2)\rangle = \frac{\hbar(2\pi)^3}{k_1} \, \delta^3(\vec{k}_1 - \vec{k}_2). \tag{3.19}$$

The position space two-point correlator is obtained by taking the inverse Fourier transform:

$$\langle\Phi(\vec{x})\Phi(\vec{y})\rangle = \int \frac{d^3k}{(2\pi)^3} \, \frac{\hbar}{k} \, e^{i\vec{k}\cdot(\vec{x}-\vec{y})} \tag{3.20}$$

$$= \frac{2\hbar}{(2\pi)^2(\vec{x}-\vec{y})^2}. \tag{3.21}$$

Consequently

$$\Delta\Phi_L = \frac{\sqrt{2\hbar}}{2\pi L}. \tag{3.22}$$

That is, quantum fluctuations become large on small scales. We shall subsequently see how fluctuations in the gravitational field exhibit a similar qualitative and quantitative behavior.

    The analysis given above is a formalization of some rather loose and wooly statements often encountered in the literature. Suppose one considers a field configuration such that the field Φ is "close" to zero. That is, Φ differs from zero by an amount that we rather crudely call $\Delta\Phi$, only on a region Ω characterized by a distance scale L. Then we can estimate

$$\Psi_0[\Phi(\vec{x})] \propto \exp\left\{-\frac{1}{2\hbar c}\int d^3x \, \Phi(\vec{x}) \, \sqrt{\nabla^2} \, \Phi(\vec{x})\right\}$$

$$\sim \exp\left\{-\frac{1}{2\hbar}\Delta\Phi \, \sqrt{\frac{1}{L^2}} \, \Delta\Phi \, L^3\right\},$$

$$\sim \exp\left\{-\frac{1}{2\hbar}(\Delta\Phi \, L)^2\right\}. \tag{3.23}$$

Then, speaking rather loosely, this implies that the characteristic size of the quantum fluctuations about the average field ($\langle \Phi \rangle = 0$) depend on the length scale being probed:

$$\Delta \Phi \sim \frac{\sqrt{\hbar}}{L}. \tag{3.24}$$

There is much more that could be said about quantum field theory and its many successes in describing physical reality. Canonical quantum field theory in the Schrödinger picture is useful because it makes manifest the deep connections between quantum field theory and ordinary quantum mechanical systems. Particle physics calculations, however, are often more usefully carried out in the Heisenberg picture, or even outside the canonical formalism altogether. For the purposes of this monograph, we will not need to delve too deeply into these fascinating but technically complicated issues.

Exercise 1: Using any of the standard quantum field theory textbooks, write down an explicit formula for the two-point correlation function. For simplicity, start by considering the symmetric two-point function. Consider both the massless and massive cases.

Exercise 2: Now write down explicit formulas for the various other two-point functions [the commutator, the advanced Green function, the retarded Green function, the time-ordered Green function (Feynman propagator)].

Exercise 3: Repeat for spin-half fermion fields.

3.2 Feynman functional integrals

The Feynman path integral is an alternative formulation of quantum mechanics equivalent to the usual canonical quantization techniques. In the path integral formalism one expresses quantum mechanical expectation values in terms of a functional integral over all possible particle paths. (This is best carried out in the Heisenberg picture. Quantum states are time independent and the time evolution of a quantum system is encoded in the operators.) One defines the expectation value $\langle \mathcal{X} \rangle$ of some quantum mechanical quantity $\mathcal{X}(q)$ by

$$\langle \mathcal{X} \rangle \propto \int \mathcal{D}[q(t)] \, \mathcal{X}(q) \, \exp\left\{ - i \int L[q(t), \dot{q}(t)] dt / \hbar \right\}. \tag{3.25}$$

The symbol $\mathcal{D}[q(t)]$ indicates an "integration measure" on the set of all possible paths in configuration space. Formally

$$\mathcal{D}[q(t)] = \prod_{t \in \Re} \{ dq(t) \}. \tag{3.26}$$

This formalism is easily extended to quantum fields:

$$\langle \mathcal{X} \rangle \propto \int \mathcal{D}[\Phi(x)] \, \mathcal{X}(\phi(x)) \, \exp\{-iS[\Phi]/\hbar\}. \qquad (3.27)$$

The mathematical community will be horrified by this bland statement, and quite justifiably so. For a relatively accessible and rigorous discussion of the full horrors involved in setting up the functional integral formalism for field theories on flat spacetime see the book by Glimm and Jaffe [115].

The integration measure is now, formally,

$$\mathcal{D}[\Phi(x)] = \prod_{x \in \Re^4} \{d\Phi(x)\}, \qquad (3.28)$$

while $S[\Phi]$ is the classical action associated with the field configuration $\Phi(t, \vec{x})$:

$$S[\Phi] = \int d^4x \, \mathcal{L}(\partial_\mu \Phi, \Phi). \qquad (3.29)$$

For the special case of the free massless scalar field,

$$\mathcal{L} = \frac{1}{2} \eta^{\mu\nu} \partial_\mu \Phi \partial_\nu \Phi. \qquad (3.30)$$

One probes the quantum fluctuations by defining

$$\Delta \Phi_L \equiv \sqrt{\langle \Phi(x)\Phi(y) \rangle} \bigg|_{x^0 = y^0 = t; \, |\vec{x} - \vec{y}| = L}. \qquad (3.31)$$

In this form, the size of the quantum fluctuations is seen to be related to the equal-time behavior of the Feynman propagator. From the general prescription enunciated above, using an integration by parts,

$$
\begin{aligned}
G_F(x, y) &\equiv \langle \Phi(x)\Phi(y) \rangle \\
&\propto \int \mathcal{D}[\Phi(z)] \, \Phi(x)\Phi(y) \, e^{-(i/2\hbar) \int (\nabla \Phi)^2 - i\epsilon \Phi^2} \\
&\propto \int \mathcal{D}[\Phi(z)] \, \Phi(x)\Phi(y) \, e^{+(i/2\hbar) \int \Phi(\Box + i\epsilon)\Phi}. \\
&\propto \int \mathcal{D}[\Phi(z)] \, \Phi(x)\Phi(y) \, e^{\int \Phi(i\Box - \epsilon)\Phi/(2\hbar)}. \qquad (3.32)
\end{aligned}
$$

Since the integral is Gaussian, it can easily be performed. (See any standard textbook on quantum field theory [154, 219, 232].) With proper care paid to the overall normalization,

$$G_F(x, y) = \hbar(\Box + i\epsilon)^{-1}(x, y) = \hbar \int \frac{d^4k}{(2\pi)^4} \frac{e^{ik(x-y)}}{k^2 + i\epsilon}. \qquad (3.33)$$

(At this elementary level of discussion, the "$i\epsilon$" prescription can be thought of as a formal trick to make the functional integral converge. At a deeper level the "$i\epsilon$" prescription is related to the causality properties of the Feynman propagator and the location of its momentum space poles. See, for example, [18, 154, 219, 232].)

Now go to equal times ($x^0 = y^0 = t$), and define $\omega = k^0$. Then

$$
\begin{aligned}
G_F(t, \vec{x}; t, \vec{y}) &= \hbar \int \frac{d\omega \, d^3k}{(2\pi)^4} \frac{e^{i\vec{k}\cdot(\vec{x}-\vec{y})}}{\omega^2 - \vec{k}^2 + i\epsilon} \\
&= \frac{2\hbar}{(2\pi)^2(\vec{x}-\vec{y})^2}.
\end{aligned}
\tag{3.34}
$$

This is a standard result that can be extracted from most quantum field theory textbooks. There are many equivalent ways of getting to the same result. Finally, one evaluates

$$
\Delta\Phi_L = \frac{\sqrt{2\hbar}}{2\pi L}.
\tag{3.35}
$$

That this agrees with the estimate extracted from canonical wave functional techniques is encouraging, and indeed necessary.

I mention in passing that the current discussion is easily generalized to the case of a massive free field. In that case (see, for example, [18, p. 389])

$$
\Delta\Phi_L = \sqrt{\frac{\hbar}{2\pi^2 L} \frac{\partial K_0(mL/\hbar)}{\partial L}}.
\tag{3.36}
$$

Here K_0 is the cylindrical Bessel function.

Again, the discussion above is a more formal presentation of a rather wooly argument often encountered in the literature: Suppose one knows that the average field is Φ_0, and suppose that one then considers a field configuration Φ that differs from Φ_0 by a quantity $\Delta\Phi$ over a spacetime region of size L. One can estimate

$$
\Delta S[\Phi] \equiv S[\Phi] - S[\Phi_0] \approx \frac{1}{2}(\Delta\Phi/L)^2 \, L^4 = \frac{1}{2}(\Delta\Phi L)^2.
\tag{3.37}
$$

Once $\Delta S[\Phi] \approx 2\pi\hbar$, the field configuration Φ interferes destructively with Φ_0, the assumed average field configuration. This suggests that the characteristic scale for coherent quantum fluctuations over a distance scale L is

$$
\Delta\Phi_L \sim \frac{\sqrt{\hbar}}{L}.
\tag{3.38}
$$

The rather superficial aspects of quantum field theory discussed so far will be quite sufficient for the time being. Understanding the vast phenomenological success of quantum field theory in describing elementary particle physics is not necessary to understanding this monograph.

Exercise:　Explicitly check that all correlation functions calculated via the path integral formalism are identical to those calculated via canonical quantization.

Chapter 4

Units and natural scales

4.1 The Planck scale

In 1899 Planck introduced the fundamental constant $\hbar$—now known as Planck's quantum of action. Planck's constant was introduced as a way of parameterizing what was at that stage an ad hoc mutilation of the spectrum of blackbody radiation, this mutilation being performed in order to avoid the ultraviolet catastrophe of classical thermodynamics. It was immediately clear to Planck that by using suitable algebraic combinations of $\hbar$, c, and G it was possible to define "natural" units of mass, length, and time (now known as the Planck mass, Planck length, and Planck time, respectively). Specifically, by taking the modern values [201]

$$\hbar = 1.054\,572\,66(63) \times 10^{-34}\ \text{J s}, \tag{4.1}$$

$$c \equiv 2.997\,924\,58 \times 10^{8}\ \text{m s}^{-1}, \tag{4.2}$$

$$G = 6.672\,59(85) \times 10^{-11}\ \text{m}^3\ \text{kg}^{-1}\ \text{s}^{-1}, \tag{4.3}$$

one may construct

$$m_P \equiv \sqrt{\frac{\hbar c}{G}} = 1.221\,047(79) \times 10^{19}\ \text{GeV}/c^2, \tag{4.4}$$

$$= 2.176\,71(14) \times 10^{-8}\ \text{kg}, \tag{4.5}$$

$$\ell_P \equiv \frac{\hbar}{m_P c} \equiv \sqrt{\frac{\hbar G}{c^3}} = 1.616\,048 \times 10^{-35}\ \text{m}, \tag{4.6}$$

$$T_P \equiv \ell_P/c \equiv \sqrt{\frac{\hbar G}{c^5}} = 5.390\,557 \times 10^{-44}\ \text{s}. \tag{4.7}$$

Thus various combinations of the fundamental constants $\hbar$, c, and G give a "unique" set of natural units with which to describe nature.

At this stage, of course, this is nothing more than algebraic numerology—the physical import of m_P, ℓ_P, and T_P is far from clear. That the Planck scale has a deep physical significance for questions of quantum gravity is one of the major themes of this book.

4.2   The electrogravitic scale

In 1897 J. J. Thomson announced the direct experimental discovery of the electron and effectively measured its charge to mass ratio. That the electric charge of any material body is quantized as an integral number of electron charges was finally settled in 1910 by the work of R. A. Millikan [183, 184]. (Please, no niggling comments about quarks.) This observation effectively adds a new fundamental constant $e^2/(4\pi\epsilon_0)$ to the laws of physics. In a manner entirely analogous to that used to obtain the Planck units it is possible to combine $e^2/(4\pi\epsilon_0)$, c, and G to construct completely classical natural units of mass, length, and time. Taking the modern values [201]

$$e = 1.602\ 10(7) \times 10^{-19}\ \text{C}, \tag{4.8}$$

$$\frac{e^2}{4\pi\epsilon_0} = 2.306\ 856 \times 10^{-28}\ \text{J m}, \tag{4.9}$$

one may construct

$$m_Q \equiv \sqrt{\frac{e^2}{4\pi\epsilon_0}\frac{1}{G}} = 1.859\ 358 \times 10^{-9}\ \text{kg}, \tag{4.10}$$

$$\ell_Q \equiv \sqrt{\frac{e^2}{4\pi\epsilon_0}\frac{G}{c^4}} = 1.380\ 435 \times 10^{-36}\ \text{m}, \tag{4.11}$$

$$T_Q \equiv \sqrt{\frac{e^2}{4\pi\epsilon_0}\frac{G}{c^6}} = 4.604\ 636 \times 10^{-45}\ \text{s}. \tag{4.12}$$

The important point is that these electrogravitic natural units are completely classical in nature, and do not involve $\hbar$ in any way, shape, or form.

That the electrogravitic scale is relatively close to the Planck scale is essentially an accident caused by the fact that the fine structure constant is relatively close to unity. Indeed the classical unit of action is

$$\mathcal{J} \equiv \frac{e^2}{4\pi\epsilon_0 c} \equiv \alpha\hbar. \tag{4.13}$$

Numerically

$$\alpha = 1/137.035\ 989\ 5(61), \tag{4.14}$$

and

$$\mathcal{J} = 7.694\,843 \times 10^{-37} \text{ J s}. \tag{4.15}$$

The electrogravitic scale is related to the Planck scale by

$$m_Q \equiv \sqrt{\alpha}\, m_P, \tag{4.16}$$

$$\ell_Q \equiv \sqrt{\alpha}\, \ell_P, \tag{4.17}$$

$$T_Q \equiv \sqrt{\alpha}\, T_P. \tag{4.18}$$

It is largely an accident of timing that the electrogravitic scale is not more widely known and recognized. Interest in these quantities was largely preempted by the prior appearance of the Planck scale.

An example of the utility of the electrogravitic scale is in interpreting the "charge radius" of the Reissner–Nordström solution:

$$r_Q = \sqrt{\frac{G}{4\pi\epsilon_0 c^4}}\, \sqrt{Q_e^2 + Q_m^2} = \ell_Q \sqrt{Z_e^2 + Z_m^2}. \tag{4.19}$$

Here Z_e and Z_m are now the dimensionless numbers $Z_e \equiv Q_e/e$ and $Z_m \equiv Q_m/e$.

4.3 Units

4.3.1 Seminatural units

Most of this book will be written using "seminatural units". I shall define $c \equiv 1$, so that $m_P = \sqrt{\hbar/G}$; $\ell_P \equiv T_P \equiv \sqrt{\hbar G}$. Equivalently $G \equiv \hbar/m_P^2 \equiv \ell_P^2/\hbar \equiv \ell_P/m_P$. For pedagogical purposes this seems to me to be the best compromise between constantly keeping track of explicit factors of the fundamental constants and quickly getting totally confused about the dimensionality of various physical quantities.

Other conventions in common use are as follows:

4.3.2 Geometrodynamic units

Define $c \equiv G \equiv 1$, so that $m_P \equiv \ell_P \equiv T_P \equiv \sqrt{\hbar}$. These units are commonly used by classical relativists but are guaranteed to drive particle physicists rapidly up the wall.

4.3.3 Quantum units

Define $c \equiv \hbar \equiv 1$, so that $m_P \equiv 1/\sqrt{G}$, while $\ell_P \equiv T_P \equiv \sqrt{G}$. These units are commonly used by particle physicists but are guaranteed to drive classical relativists rapidly up the wall.

4.3.4   Planck units—completely natural units

Define $c \equiv \hbar \equiv G \equiv 1$, so that $m_P \equiv \ell_P \equiv T_P \equiv 1$. This system of units serves to keep both classical relativists and particle physicists happy. This system also serves to keep both classical relativists and particle physicists confused since it is essentially impossible to use dimensional analysis to check results for consistency while using this system. This system works in the first place because for suitable exponents (α, β, γ) and (a, b, c) almost all interesting physical quantities X are dimensionless multiples of

$$[X] \equiv m_P^\alpha \, \ell_P^\beta \, T_P^\gamma \equiv \hbar^a \, c^b \, G^c. \tag{4.20}$$

[Please, no niggling comments about thermodynamics (Boltzmann's constant), chemistry (moles), luminous intensity (candelas), etc.]

Part II

History

Chapter 5

The Einstein–Rosen bridge

Wormhole physics can be traced back at least as far as the 1916 paper by Flamm [84]. Further speculations can be gleaned from the 1928 philosophical musings of Weyl [282, 283]. Serious calculations date from the 1935 paper by Einstein and Rosen [77]. Of course, the word "wormhole" had not yet been coined, so Einstein and Rosen phrased their discussion in terms of a "bridge" across a double-sheeted physical space.

Einstein had for a considerable time been very disturbed by the inherent dichotomy between the notion of a field theory and the notion of a particle. Even at the classical level, he was very perturbed by the question of just how a singular particle-like object might be matched up to the notion of a continuous field theory such as general relativity. Quote [77, p. 73]:

> [Some] writers have occasionally noted the possibility that material particles might be considered as singularities of the field. This point of view, however, we cannot accept at all.

In their 1935 paper "The particle problem in general relativity" [77], Einstein and Rosen were attempting to build a geometrical model of a physical elementary "particle" that was everywhere finite and singularity free [77, p. 73]:

> These solutions involve the mathematical representation of physical space by a space of two identical sheets, a particle being represented by a "bridge" connecting these sheets.

The particular model they constructed must be considered a failure, but the way that it fails is interesting and presages many of the ideas to be encountered later in this book.

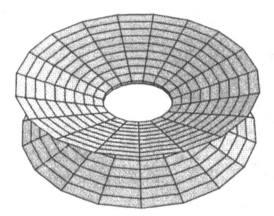

Figure 5.1: Schematic representation of an Einstein–Rosen bridge. This is a black hole in disguise. The region outside the event horizon is covered twice, while the region inside the event horizon is discarded.

Though Einstein and Rosen discussed two specific types of "bridges", neutral and "quasicharged"; their discussion can and will be easily generalized.

5.1   The neutral bridge

The uncharged Einstein–Rosen bridge is nothing more nor less than the observation that a suitable coordinate change seems to make the Schwarzschild singularity disappear.

Now, at the time Einstein and Rosen were writing, the notions of "coordinate singularity" and "physical singularity" had not yet been cleanly separated in physicists' minds. Nor was there any deep understanding of the behavior of the Schwarzschild geometry in the vicinity of the event horizon. To many physicists, the horizon *was* the singularity.

In modern language the discovery of Einstein and Rosen reduces to the observation that some coordinate systems (such as, for instance, isotropic coordinates) naturally cover only the two asymptotically flat regions of the maximally extended Schwarzschild spacetime. The interior region containing the Schwarzschild curvature singularity is not covered by either the Einstein–Rosen or isotropic coordinate systems.

Consider the ordinary Schwarzschild geometry. Adopt geometrodynamic units. In Schwarzschild coordinates

$$ds^2 = -(1 - 2M/r)\,dt^2 + \frac{dr^2}{1 - 2M/r} + r^2 d\Omega^2. \tag{5.1}$$

By making the coordinate change $u^2 = r - 2M$, this can be put into the Einstein–Rosen form

$$ds^2 = -\frac{u^2}{u^2 + 2M}dt^2 + 4(u^2 + 2M)du^2 + (u^2 + 2M)^2\,d\Omega^2, \tag{5.2}$$

with $u \in (-\infty, +\infty)$. This coordinate change discards the region containing the curvature singularity $r \in [0, 2M)$, and twice covers the asymptotically flat region, $r \in [2M, +\infty)$. The region near $u = 0$ is interpreted as a "bridge" connecting the asymptotically flat region near $u = +\infty$ with the asymptotically flat region near $u = -\infty$.

To justify this appellation, consider a spherical surface defined by taking the u coordinate to be a constant. The area of this surface is $A(u) = 4\pi(2M + u^2)^2$. This area is a minimum at $u = 0$, with $A(0) = 4\pi(2M)^2$. One defines the narrowest part of the geometry to be the "throat", while the region nearby is called the "bridge", or in modern terminology, the "wormhole".

Note that the Einstein–Rosen construction does not work if $M < 0$. The construction requires the existence of a horizon to set up the coordinate transformation. The negative mass Schwarzschild solution has a naked singularity, there is no horizon, and the "bridge" construction fails. (Aside: Recall that a naked singularity is any curvature singularity not surrounded by an event horizon. For more precise statements, see various discussions of the "cosmic censorship conjecture", for example, [275, pp. 299–308].)

The net result of these observations is that the neutral "Einstein–Rosen bridge" (also called the "Schwarzschild wormhole") is *identical* to a *part* of the maximally extended Schwarzschild geometry. If you discover an Einstein–Rosen bridge, do not attempt to cross it, you will die. You will die just as surely as by jumping into a black hole. You will die because you *are* jumping into a black hole. The Einstein–Rosen coordinate u is a bad coordinate at the horizon. Attempting to cross the horizon, say from $u = +\epsilon$ to $u = -\epsilon$, will force one off the u coordinate patch and into the curvature singularity.

(This is not supposed to be obvious. A full and careful technical discussion of this phenomenon requires one to invoke the Kruskal–Szekeres maximal analytic extension of the Schwarzschild solution. See, for example, [186, pp. 833–835], or [275, pp. 148–156].)

Exercise 1: (Easy) Read the Einstein–Rosen article [77].

Exercise 2:   Read any of the standard textbooks to become familiar with the notions of Kruskal diagrams, Penrose diagrams, the Kruskal–Szekeres construction, and maximal analytic extension.

5.2   The "quasicharged" bridge

The "quasicharged" Einstein–Rosen bridge is also interesting. Start with the Reissner–Nordström geometry (an electrically charged black hole of charge Q and mass M). In Schwarzschild coordinates

$$ds^2 = -(1 - 2M/r + Q^2/r^2)\, dt^2 + \frac{dr^2}{1 - 2M/r + Q^2/r^2} + r^2 d\Omega^2. \quad (5.3)$$

To make the bridge construction work Einstein and Rosen found that they had to make a completely *ad hoc* mutilation of the theory—they found that they had to reverse the sign of the electromagnetic stress-energy tensor (so that the energy density in the electromagnetic field became negative). The mutilated geometry is not quite the Reissner–Nordström geometry. In Schwarzschild coordinates

$$ds^2 = -(1 - 2M/r - \epsilon^2/r^2)dt^2 + \frac{dr^2}{1 - 2M/r - \epsilon^2/r^2} + r^2 d\Omega^2. \quad (5.4)$$

Taking $M = 0$ the metric reduces to

$$ds^2 = -(1 - \epsilon^2/r^2)dt^2 + \frac{dr^2}{1 - \epsilon^2/r^2} + r^2 d\Omega^2. \quad (5.5)$$

Thus, with $M = 0$, the coordinate change $u^2 = r^2 - \epsilon^2/2$ results in

$$ds^2 = -\frac{u^2}{u^2 + \epsilon^2/2} dt^2 + du^2 + (u^2 + \epsilon^2/2)^2 d\Omega^2. \quad (5.6)$$

This is a very peculiar geometry. It represents a massless, quasicharged object, whose energy density is everywhere *negative*. There is still a horizon at $r = \epsilon$, $u = 0$. It is this object that Einstein and Rosen wished to interpret as an "electron".

The reason for the mutilation of the theory is that for $M = 0$ the Reissner–Nordström solution is a naked singularity. It has no horizon, so no bridge construction is possible. The maiming of the theory, $Q^2 \to -\epsilon^2$, is made merely in order to avoid the appearance of a naked singularity. (It is amusing to note that this modification, $Q^2 \to -\epsilon^2$, corresponds to assigning a purely imaginary value to the electric charge.)

In fact this problem arises for all $M < |Q|$. Since the mass of a physical electron is $m_e = 9.109\,389\,7(54) \times 10^{-31}$ kg, while in geometrodynamic units

$|Q| = |e| = m_Q \approx 1.859\,358 \times 10^{-9}$ kg, one has $m_e/|e| \approx 10^{-22}$. It is clear that allowing for the effect of the nonzero electron mass is not sufficient to fix the problem.

This is a special case of a more general result: if one asks "What is the gravitational field of a *single* isolated elementary particle?", one gets very strange results. Since elementary particles are characterized by their mass, electric charge, and spin angular momentum, one might be tempted to use the Kerr–Newman geometry (the generalization of the Reissner–Nordström geometry to a rotating system) as a model for the gravitational field of an elementary particle. If you insist on doing this then *every* elementary particle currently known to particle physics is modeled by a naked singularity. The sole exception to this general rule is the neutral Higgs scalar (which has not yet been seen experimentally). The significance, if any, of this observation is unknown. Worse, if one takes the Kerr–Newman geometry seriously, then particles with spin are modeled by a rotating ring singularity whose radius is of order the particle's Compton wavelength $R \approx \hbar/(mc)$. Thus a naive application of general relativistic ideas seems to predict internal structure for elementary particles. This naive suggestion is in violent conflict with the experimental situation—direct probes of the internal structure of elementary particles such as the electron show that these systems appear to be pointlike down to energy scales of order several TeV, corresponding to distance scales of order 10^{-20} m. This implies that the electron appears pointlike down to distances of order 10^{-8} times its Compton wavelength. (A micro review regarding limits on quark and lepton substructure appears in [201, IX.12–IX.15].) Again, the deeper significance, if any, of this observation is unknown. I shall have more to say on this topic later.

Exercise 1: (Easy) Show that in geometrodynamic units electric charge has the same dimensions as length. Evaluate the charge on the electron in metres. Evaluate the charge on the electron as a multiple of the Planck length. The dimensionless number

$$\frac{e_{\text{geometrodynamic}}}{\ell_P} \tag{5.7}$$

should be familiar to you. What is its significance?

Exercise 2: (Easy) Show that in geometrodynamic units mass has the same dimensions as length. Evaluate the mass of the electron in metres. Evaluate the mass of the electron as a multiple of the Planck length. Is there any particular significance to the dimensionless number

$$\frac{(m_e)_{\text{geometrodynamic}}}{\ell_P}? \tag{5.8}$$

Exercise 3: (Easy) Evaluate (e/m_e) in geometrodynamic units. What are the dimensions of $(e/m_e)_{\text{geometrodynamic}}$? Check that this is consistent with the discussion given above.

5.3   The generalized bridge

With the two examples discussed by Einstein and Rosen now in hand, it is comparatively easy to generalize the bridge construction. Merely consider an arbitrary symmetric geometry that possesses an event horizon, one that may or may not have some unspecified form of matter surrounding the event horizon. As we have already discussed, the metric can always (without loss of generality) be put into the form

$$ds^2 = -e^{-\varphi(r)}[1 - b(r)/r]\,dt^2 + \frac{dr^2}{1 - b(r)/r} + r^2 d\Omega^2. \tag{5.9}$$

The horizon occurs at $r = r_H$, where r_H is defined by the equation $b(r_H) = r_H$. Now introduce the coordinate u by setting $u^2 = r - r_H$. Then one has the exact result

$$ds^2 = -e^{-\varphi(r_H + u^2)} \frac{r_H + u^2 - b(r_H + u^2)}{r_H + u^2} dt^2$$

$$+ 4\frac{r_H + u^2}{r_H + u^2 - b(r_H + u^2)} u^2 du^2 + (r_H + u^2)^2 d\Omega^2. \tag{5.10}$$

The region near $u = 0$ is the bridge connecting the asymptotically flat region near $u = +\infty$ to the asymptotically flat region near $u = -\infty$. Near the bridge/horizon one has $r \approx r_H$, and $u \approx 0$, so that

$$ds^2 \approx -e^{-\varphi(r_H)} \frac{u^2[1 - b'(r_H)]}{r_H} dt^2 + 4\frac{r_H + u^2}{1 - b'(r_H)} du^2 + (r_H + u^2)^2 d\Omega^2. \tag{5.11}$$

This metric is qualitatively of the same form as the neutral and "quasicharged" Einstein–Rosen bridges. This is most easily seen by introducing constants A and B and rewriting the above metric as

$$ds^2 \approx -A^2 u^2 dt^2 + 4B^2(u^2 + r_H)du^2 + (u^2 + r_H)^2 d\Omega^2. \tag{5.12}$$

This shows that the key ingredient of the bridge construction is merely the existence of a horizon.

For completeness, note that far away from the bridge $u \to \pm\infty$. Asymptotic flatness implies that $\varphi(r) \to 0$, and $b(r) \to 2m$, so one has

$$ds^2 \approx -\frac{r_H - 2m + u^2}{r_H + u^2} dt^2 + \frac{4(r_H + u^2)u^2}{r_H - 2m + u^2} du^2 + (r_H + u^2)^2 d\Omega^2, \tag{5.13}$$

which is again very similar to the canonical Einstein–Rosen form given above.

However, I should reiterate this very important point: The Einstein–Rosen bridge is only a coordinate artifact arising from choosing a rather special coordinate patch. This coordinate patch is defined to double-cover the asymptotically flat region exterior to the black hole event horizon. Coordinate artifacts will not protect one from an untimely demise should one be so foolish as to attempt crossing an Einstein–Rosen bridge.

Exercise: Read (and understand) the discussion in Misner, Thorne, and Wheeler regarding the fatal futility of attempting to cross an Einstein–Rosen bridge. See [186, pp. 836–840].

Chapter 6

Spacetime foam

6.1 Wheeler wormholes

After the pioneering 1935 work of Einstein and Rosen, the field lay largely dormant for a good twenty years. Interest in systems of this nature was rekindled in by Wheeler in 1955. The closing comments of his paper on "geons" indicate that Wheeler was beginning to be interested in topological issues in general relativity [284, pp. 534–536].

Geons are hypothesized unstable but long lived solutions to the combined Einstein–Maxwell field equations (gravity + electromagnetism) [284]. Indeed the word "geon" was coined by Wheeler to denote a "gravitational-electromagnetic entity". (The word "kugelblitz" also had a brief vogue.) In modern language the geon might best be thought of as a hypothetical "unstable gravitational–electromagnetic quasisoliton". In his 1955 paper Wheeler observed that [284, p. 534]:

> One can consider a metric which on the whole is nearly flat except in two widely separated regions, where a double connectedness comes into evidence as symbolized in Fig. ...

The figure referred to is reproduced herein [Figure (6.1)]. It is notable that this is the first diagram of what we would now call a wormhole to appear in the scientific literature.

Furthermore, Wheeler's observation can be used as the basis for building a model of nonzero classical charge with an everywhere zero charge density [284, p. 534]:

> The general divergence free electromagnetic disturbance holding sway in the space around one of these "tunnel mouths" will send forth lines of force into the space, and appear to have a charge. However an equal number of lines of force must enter

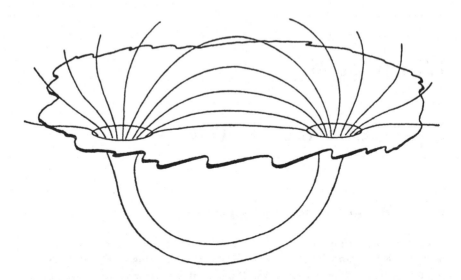

Figure 6.1: Wheeler's schematic picture of a wormhole: "Schematic representation of lines of force in a doubly-connected space. In the upper continuum the lines of force behave much as if the tunnel mouths were seats of equal and opposite charges."

> into the region of disturbance from the tunnel. Consequently the other mouth of the tunnel must manifest an equal and opposite charge.

It is, at this stage, possible to head in two different directions. On the one hand, assuming the existence of such tunnel configurations one could investigate their classical dynamics. On the other hand, investigation of quantum gravitational processes that might give rise to such configurations is also of interest. Wheeler pursued both directions of investigation. The classical route led Wheeler, together with Misner, to their opus minor "Classical physics as geometry: gravitation, electromagnetism, unquantized charge, and mass as properties of curved empty space" [187]. The quantum route led to the extraordinarily tenacious concept of "spacetime foam". The present discussion will stick to the classical analysis. We leave Wheeler's 1955 paper with his parting observation [284, p. 536]:

> One's interest in following geon theory down into the quantum domain will depend on one's considered view of the rela-

tionship between very small geons and elementary particles.

The 1957 paper by Misner and Wheeler [187] was a *tour de force* wherein the Riemannian geometry of manifolds of nontrivial topology was investigated with a view to explaining *all* of classical physics. This very ambitious project was one of the very first uses of abstract topology, homology, cohomology, and differential forms in physics. Their point of view is best summarized by their phrase [187, p. 526]: "Physics is geometry." It is this paper that first introduces the word "wormhole" to the scientific community [187, p. 532]:

> [T]here is a net flux of lines of force through what topologists would call a handle of the multiply–connected space and what physicists might perhaps be excused for more vividly terming a "wormhole".

This paper is also the source of such *mantras* as "charge without charge" and "mass without mass" [187, p. 534]. Ultimately the aim was to use the *source-free* Maxwell equations, coupled to Einstein gravity, with the seasoning of nontrivial topology, to build models for classical electric charges and all other particle like entities in classical physics.

It is now known that this classical conception of "charge without charge" cannot in fact work as originally conceived—classically the tunnels will collapse to form black holes, and the otherwise interesting topology is hidden behind event horizons. This is the content of the classical "topological censorship theorem" that will be discussed in a subsequent chapter. Quantum effects *might* allow one to evade topology censorship; details will be discussed in due course.

6.2 Metric fluctuations

"Quantum gravity" is a noise, a confusion, an agony in manifold fits. (Apologies to J. Steinbeck and C. L. Dodsgon.) Nevertheless, there are a few qualitative and even quantitative comments that one might be willing to make with some degree of confidence. For instance, discussing the quantum fluctuations in the spacetime metric, Wheeler states [284, p. 514]:

> [T]he fluctuation in a typical gravitational potential is

$$\Delta g \sim (\hbar G/c^3)^{1/2}/L.$$

These fluctuations will be inappreciable in comparison to typical average values of the metric $g \sim 1$, so long as the distances, L, under consideration are substantial in comparison with [the Planck length].

In a later publication [285, p. 604]:

> Gravitational field fluctuations are concluded to have qual-
> itatively new consequences at distances of order $(\hbar G/c^3)^{1/2} = 1.6 \times 10^{-33}$ cm.

What is the origin of such claims?

Ultimately, such claims are based on the fact that we believe we understand the quantization of linearized (weak field) Einstein gravity. Rephrase
the classical linearized gauge fixed Einstein equations as

$$\Box h_{\mu\nu} = -\frac{16\pi\ell_P^2 c}{\hbar}\left(T_{\mu\nu} - \frac{1}{2}T\eta_{\mu\nu}\right) + O(h^2). \tag{6.1}$$

These equations can be derived from an action principle

$$S_L = -\frac{\hbar}{32\pi\ell_P^2}\int d^4x\, h^{\mu\nu}\Box h_{\mu\nu} + c\int d^4x\, h^{\mu\nu}\left(T_{\mu\nu} - \frac{1}{2}T\eta_{\mu\nu}\right) + O(h^3). \tag{6.2}$$

This classical linearized gauge fixed action can also be extracted directly
from the Einstein–Hilbert action by expanding the Ricci scalar to *second*
order in $h_{\mu\nu}$. Proceeding in that manner is more fundamental but also
somewhat more tedious than the present discussion.

Since this linearized action is quadratic in h, quantization can immediately be carried out by analogy with the free scalar field . (The stress-energy
tensor is, for the time being, considered to be a fixed classical external
field. This situation is in complete analogy with the quantization of the
electromagnetic field in the presence of a fixed classical current. See, for
example, [154, pp. 163–176].) One is immediately led to consider two-point
correlation functions such as

$$\langle h(x)h(y)\rangle, \qquad \langle h_{\mu\nu}(x)h(y)\rangle, \qquad \text{and} \qquad \langle h_{\mu\nu}(x)h_{\sigma\rho}(y)\rangle. \tag{6.3}$$

When the externally imposed stress-energy is zero, the first of these two-
point functions, the trace–trace correlator, is immediately calculable in
terms of the scalar-field two-point function [see equation 3.21]:

$$\langle h(x)h(y)\rangle = \frac{16\pi\ell_P^2}{\hbar}\langle \Phi(x)\Phi(y)\rangle. \tag{6.4}$$

At equal times this reduces to

$$\langle h(t,\vec{x})h(t,\vec{y})\rangle = \frac{16\pi\ell_P^2}{\hbar}\frac{2\hbar}{(2\pi)^2(\vec{x}-\vec{y})^2} = \frac{8}{\pi}\frac{\ell_P^2}{(\vec{x}-\vec{y})^2}. \tag{6.5}$$

Thus, at a distance scale L, the quantum fluctuations in $h(x)$ are

$$(\Delta h)_L \equiv \sqrt{\langle h(t,\vec{x})h(t,\vec{y})\rangle}\Big|_{|\vec{x}-\vec{y}|=L} = \sqrt{\frac{8}{\pi}}\frac{\ell_P}{L}. \tag{6.6}$$

The tensor–trace correlator is somewhat more complicated. One can use the symmetries of the background Minkowski space to deduce that this correlator must be of the form

$$\langle h_{\mu\nu}(x)h(y)\rangle = \frac{16\pi\ell_P^2}{\hbar}\langle\Phi(x)\Phi(y)\rangle\left\{A\,\eta_{\mu\nu} + B\,\frac{z^\mu z^\nu}{z^2}\right\}. \qquad (6.7)$$

Here we have defined $z = x - y$. By translation invariance, the correlator can depend only on this difference z. The coefficients A and B are dimensionless constants. The expression in braces is the most general dimensionless tensor that can be constructed solely out of the flat Minkowski metric $\eta_{\mu\nu}$ and the relative separation z, which also has the right symmetry properties.

Exercise: (Relatively easy) Determine the dimensionless coefficients A and B. Hints: First normalize using the trace–trace correlator—taking a trace on $\mu\nu$ implies $4A + B = 1$. Second, use the Hilbert gauge condition, $\partial_\mu(h^{\mu\nu} - \frac{1}{2}h\,\eta^{\mu\nu}) = 0$. Apply this to the tensor–trace correlator to deduce

$$\partial^\mu\left(\langle h_{\mu\nu}(z)h(0)\rangle - \frac{1}{2}\langle h(z)h(0)\rangle\,\eta^{\mu\nu}\right) = 0. \qquad (6.8)$$

Now just perform the differentiation to get a second linear equation relating A and B.

Finally, the tensor–tensor correlator is considerably more complicated. The symmetries of the background Minkowski space allow one to deduce that this correlator must be of the form

$$\langle h_{\mu\nu}(x)h_{\sigma\rho}(y)\rangle = \frac{16\pi\ell_P^2}{\hbar}\langle\Phi(x)\Phi(y)\rangle$$
$$\times\left\{A\,\eta_{\mu\nu}\,\eta_{\sigma\rho} + B\left(\eta_{\mu\nu}\,\frac{z^\sigma z^\rho}{z^2} + \frac{z^\mu z^\nu}{z^2}\,\eta_{\sigma\rho}\right) + C\,\frac{z^\mu z^\nu}{z^2}\,\frac{z^\sigma z^\rho}{z^2}\right\}. \qquad (6.9)$$

Again, $z = x - y$. The expression in braces is the most general dimensionless tensor that can be constructed solely out of the flat Minkowski metric $\eta_{\mu\nu}$ and the relative separation z, which respects the symmetry properties of the tensor–tensor correlator.

Exercise: (Rather tedious) Determine the dimensionless coefficients A, B, and C. These can be deduced by a more tedious variation on the theme of the previous exercise. For instance, taking a trace on $\mu\nu$ and $\sigma\rho$ implies $16A + 8B + C = 1$. Complete the exercise by applying the Hilbert gauge condition.

Qualitatively (that is, ignoring the messy tensorial structure), the weak field metric perturbations can be estimated as

$$(\Delta h_{\mu\nu})_L \approx \sqrt{\frac{16\pi\ell_P^2}{\hbar}}\,\frac{\sqrt{2\hbar}}{2\pi L}. \qquad (6.10)$$

Expressed in terms of the metric components themselves,

$$(\Delta g_{\mu\nu})_L \approx \sqrt{\frac{8}{\pi}\frac{\ell_P}{L}}. \qquad (6.11)$$

It is convenient and conventional to simply omit the numerical factor and write

$$(\Delta g_{\mu\nu})_L \sim \frac{\ell_P}{L}. \qquad (6.12)$$

This argument is sufficient to indicate the overwhelming importance of the Planck scale in quantum gravitational physics. At distances below the Planck length quantum fluctuations are large—so large that linearized theory breaks down irretrievably and the quantum physics of full nonlinear Einstein gravity must be faced.

A naive reading of this result would seem to indicate that sufficiently small-scale vacuum fluctuations ($L < \ell_P$) might bootstrap themselves into Planck scale black holes, thereby presumably curdling spacetime. It seems difficult to reconcile such a picture, a curdled froth of quantum black holes popping into and out of existence via vacuum fluctuations, with the observed relatively benign behavior of spacetime. It is far from clear, for instance, whether or not a relatively low-energy photon could traverse such a curdled region without being scattered all over creation. (See [222, 223]. For a different point of view, however, see Wheeler's comments in [285, p. 612] and the more recent comments in [89].) A better, more reliable, reading of this result is that quantum effects make classical gravity break down once $L < \ell_P$.

A few more words about the quantization of linearized gravity are in order. It is possible to put the linearized action into a form more suitable for discussing quantum field theory by introducing the variable

$$f_{\mu\nu} = \sqrt{\frac{\hbar}{16\pi\ell_P^2}}\,h_{\mu\nu} = \frac{m_P c}{\sqrt{16\pi\hbar}}\,h_{\mu\nu} = \kappa^{-1}\,h_{\mu\nu}. \qquad (6.13)$$

Then

$$\begin{aligned}
S_L &= -\frac{1}{2}\int d^4x\, f^{\mu\nu}\Box f_{\mu\nu} + \kappa\int d^4x\, f^{\mu\nu}\left(T_{\mu\nu} - \frac{1}{2}T\,\eta_{\mu\nu}\right) \\
&\quad + O\big((\kappa f)^3\big).
\end{aligned} \qquad (6.14)$$

Ignore, for now, the $O(f^3)$ contributions to the action. In the absence of matter, the model reduces to the quantum field theory of a free massless tensor field (in fact, a free massless spin-2 field). In the presence of matter, the remaining pieces of the linearized action describe a quite ordinary nonrenormalizable quantum field theory. The theory is known to be nonrenormalizable because of the appearance of the dimensional coupling constant $\kappa = \sqrt{16\pi\hbar}/(m_{P}c)$. (See, for instance, [154, pp. 379–382].) In principle, this field theory is no more mysterious than the old "intermediate vector boson" theories that were used to describe the weak interactions before the advent of gauge theories and spontaneous symmetry breaking via the Higgs mechanism. (See, for example, [249, pp. 5–12].) Though the theory, in view of its nonrenormalizability, is known to be "diseased" (it is, at best, an effective theory), the various unpleasant features are not expected to show up until energies exceed the Planck mass and/or distances are smaller than the Planck length. This result is completely analogous to the behavior of the more well-known nonrenormalizable intermediate vector boson theories of the weak interactions. In those theories the diseases attributable to nonrenormalizability do not manifest themselves until the energy scale is greater than that specified by the dimensional Fermi coupling constant. (See, for example, [249, pp. 4–12].)

Furthermore, from our experience with various quantum field theories we have built up considerable insight as regards to how to take the classical limit. For instance, in quantum electrodynamics the classical Coulomb law for the attraction of two charged particles can be rephrased in terms of the exchange of virtual photons. Likewise, in quantized linearized gravity one can rephrase the classical gravitational attraction of two massive objects in terms of the exchange of virtual gravitons. (Definition: A graviton is the gravitational analog of a photon. It is the quantum field theoretical particle that results from quantizing the gravitational field.) The fact that we can reproduce Newton's law of universal gravitation in this manner is the best (albeit indirect) experimental evidence we currently have for the existence of gravitons.

It is important to realize precisely what *cannot* be done with quantized linearized gravity.

1. One cannot possibly recover the maximally extended Schwarzschild solution. There are two reasons for this. If the $O(f^3)$ terms in the Lagrangian are neglected, then gravitons do not self-interact, and nonlinearities are explicitly excluded. Even if the $O(f^3)$ and higher-order pieces are retained, one is intrinsically performing a perturbation expansion around flat space. But the topology of flat Minkowski space is radically different from the topology of the maximally extended Schwarzschild solution. There is no way to get from one to the other perturbatively. (For example, see [207].) If one is not worried about

issues of maximal extension the the region outside the event horizon can be approximated with arbitrary accuracy.

2. One cannot probe energy scales above the Planck mass. In the absence of matter, this is because quantum fluctuations become large and linearized theory is no longer appropriate. In the presence of matter, one has the additional complications due to nonrenormalizability and the attendant violations of unitarity.

Despite the caveats above, Wheeler's qualitative description of the quantum geometry of spacetime seems safe [285, p. 607]:

> On the atomic scale the metric appears flat, as does the ocean to an aviator far above. The closer the approach, the greater the degree of irregularity. Finally, at distances of the order of ℓ_P, the fluctuations in the typical metric component, $g_{\mu\nu}$, become of the same order as the $g_{\mu\nu}$ themselves.

Once the metric fluctuations become nonlinear and strongly interacting, the analogy with the surface of the ocean suggests that the metric fluctuations might "break" in a manner analogous to the breaking of ocean waves. This might be expected to endow spacetime with a "foamlike" structure. It is this suggestion that Wheeler picturesquely describes by the sentence [287, p. 264]:

> [S]pace "resonates" between one foamlike structure and another.

More succinctly, one often encounters the phrase "spacetime foam" which refers to Wheeler's suggestion that the geometry (and topology) of space might be constantly fluctuating.

Warning: There is still considerable disagreement on these issues among various practitioners in the field. For example, recent work by Redmount and Suen [222, 223] calls into question the very basis of the notion of spacetime foam.

6.3 Two "quick and dirty" arguments

Here are two "quick and dirty" arguments, avoiding the subtleties of quantum field theory, that also indicate the significance of the Planck scale.

Consider a quantum particle that is confined to a region of spatial extent L. Then Heisenberg's uncertainty principle implies

$$\Delta p\, L \sim \hbar. \tag{6.15}$$

On the other hand, special relativistic effects imply

$$\Delta E \sim \sqrt{m_0^2 c^4 + (\Delta p\, c)^2} \sim c\Delta p. \qquad (6.16)$$

The Newtonian gravitational potential ϕ near the edge of the confinement region is then of order

$$\phi \sim \frac{Gm_{\text{tot}}}{L} \sim \frac{G(\Delta E/c^2)}{L} \sim c^2 \left(\frac{G\hbar}{c^3 L^2}\right) \sim c^2 \left(\frac{\ell_P}{L}\right)^2. \qquad (6.17)$$

Reexpressing this in terms of the approximate change in the spacetime metric, one has

$$\Delta g_{\mu\nu} \sim \left(\frac{\ell_P}{L}\right)^2. \qquad (6.18)$$

Note: the quantum field theory estimates considered previously produced $\Delta g \sim (\ell_P/L)$, while the present estimate is $\Delta g \sim (\ell_P/L)^2$. These are different physical systems, so a difference in the exponents is not unexpected. The difference is that in one case we are looking at correlations in quantum fluctuations, while in the other case we are looking at the gravitational effects of confining a physical particle to a small region of space. Wheeler has phrased this in terms of the difference between dynamic and static configurations, respectively [285, p. 606]. One could also view this as the difference between virtual gravitons versus real physical particles. Either style of argument implies $\Delta g \sim 1$ at the Planck scale.

The second "quick and dirty" argument proceeds as follows. To any particlelike object of mass M it is possible to assign two fundamental length scales: the Schwarzschild radius and the Compton wavelength.

$$r_s = \frac{2GM}{c^2}; \qquad r_c = \frac{\hbar}{Mc}. \qquad (6.19)$$

For all elementary particles, whose physics is dominated by quantum processes, $r_s \ll r_c$. For systems whose physics dominated by classical gravitation $r_s \gg r_c$. Systems with $r_s \sim r_c$ (which implies $M \sim m_P$) seem, loosely, to be in the transition region between the quantum and the classical regimes. (See, for example, [208, pp. 348–373].)

All these arguments, of varying levels of rigor, lead to agreement on at least one issue—quantum fluctuations of the gravitational field are expected to be important and nontrivial once $L \lesssim \ell_P$ or $E \gtrsim m_P$.

6.4 Topology change?

A critical aspect of the spacetime foam picture has been glossed over in the preceding discussion. What happens when the metric fluctuations become large? While there is no reasonable disagreement that the *geometry* of spacetime undergoes quantum fluctuations, the question of whether or not the *topology* of spacetime undergoes quantum fluctuations is considerably more subtle: Does the topology of spacetime fluctuate? For that matter, what does it mean for the topology of spacetime to fluctuate?

Wheeler has argued forcefully in favor of topology change. For instance, in describing the large metric fluctuations expected at and below the Planck length Wheeler asserts [287, p. 263]:

> Values of Δg comparable to unity and larger indicate changes in geometry so drastic that the word "curved space" is hardly adequate to describe them, "changes in topology" seems a more reasonable description.

The logical structure of this argument is less than watertight. (Wheeler's comment is a plausibility argument. It is not, and is not intended to be, a formal logical deduction. Keep this point in mind.)

Furthermore, arguing from the Feynman functional integral formalism, Wheeler states [284, p. 535]:

> Because it is the essence of quantum mechanics that *all* field histories contribute to the probability amplitude, the sum ... not only may contain doubly and multiply connected metrics; it must do so.

There is some room for disagreement on this point.

Wheeler's position is summarized by the ringing conclusion [284, p. 535]:

> General relativity, quantized, leaves no escape from topological complexities ...

Many of the researchers studying quantum gravity have uncritically accepted the notion that the "spacetime foam" picture automatically leads to topology-changing quantum amplitudes, and further, to interference effects between spacetimes of different topologies. I feel that a certain amount of caution and circumspection is in order. Warning: There are two very separate issues here. They are often confused.

1. In a fixed spacetime, does the topology of space change as a function of time?

2. Do different spacetime topologies interfere quantum mechanically?

Even if one lets the topology of space depend on time it is a much more radical proposal to assert the second possibility. It is logically possible to accept possibility 1 without asserting the reality of possibility 2. Papers arguing against the topology of space changing as a function of time include [67, 5, 6, 265, 266]. With regard to the second possibility, DeWitt has argued [66, pp. 394–395]:

> [T]he notion that *alternative topologies* for spacetime may be alternative dynamical possibilities for a *given* universe makes no sense. ... There exists no concept analogous to homotopy which would allow one ... to follow a transition from one topology to another, or to superpose "amplitudes" for different topologies. I therefore view all attempts to do just this as meaningless.

This topic is the first place where we shall encounter serious disagreement between various practitioners in the field. The disagreements span all aspects of the issue, from basic definitions to highly technical aspects. I shall discuss a few issues, temporarily stepping outside the historical perspective.

It is surprisingly difficult to even get agreement as to what sort of process would constitute a true change in topology. For the time being, limit discussion to the classical situation. Classically, spacetime is modeled by a (single) Lorentzian manifold. This manifold, whatever it is, is fixed, and *spacetime* (and the topology of spacetime) cannot be said to change in any meaningful way. Working within a single Lorentzian manifold, if one has some natural definition of time, one can use this time function to slice the spacetime up into a set of hypersurfaces which might be thought of as "space". One can certainly ask whether or not the topology of these three-dimensional slices of space change as a function of "time". If one adopts such an approach there are many technical questions to be settled. (See, for instance, Harris [125].) What class of time functions is acceptable? Does the topology of space change in any intrinsic way, or are apparent changes in the topology of space merely reflections of the rather arbitrary nature of setting up the time function? If one views the existence of the Lorentzian metric as paramount, then the very existence of that Lorentzian metric imposes topological restrictions on the spacetime. Considerations of this nature have lead to a number of interesting theorems about topological constraints on the *classical* evolution of general relativistic spacetimes. Very roughly speaking:

1. In causally well-behaved classical spacetimes the topology of space does not change as a function of time.

2. In causally ill-behaved classical spacetimes the topology of space can sometimes change.

6.5   Topology change: Classical theorems

One set of classical theorems proceeds from the assumed existence of an everywhere Lorentzian metric to derive as many restrictions as possible. I quote some of these theorems without proof. (Technical details may be found in the cited literature. Note that because we understand classical general relativity well enough to have it formalized as a particular mathematical structure we have the ability to prove theorems based on this mathematical structure. This is a luxury that is sorely missed in the quantum realm.)

I shall begin setting the stage with an abstract mathematical definition:

Definition 5 *A manifold M admits a Lorentzian metric if and only if it admits a smooth nondegenerate rank 2 tensor field whose signature is everywhere $(-, +, +, +)$.*

This simply formalizes the notion of a Lorentzian manifold. For the time being, all manifolds will be assumed to be Hausdorff. (Towards the end of this monograph I shall have a few words to say about non-Hausdorff manifolds.)

Theorem 1 *A manifold M admits a Lorentzian metric if and only if it is (a) paracompact and (b) its double-cover $D(M)$ admits an everywhere nonvanishing continuous vector field.*
(Geroch, [105, 106], Geroch and Horowitz, [107])

The existence of this nonvanishing continuous vector field on the double-cover implies certain topological restrictions on the existence of Lorentzian manifolds. (For instance, the double-cover is constrained to have Euler characteristic zero.)

Since the above existence theorem is so basic, I will give an alternative formulation.

Definition 6 *A direction field $d(x)$ on a manifold M assigns to each point x a pair of equal but opposite vectors in the tangent space $T_x(M)$. That is, $x \to d(x) = \{V(x), -V(x)\}$.*

Given a continuous vector field $V(x)$, the construction of a continuous direction field is trivial. On the other hand, given a continuous direction field $d(x)$, an attempt at constructing an underlying vector field $V(x)$ need not yield a continuous result. While one can always construct a vector field $V(x)$ this vector field may have sign-reversal discontinuities. Another way of saying this is that a direction field is a vector field that is continuous only up to overall sign-reversal. Using this definition, the existence theorem for Lorentzian metrics can be recast as

Theorem 2 *A manifold $\mathcal{M}$ admits a Lorentzian metric if and only if (a) it is paracompact and (b) it admits an everywhere nonvanishing continuous direction field.*
(Geroch, [105, 106], Geroch and Horowitz, [107])

Exercise: Prove either version of the existence theorem. Hints: Use paracompactness to deduce the existence of a Euclidean signature metric g_E (there will be many such Euclidean metrics). Now consider the object

$$(g_L)^{\mu\nu} \equiv (g_E)^{\mu\nu} - 2\,\frac{d^\mu d^\nu}{(g_E)_{\alpha\beta}\,d^\alpha d^\beta}. \tag{6.20}$$

Show that g_L is a Lorentzian metric. It is of course in no sense unique. This construction will provide a Lorentzian metric for any choice of Euclidean metric and any choice of direction field. Note that with this choice of construction for g_L the direction field is timelike.

Unless further hypotheses are added, the constraints imposed by the existence of a Lorentzian metric are rather mild. For instance:

Theorem 3 *Given any two compact three-dimensional manifolds (Σ_1 and Σ_2), there exists a compact Lorentzian geometry that interpolates from one three-dimensional manifold to the other and such that Σ_1 and Σ_2 are both spacelike.*
(Geroch, [105, 106], Sorkin [243], Horowitz [143], Borde [24])

This is often rephrased as the more technical statement

Theorem 4 *Any two compact three-manifolds are Lorentz cobordant.*

One of the standard subsidiary hypotheses often invoked is

Definition 7 *A Lorentzian spacetime is time-orientable if and only if there exists a globally defined everywhere nonzero continuous timelike vector field.*

This simply means that in time-orientable spacetimes it is possible to divide the light cones into past and future components—and to do so in a globally consistent manner.

Once one imposes some causality constraints on the spacetime, one can begin to deduce interesting constraints on the topology. Indeed:

Theorem 5 *If a compact Lorentzian spacetime $\mathcal{M}$ interpolates between two compact spacelike surfaces (Σ_1 and Σ_2), and if $\mathcal{M}$ is time-orientable and contains no closed timelike curves, then Σ_1 and Σ_2 are diffeomorphic and further $\mathcal{M}$ is topologically $\Sigma_1 \times [0,1]$.*
(Geroch, [105, 106])

Note that to obtain this theorem the interpolating spacetime is assumed to be *compact*. This is a rather strong topological assumption. Physically, this asserts that there are no "holes" or "rips" in spacetime that allow information to "come in from infinity" [129, 130]. Recently, Hawking has proved the stronger result:

Theorem 6 *In a Lorentzian spacetime, if there is a timelike tube connecting spacelike surfaces of different topology, then the interior of the timelike tube contains closed timelike curves.*
(Hawking, [129, 130])

Roughly speaking: topology change in a bounded region of spacetime entails causality violations. By assuming nice causality properties, even stronger statements can be made:

Theorem 7 *A manifold M admits a time-orientable Lorentzian metric if and only if (a) it is paracompact and (b) it admits an everywhere nonvanishing vector field.*
(Geroch, [105, 106], Geroch and Horowitz, [107])

Given the first quoted theorem on the existence of Lorentzian metrics, and the definition of time-orientability, this result is automatic. Another useful causality condition is

Definition 8 *A Lorentzian spacetime is stably causal if and only if there exists a globally defined scalar function $\tau(x)$ such that $\nabla \tau$ is everywhere timelike.*
(Hawking and Ellis, [132, p. 198])

Theorem 8 *Any stably causal Lorentzian spacetime can be globally foliated by spacelike hypersurfaces. If all of these spacelike hypersurfaces are compact, then they are all diffeomorphic and spacetime has the topology $M \sim \Re \times \Sigma$. That is, $M \sim$ "time" $\times$ "space".*
(Hawking and Ellis, [132, p. 201])

A minor variant on the Hawking topology change theorem is

Theorem 9 *In any stably causal Lorentzian spacetime, if there is a timelike tube connecting spacelike hypersurfaces Σ_1 and Σ_2, then these hypersurfaces must have the same topology.*

Roughly speaking: Causality forbids "localized" changes in topology. The strongest causality condition in common use is

Definition 9 *A Lorentzian spacetime is globally hyperbolic if and only if the scalar wave equation for a point source has a unique solution which vanishes outside the forward light cone.*
(Hawking and Ellis, [132, p. 206])

There are many other equivalent definitions of global hyperbolicity (for instance, the existence of a global Cauchy surface).

Theorem 10 *Any globally hyperbolic Lorentzian spacetime can be globally foliated by spacelike hypersurfaces. All of these spacelike hypersurfaces are diffeomorphic and spacetime has the topology $M \sim \Re \times \Sigma$. That is, $M \sim$ "time" $\times$ "space".*
(Hawking and Ellis, [132, p. 212])

One possible way around these theorems is by abandoning the notion that spacetime is everywhere Lorentzian. The idea is that spacetime might be "almost-everywhere" Lorentzian, with the nonexistence or degeneracy of the Lorentzian metric being (presumably) confined to a region of measure zero. For instance, one can rephrase the Hawking topology change theorem given above as

Theorem 11 *In a manifold with a somewhere degenerate Lorentzian metric, if there is a timelike tube T connecting spacelike hypersurfaces (Σ_1, Σ_2) of different topology and the interior of the tube $[\text{int}(T \cup \Sigma_1 \cup \Sigma_2)]$ does not contain any closed timelike curves, then the metric must be degenerate somewhere in the interior of the tube.*

This in essence implies abandoning the strong equivalence principle [126, 143, 243].[1] (Since there are now points in spacetime where the manifold is *not* locally isomorphic to Minkowski space.) This option makes me personally more than a little queasy. The equivalence principle is the part of classical general relativity that is most amenable to experimental test. The weak equivalence principle (the uniqueness of free fall) has been tested to fantastic accuracy by Eötvös type experiments. (See Eötvös, Pektar, and Fekete [78], Roll, Krotlov, and Dicke [224], Braginskii and Panov [27], and Adelberger and co-workers [3, 1, 2, 242].) It is not clear to me (or to anyone else for that matter) whether or not the violations of the strong equivalence principle attendant on adopting an "almost-everywhere Lorentzian" model of spacetime would have measurable experimental consequences.

Research Problem 1 *One might have reason to hope that the experimental consequences of these "almost-everywhere Lorentzian" metrics might be suppressed by suitable powers of the Planck mass. Prove it! Better yet, disprove it!*

---

[1]Definitions: Strong equivalence principle — Spacetime is an everywhere Lorentzian manifold. Freely falling particles follow geodesics of the metric. This implies that local physics (small laboratories) can always be analyzed using special relativity. Weak equivalence principle — All freely falling particles follow the same trajectories independent of their internal composition.

Summary: Working at the classical level, and provided that one does not immediately abandon causality, the case against topology change is rather strong. (On the other hand, Borde has recently provided a nice readable account of what happens when causality constraints are relaxed [24].)

6.6   Topology change: Quantum arguments

At the quantum level, the situation becomes decidedly murkier. In the absence of any generally agreed upon framework for quantum gravity it is impossible to prove general theorems simply because one does not know the general hypotheses. Typically, one argues by analogy, or proves very specific theorems that might be valid only within one very specific interpretation of what quantum gravity might be.

6.6.1   Canonical quantization

The canonical quantization of gravity is a subject fraught with conceptual and calculational difficulties. A key paper is that by DeWitt [61]. Crudely speaking, one slices spacetime up into space and time, and adopts the ADM decomposition of the metric. After suitable gauge fixing, canonical quantization of the physical degrees of freedom is attempted.

Where, in this framework, is there place for topology change?

To set up the canonical framework in the first place, one has had to *assume* a global foliation of spacetime by spacelike hypersurfaces. (This is true in both the Heisenberg and Schrödinger pictures. It is easier to see what is going on in the Heisenberg picture.) Thus one assumes stable causality from the outset. (For some technical discussions of spacetime foliations, see [13, 104, 181].) Now in canonical quantization, the configuration space of the quantum theory is the same as the configuration space of the classical theory. Thus those classical topology change theorems that do not make explicit use of the equations of motion (the Einstein field equations) will carry over to canonically quantized gravity. The Hawking topology change theorem is thus enough to show that the topology of space cannot change in canonically quantized gravity.

Theorem 12 *Canonical quantization of the gravitational field is incompatible with topology change.*

To avoid this conclusion, one will have to do violence to classical general relativity. Since one needs both a global time function and a Hamiltonian in order to set up the canonical quantization procedure, the place to mutilate the theory seems to be in the assumption of the existence of an everywhere Lorentzian metric. The objections of the previous section then come into play.

6.6.2 Functional integral quantization

In the Feynman functional integral quantization of gravitation one formally writes

$$\langle \mathcal{X} \rangle \propto \int \mathcal{D}[g(x)] \, \mathcal{X} \, \exp\{-iS[g]/\hbar\}. \qquad (6.21)$$

The integral is to run over "all" metrics $g(x)$, while $S[g]$ is the classical action.

The critical issue, of course, is just what should be included in the set of "all" metrics. All globally hyperbolic Lorentzian metrics? All stably causal Lorentzian metrics? Or all Lorentzian metrics without further qualifications? Should the metrics integrated over be smooth or at least nondegenerate? Should one fix the manifold structure once and for all and only integrate over metrics keeping the manifold structure fixed, or should one include a sum over the various different topological classes of Lorentzian manifolds?

The problem becomes more acute if one "Wick rotates" to Euclidean signature. Now for quantum field theory on flat spacetime Wick rotation is in no sense a generalization of the theory—merely a reformulation. Causality properties of the Lorentzian signature quantum field theory (as embodied in Feynman's $i\epsilon$ prescription) underlie and justify the Wick rotation procedure. The resulting Euclidean quantum field theory inherits the property of Osterwalder–Schrader positivity (OS positivity) from the causal structure of the underlying Lorentz invariant quantum field theory [115].

When attempting to "Wick rotate" gravity this entire structure is absent. One loses both the $i\epsilon$ prescription in Lorentzian signature and OS positivity in Euclidean signature. One formally writes

$$\langle \mathcal{X} \rangle \propto \int \mathcal{D}[g_E(x)] \, \mathcal{X} \, \exp\{-S[g_E]/\hbar\}. \qquad (6.22)$$

On the one hand, if one takes the manifold structure (and in particular the spacetime topology) as fixed, this is no great generalization. In this case the Euclidean metrics one integrates over must all, by construction, be compatible with the topological constraint of the existence of a Lorentzian metric on the manifold.

On the other hand, if one views quantization as including a sum over topologically inequivalent manifolds, should one now include only those manifolds compatible with the existence of a Lorentzian metric? Or should one sum over all topologically inequivalent manifolds without restriction? (See, for instance, [128].)

Depending on the answers one chooses for the preceding questions one is led to at least ten different candidate models for what it means to quantize gravity via functional integrals.

Exercise: Remaining within the functional integral formalism, explicitly list as many different candidate models for quantum gravity as you can think of. Can you kill any of these candidate models by appeal to experiment?

6.6.3   Quantum tunneling

It is often claimed[2] that quantum gravity induces transitions in the topology of space in the same way that quantum mechanics induces tunneling between classically disjoint regions of phase space. This argument is grossly misleading.

Consider some nonrelativistic system whose classical configuration space is given by some n-dimensional manifold $\mathcal{M}$. The n coordinates of points on the manifold are just the generalized coordinates of classical Lagrangian mechanics. To set up a Lagrangian for the system, the configuration space must be given a Riemannian metric. Then

$$L(q, \dot{q}) = \frac{1}{2} g_{ij} \, \dot{q}^i \dot{q}^j - V(q). \qquad (6.23)$$

The classical configuration space is by definition a connected n-dimensional Riemannian manifold.

The classical phase space is now the cotangent bundle $T^*(\mathcal{M})$, and the classical Hamiltonian is

$$H(p_i, q^j) = \frac{1}{2} g^{ij} \, p_i p_j + V(q). \qquad (6.24)$$

The constant energy hypersurfaces $\Sigma(E)$ are $(2n - 1)$-dimensional submanifolds of the phase space $T^*(\mathcal{M})$. (Technical point: Actually, they are in general varieties rather than manifolds.) Though the phase space $T^*(\mathcal{M})$ is by definition connected, the constant energy hypersurfaces $\Sigma(E)$ may be disconnected. The energetically accessible region of configuration space may be defined by projection

$$\mathcal{A}(E) \equiv \pi[\Sigma(E)]. \qquad (6.25)$$

More prosaically

$$\mathcal{A}(E) \quad \equiv \quad \left\{ q^i \in \mathcal{M} \,\middle|\, \exists p_j : H(p, q) = E \right\}$$
$$= \quad \left\{ q^i \in \mathcal{M} \,\middle|\, V(q) \leq E \right\}. \qquad (6.26)$$

For each energy E the energetically accessible region is in general a disjoint collection of n-dimensional submanifolds of $\mathcal{M}$ with $(n - 1)$-dimensional boundaries.

---

[2]The guilty parties know who they are.

- Classically a system of fixed energy E is constrained to remain within whichever of the disconnected components of $\mathcal{A}(E)$ it initially resides in.

- Quantum mechanically, there is a finite probability for the system to "tunnel" through the classically forbidden region (tunnel under the potential barrier). This quantum tunneling process allows quantum systems to explore energetically inaccessible regions of the classical configuration space. Quantum tunneling does *not* permit a quantum system to leave the classical configuration space, nor does it modify or mutilate of the classical configuration space. The set of kinematically allowed trajectories is not changed by quantization.

The lessons with regard to topology change in quantum gravity are instructive. Topology change in classical general relativity is not merely an energetically forbidden process, rather (assuming suitable causality properties and an everywhere Lorentzian metric) it is a kinematically forbidden process. Topology-changing geometries are (modulo the sort of technical fiddles discussed previously) simply not included in the class of kinematically allowed trajectories of ordinary classical general relativity. Dynamical information (such as the Einstein field equations) is not needed to arrive at this conclusion.

Blandly asserting, without further qualification, that quantum gravity permits and requires topology change via the quantum tunneling mechanism makes as much sense as asserting that quantum mechanics permits and requires the electron in a hydrogen atom to tunnel to negative values of the radius.

To circumvent this argument, one would have to modify classical general relativity, thereby modifying the classical configuration space. More precisely, one has to modify the set of kinematically allowed trajectories. This may be achieved, for instance, by adopting the use of "almost-everywhere Lorentzian" spacetimes, or by abandoning causality (permitting closed causal loops). Alternatively, one can simply start out in Euclidean signature and simply agree not to ask awkward questions.

Summary: The oft repeated claim that quantum tunneling permits and requires topology change is somewhat misleading:

- Quantum tunneling does not magically permit changes in the topology of space.

- Quantum tunneling permits topology change if and only if changes in the topology of space are already kinematically allowed at the classical level.

- Quantum physics does not require topology change. If you want the topology of space to change, you must explicitly put it into your classical model for gravity.

- Topology change in quantum gravity is a choice, not a logical necessity.

Some readers may feel that I am belaboring the obvious—others may feel that my comments regarding topology change are far too strong. Beginning students should be made aware of the fact that there are some nontrivial choices to be made regarding these issues.

6.6.4   Topology change without topology change?

It is amusing and interesting to note that, even if the topology of space is fixed once and for all, even if topology change is utterly and totally forbidden, it is nevertheless possible for geometrical effects to mimic the effects of a topology change [264]. This comes about because topology is a mathematical abstraction. Physical probes are not necessarily sensitive to this mathematical abstraction called topology—physical probes typically couple to geometrical features of spacetime.

For instance, a physical probe of energy E is intrinsically limited in its resolving power by the Heisenberg uncertainty principle. A physical probe of energy E cannot probe structural details at distance scales below

$$R \sim \frac{\hbar c}{E}. \tag{6.27}$$

In particular, if a geometry possesses a thin neck of radius R, then such structure can only be resolved by probes of energy

$$E > E_{\text{resolve}} = \hbar c / R. \tag{6.28}$$

This suggests that it might be useful to introduce the notion of an energy-dependent coarse-grained physicists' topology in contradistinction to the mathematicians' notion of topology.

All of the topology change theorems discussed previously make precise mathematical statements about the mathematicians' topology. Unless one uses dynamical information such as the Einstein field equations, there are no general theorems that can be derived about the physicists' topology. In particular, there is nothing to stop regions of space developing small necks that shrink to "umbilical cords" of sufficiently small size to effectively change the topology for all practical purposes [264]. (See figure 6.2.) This situation is best summarized by the positively Wheeleresque mantra "topology change without topology change".

Example: Consider a two-dimensional space that contains a single worm-hole. The mathematicians' topology is that of the punctured torus $T^2 - 0$. Let the minimum circumference of the wormhole throat be C. This can be defined by looking for minimum length closed geodesics that represent the nontrivial homotopy classes of the topology. From the physicists' point of view the high-energy (short wavelength, $\lambda \ll C$) physicist's topology is identical to the mathematicians' topology. However, at low energies (long wavelength, $\lambda \gg C$) physical probes will fail to penetrate the throat and, for all practical purposes, the topology is effectively just $\Re^2$. This defines the low-energy physicists' topology to be $\Re^2$.

6.6.5 Summary

Classically, topology change in ordinary general relativity is essentially forbidden (given reasonable and mild causality assumptions). Mutilating general relativity by admitting "almost-everywhere Lorentzian" spacetimes circumvents the topology change theorems but does so at a very high price—the strong equivalence principle is abandoned. (One might *hope* that experimental consequences of this equivalence principle violation be confined to high energy.)

Quantum mechanically the possibility of topology change is still a rather dubious proposal. Whether or not topology change is allowed in a particular model for quantum gravity depends on the technical details of the particular model. No blanket statements can be made.

Even if (mathematical) topology change is completely and utterly forbidden the use of a finite resolution physicists' topology can lead to "topology change without topology change".

My own views on this matter (a minority opinion) are that the best bet would be to take a conservative stance: to fix the topology of spacetime once and for all and to functionally integrate only over stably causal (or even globally hyperbolic) metrics. If this prescription proves inadequate I would be willing to contemplate the issue of "almost-everywhere Lorentzian" spacetimes—but I would be much happier if reliable estimates of their effects on equivalence principle experiments could be carried out. I view the promiscuity inherent in the more radical Euclidean signature proposals with considerable alarm.

Disagreement on all of these issues is pandemic to the field.

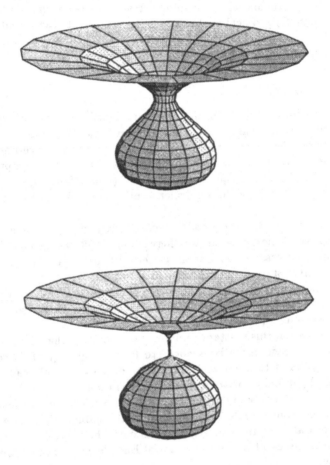

Figure 6.2: Topology change without topology change: Effective topology change. An "umbilical cord" at all times connects the "baby universe" to its parent.

Chapter 7

The Kerr wormhole

7.1 Kerr geometry

The discovery in 1963 of the Kerr [162] solution to the vacuum Einstein field equations led to the discovery of a new type of wormhole [132, 186, 275]. The Kerr metric corresponds to a rotating object possessing both mass M and angular momentum L.

Roughly speaking, if one enforces the vacuum field equations of general relativity, and one takes a point mass and spins it about some axis, then the object cannot remain a point. Instead the mass must redistribute itself as a rotating ring of material. This ring has "radius"

$$a = L/Mc. \tag{7.1}$$

The doughnut hole in the middle of the ring is a doorway to another universe. This grossly oversimplified picture should be taken with several kilograms of salt.

A careful discussion of the global geometry of the maximally extended Kerr spacetime can be found in Hawking and Ellis [132, pp. 161–168]. For simplicity, adopt geometrodynamic units so that both a and M have units of length, that is:

$$M_{\text{geometrodynamic}} = M_{\text{physical}} \, (\ell_P/m_P). \tag{7.2}$$

The global geometry of the maximally extended spacetime depends critically on whether $M > a$, $M = a$, or $M < a$.

7.1.1 Case $a < M$

If $a < M$ the Kerr solution corresponds to a rotationally distorted black hole. Many of the qualitative features of the Schwarzschild black hole survive (such as the existence of an event horizon). Other features (such as the existence of an ergoregion) are completely new.

[Definitions: The ergosphere is the surface where the time translation Killing vector $K^\mu \equiv (\partial/\partial t)^\mu = (1,0,0,0)$ is null, $g_{\mu\nu}K^\mu K^\nu = 0$. The ergoregion is the region between the event horizon and the ergosphere. In the case of zero rotation the ergosphere coincides with the event horizon and the ergoregion shrinks to zero volume.]

In Boyer–Lindquist coordinates (t,r,θ,ϕ) the Kerr metric is:

$$ds^2 = -dt^2 + \frac{2Mr}{\rho^2}(dt - a\sin^2\theta\, d\phi)^2$$
$$+\rho^2\left(\frac{dr^2}{\Delta} + d\theta^2\right) + (r^2 + a^2)\sin^2\theta\, d\phi^2. \qquad (7.3)$$

Here

$$\rho^2(r,\theta) \equiv r^2 + a^2\cos^2\theta, \qquad (7.4)$$
$$\Delta(r) \equiv r^2 - 2Mr + a^2. \qquad (7.5)$$

The inner and outer horizons occur at the two zeros of $\Delta(r)$, that is, at

$$r_\pm \equiv M \pm \sqrt{M^2 - a^2}. \qquad (7.6)$$

These horizons are qualitatively similar to the inner and outer horizons of the Reissner–Nordström geometry. (The outer horizon is an event horizon; the inner horizon is a Cauchy horizon.)

A construction along the lines of the Einstein–Rosen bridge is possible. (To do this, work with the outermost horizon only.) As in the case of the Schwarzschild and Reissner–Nordström solutions such a construction is useless for the purpose of safe transit. Any attempt at crossing the event horizon will lead (in very short order) to the traveler's death. While it might now appear to be possible to avoid falling into the central singularity by jumping through the hoop into the universe next door, such a trip would still be fatal as the intrepid traveler is cooked by the blueshift instability associated with the inner horizon. Loosely speaking, infalling radiation is given an infinite blueshift as it accumulates at the inner event horizon. Since details are not germane to the main thrust of this monograph I shall not explore these issues in more detail. See, for instance, standard textbook discussions in [132, 186, 275], or more recent work in [153, 212, 180].

7.1.2   Case $a = M$

This marginal case describes an extremal black hole. It is not of immediate concern.

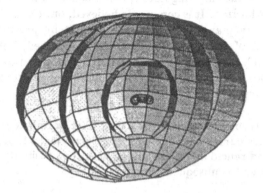

Figure 7.1: Cut away diagram of the Kerr black hole for $a < M$. Proceeding from the exterior, one encounters the ergosphere, event horizon (outer horizon), Cauchy horizon (inner horizon), and finally the ring singularity.

7.1.3 Case $a > M$

For $a > M$ the event horizon vanishes, and one is left dealing with a naked singularity. The maximal extension of the geometry is now pleasingly simple. It consists of two asymptotically flat regions, each of topology $\Re^4$, that are carefully mutilated in the following manner. In each constant t spatial slice, remove the disk $z = 0$, $x^2 + y^2 \le a^2$. Then join the top of the excised disk in universe 1 to the bottom of the excised disk in universe 2, and the top of the excised disk in universe 2 to the bottom of the excised disk in universe 1. The ring singularity is located at $x^2 + y^2 = a^2$, $z = 0$. The metric in Kerr–Schild coordinates is

$$ds^2 = - dt^2 + dx^2 + dy^2 + dz^2$$
$$ + \frac{2Mr}{r^4 + a^2 z^2} \left[dt + \frac{z dz}{r} + \frac{r(x dx + y dy) - a(x dy - y dx)}{r^2 + a^2} \right]^2 , \tag{7.7}$$

where

$$r^2 = \frac{(x^2 + y^2 + z^2 - a^2) \pm \sqrt{(x^2 + y^2 + z^2 - a^2)^2 + 4a^2 z^2}}{2} . \tag{7.8}$$

This represents a mass M, with angular momentum $L = Ma$ pointing along the z axis.

Unfortunately, the ring singularity, being naked, is not decently hidden behind an event horizon. It is commonly believed, but certainly not proved, that under certain plausible conditions it might be possible to prove a "cosmic censorship" theorem. It is commonly believed, but not proved, that naked singularities cannot be physically constructed from reasonable initial conditions. Suffice it to say that naked singularities make most physicists nervous.

Another disturbing feature of the naked Kerr wormhole is the existence of closed timelike curves near the ring singularity [31, 54, 53, 40, 41]. The quite radical and rancid issues associated with the possibility of time travel are to be deferred to subsequent chapters.

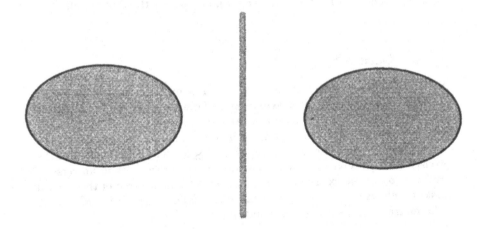

Figure 7.2: The Kerr wormhole. Maximal extension of the Kerr spacetime for $a > M$. The ring singularity is now naked. There are two asymptotically flat regions. The disk spanned by the ring singularity is to be removed from each asymptotically flat region, and top and bottom faces of this disk are to be identified between the two asymptotically flat regions. The disk can be thought of as the gateway from one asymptotically flat region to another.

7.2 Kerr–Newman geometry

The Kerr–Newman geometry possesses mass M, electric charge Q, and angular momentum $L = Ma$. For charge zero it reduces to the Kerr geometry, while for zero angular momentum it reduces to the Reissner–Nordström geometry. The geometry is qualitatively similar to the Kerr geometry. The dividing line between black hole and naked singularity is

$$M = \sqrt{a^2 + Q^2}. \tag{7.9}$$

Above this mass a naked wormhole forms. It exhibits all the diseases of the naked Kerr wormhole and need concern us no further.

7.3 Neutrinos—a note

The current limit on the electron neutrino mass is $m_\nu c^2 \leq 7.3$ eV. (See, for example, [201].) The neutrino's charge is zero, and its intrinsic spin angular momentum is $L = \frac{1}{2}\hbar$. If one insists on trying to model the gravitational field of a single neutrino by the Kerr geometry, then the radius of the ring singularity is

$$a \equiv \frac{L}{m_\nu c} = \frac{\hbar}{2m_\nu c} \geq 1.5 \times 10^{-8} \text{ m} = 150 \text{ Å}. \tag{7.10}$$

Now there are quite a few things that the particle physics community is very sure about. One of them is that neutrinos are not large objects 300 (or more) Å across—neutrinos are certainly not gateways to another universe 300 (or more) Å wide. All experiments to date show neutrinos to be pointlike objects at least to energy scales of order several TeV, that is, distance scales of order 10^{-20} metres. This is clear and conclusive evidence that one cannot blindly apply classical general relativity to elementary particle physics.

Lesson: Do not believe the Kerr geometry for $M \ll m_P$. More generally, for an elementary particle of spin s,

$$L = \sqrt{s(s+1)}\hbar. \tag{7.11}$$

A naked singularity forms once

$$M < \sqrt{L} = \sqrt[4]{s(s+1)}m_P. \tag{7.12}$$

One should probably not believe the Kerr solution for elementary particles lighter than

$$M < \sqrt[4]{s(s+1)}m_P. \tag{7.13}$$

Even for spin zero, where the argument in terms of a putative naked singularity is vitiated, one should probably not trust the Schwarzschild solution for objects lighter than a Planck mass.

The implications of these comments with respect to the motion of elementary particles in an external gravitational field is far from clear. For a "test mass" of bulk matter $m \gg m_P$ in an external gravitational field generated by an object of mass $M \gg m$, the Einstein–Hoffman–Infeld analysis [76] shows that Einstein field equations by themselves imply that the test mass follows a geodesic of the spacetime geometry determined by the mass M. (See the discussion by Wald [275, p. 74]. A more traditional point of view is presented by Synge [247, pp. 246–252].) For test masses $m \ll m_P$, the fact that naive application of the Einstein field equations leads to unphysical results indicates that this analysis seems to break down. While ordinary classical general relativity certainly predicts that elementary particles follow spacetime geodesics, we have lost one of the nice features of general relativity: the derivation of geodesic motion from the field equations.

(Technical point: I am suppressing subtleties due to the spin–Riemann interaction as they are not germane to the point I am trying to make.)

Whether or not individual elementary particles follow geodesics of the spacetime manifold is an experimental question. There are several arguments based on experimental tests of the universality of free fall (the weak equivalence principle) that strongly suggest this. Direct experiments amount to isolating an individual elementary particle and dropping it. Direct experiments are difficult and currently inconclusive. (See, for example, [179, 2].)

Chapter 8

The cosmological constant

Einstein introduced his cosmological constant Λ in an attempt to generalize his original field equations. The modified field equations are

$$G_{\mu\nu} = 8\pi G\, T_{\mu\nu} - \Lambda g_{\mu\nu}. \qquad (8.1)$$

Note that Λ has dimensions $[L^{-2}]$. By redefining

$$T_{\text{total}}^{\mu\nu} \equiv T_{\text{ordinary}}^{\mu\nu} - \frac{\Lambda}{8\pi G} g^{\mu\nu}, \qquad (8.2)$$

one can regain the original form of the field equations, $G_{\mu\nu} = 8\pi G\, T_{\mu\nu}$, at the cost of introducing a vacuum energy density and vacuum stress-energy tensor

$$\rho_\Lambda = +\frac{\Lambda}{8\pi G}; \qquad T_\Lambda^{\mu\nu} = -\rho_\Lambda\, g^{\mu\nu}. \qquad (8.3)$$

Note that ρ_Λ has the appropriate dimensions, $[M][L^{-3}]$. In this monograph I shall always interpret the cosmological constant, if present, as part of the total stress-energy.

One of the truly remarkable features of attempts to unify quantum concepts with gravity is that it is possible to make a superficially reasonable estimate of the size of the cosmological constant from first principles. The most remarkable feature of this estimate is that it is utterly, spectacularly, and grossly wrong. *Ab initio* estimates of the cosmological constant seem to imply that the universe should be about one Planck length in diameter. This is in violent disagreement with the observational evidence.

8.1   Zero point energy

To see how the *ab initio* estimate is made, start by considering the quantum simple harmonic oscillator. As every undergraduate knows, the energy eigenvalues are

$$E_n = \hbar\omega \left(n + \frac{1}{2} \right). \tag{8.4}$$

The relevant point to focus on is the zero point energy (ZPE)

$$E_0 = \frac{1}{2}\hbar\omega. \tag{8.5}$$

It is to be emphasised that the zero point energy is real and physical, and not just an artifact of the mathematics. For instance, allowing for zero point energy contributions is an essential part of molecular dynamics calculations. The zero point energy of the quantized electromagnetic field is also responsible for the Lamb shift. See, for example, [185, pp. 82–86]. (That book also discusses many other physical applications of the zero point energy.) For a complicated collection of harmonic oscillators one simply adds over all the oscillators:

$$E_{\text{ZPE}} = \frac{1}{2}\hbar \sum_j \omega_j. \tag{8.6}$$

In particular, consider a quantum field theory describing an elementary particle of mass m residing in flat Minkowski space. Put everything in a box of volume V with periodic boundary conditions. This quantum field theory is equivalent to a collection of harmonic oscillators. In terms of the three-dimensional wave vector k, the frequencies of these oscillators are given by

$$\omega_j = \sqrt{m^2 + k_j^2}. \tag{8.7}$$

(Technical point: I have suppressed factors of c and the m appearing above actually denotes $mc^2/\hbar$, but there is little point to keeping track of these quibbles.)

By going to the k-space continuum limit (this should be familiar from elementary solid state considerations) one has

$$E_{\text{ZPE}} = \frac{1}{2}\hbar \int_0^\infty \sqrt{m^2 + k^2} \, \frac{d^3k}{(2\pi)^3} \, V. \tag{8.8}$$

So the energy density in zero point fluctuations is

$$\rho_{\text{ZPE}} = \frac{1}{2}\hbar \int_0^\infty \sqrt{m^2 + k^2} \, \frac{d^3k}{(2\pi)^3}. \tag{8.9}$$

This formal result is perhaps a little abstract. (In fact, the integral diverges and the naive zero point energy is infinite, more on this point later.) While experimentalists have not been able to measure this zero point energy directly, differences in the zero point energy density are certainly real and measurable. Such differences in zero point energy density underlie the Casimir effect, a topic we will have occasion to further pursue.

Psycho-ceramics warning: There are more than a few dedicated students of psycho-ceramics (crackpots) who pursue the formalism of the zero point energy to the exclusion of empirical reality. An awful lot of gibbering nonsense has been written about the zero point energy. Proceed at your own risk. Periodic sanity checks are essential.

The zero point fluctuations also give rise to a zero point pressure. Rather than attempting to calculate this pressure *ab initio*, one can rely on the symmetry of flat Minkowski space to deduce that

$$T^{\mu\nu}_{\text{ZPE}} = -\rho_{\text{ZPE}} \, g^{\mu\nu} = -\frac{1}{2}\hbar \int_0^\infty \sqrt{m^2 + k^2} \, \frac{d^3k}{(2\pi)^3} \, g^{\mu\nu}. \qquad (8.10)$$

Thus the effect of the zero point fluctuations is equivalent to inducing a cosmological constant

$$\Lambda_{\text{ZPE}} \equiv 8\pi G \rho_{\text{ZPE}} = +4\pi G\hbar \int_0^\infty \sum_p g_p \sqrt{m_p^2 + k^2} \, \frac{d^3k}{(2\pi)^3}. \qquad (8.11)$$

The sum occurring here runs over all particle species in the cosmos, suitably weighted by a degeneracy factor, g_p, that depends on spin and mass. (In particular, g_p is positive for bosons, but negative for fermions.) This induced cosmological constant is of course infinite. If we were doing solid state physics we would know how to proceed—the momentum integral should run only over the first Brillouin zone and should effectively be cut off by the lattice spacing. In gravity physics no such lattice-based cutoff is available. Instead one has to argue that the integral should be cut off by whatever new physics comes in above the Planck scale ($k_{\text{max}} \approx 1/\ell_P$). If this is the case we expect

$$\rho_{\text{ZPE}} = \lambda_{\text{ZPE}} \frac{m_P}{\ell_P^3}; \qquad \Lambda_{\text{ZPE}} = \lambda_{\text{ZPE}} \frac{8\pi G m_P}{\ell_P^3} = \lambda_{\text{ZPE}} \frac{8\pi}{\ell_P^2}. \qquad (8.12)$$

Here λ_{ZPE} is a dimensionless number that one expects to be of order unity.

Exercise 1: For an elementary physical example of this phenomenon that does not exhibit any infinities, go to solid state physics and consider a crystal. The sound waves in a crystal are quantized as phonons, in exactly the same way as the electromagnetic waves in QED are quantized as photons. Calculate the energy density on the zero point fluctuations of the phonon field. Show that for a crystal with lattice spacing a one has

$$\rho_{\text{ZPE}}(\text{phonons}) = \lambda_{\text{ZPE}}(\text{phonons}) \, \frac{\hbar \, v_s}{a^4}. \qquad (8.13)$$

Here v_s is the speed of sound in the crystal, while $\lambda_{\text{ZPE}}(\text{phonons})$ is a dimensionless number that is now calculable and can be shown to be of order unity. By writing down the dispersion relation for phonons, and integrating only over the first Brillouin zone, calculate $\lambda_{\text{ZPE}}(\text{phonons})$ for the various standard Bravais lattices. Similarly, evaluate the contributions of the zero point fluctuations in the phonon field to the pressure. Show that

$$p_{\text{ZPE}}(\text{phonons}) \propto \rho_{\text{ZPE}}(\text{phonons}). \qquad (8.14)$$

The proportionality constant is again of order unity, and depends on the particular Bravais lattice. Finally, numerically evaluate the zero point energy density and zero point pressure for typical lattice spacings. How significant is this phenomenon compared with other solid state effects?

Exercise 2: Returning to the realm of particle physics, write the zero point energy density in the form

$$\rho_{\text{ZPE}} = \frac{1}{2} \hbar \int_0^\infty \sum_p g_p \, \sqrt{m_p^2 + k^2} \, \frac{d^3 k}{(2\pi)^3}. \qquad (8.15)$$

Now suppose that the various species of elementary particle in the cosmos conspire to enforce

$$\sum_p g_p = 0; \qquad \sum_p g_p \, m_p^2 = 0; \qquad \sum_p g_p \, m_p^4 = 0. \qquad (8.16)$$

Show that in this case the zero point energy density is finite and equal to

$$\rho_{\text{ZPE}} = \frac{1}{(2\pi)^2} \, \hbar \sum_p g_p \, m_p^4 \, \ln(m_p^4/\mu^4). \qquad (8.17)$$

Here μ is an arbitrary mass scale. You should further check that the zero point energy density is independent of μ and explicitly evaluate the weighting factors g_p as functions of spin (and mass). After thinking about this for a while, you may wish to check the similar discussion by Pauli [202, p. 33]. How physically reasonable are these finiteness constraints? [Hint: think supersymmetry (SUSY).]

8.2 Observational limit

To understand the observational limit on the cosmological constant, a crude but useful model for the overall structure of the universe is the Friedmann–Robertson–Walker (FRW) cosmology. The relevant metric is [186, 275]

$$ds^2 = -dt^2 + R(t)^2 \left[\frac{dr^2}{1 - \kappa r^2} + r^2(d\theta^2 + \sin^2\theta\, d\varphi^2) \right]. \tag{8.18}$$

Here $R(t)$ is the time dependent "size of the universe" (more properly— the scale factor), while $\kappa = +1, 0, -1$ depending on whether the constant time spatial slices are closed, flat, or open, respectively. The Einstein field equations, including the cosmological constant, reduce to the two equations

$$\left(\dot{R}/R \right)^2 = \frac{8\pi G\rho + \Lambda}{3} - \frac{\kappa}{R^2}. \tag{8.19}$$

$$\frac{d}{dt}\left(\rho R^3 \right) = -p \frac{d}{dt}\left(R^3 \right). \tag{8.20}$$

Now $\dot{R}/R$ is observationally measurable, albeit imprecisely. It is the Hubble parameter. Estimates for the present-day value[1] of the Hubble parameter are [201, p. III.2] [2]

$$\begin{aligned}
H_0 &\equiv \left(\dot{R}/R \right)_0 \\
&= h_0\, (100\ \text{km s}^{-1}\ \text{Mpc}^{-1}); \qquad h_0 \in (0.4, 1) \\
&= h_0\, (0.97781 \times 10^{10}\ \text{yr})^{-1} \\
&= h_0\, (5.69 \times 10^{60}\ T_P)^{-1}.
\end{aligned} \tag{8.21}$$

Estimates of the present day scale factor are rather crude and amount to

$$R_0 \approx \frac{c}{H_0} = \frac{3\ \text{Gpc}}{h_0} \approx \frac{6 \times 10^{60} \ell_P}{h_0}. \tag{8.22}$$

The critical density is defined as that density which would marginally close the universe ($\kappa = 0$), assuming that the cosmological constant is zero, that is,

$$\begin{aligned}
\rho_c &\equiv \frac{2H_0^2}{8\pi G} \\
&= h_0^2\, (2.775\,366\,273 \times 10^{11}\ M_\odot\ \text{Mpc}^{-3}) \\
&= h_0^2\, (1.878\,82 \times 10^{-26}\ \text{kg m}^{-3}) \\
&= h_0^2\, (3.64290 \times 10^{-123}\ m_P \ell_P^{-3}).
\end{aligned} \tag{8.23}$$

---

[1]The subscript zero denotes present-day values. The Hubble parameter, density, and pressure are of course functions of the age of the universe.

[2]Notation: Mpc, megaparsec; Gpc, gigaparsec. $M_\odot$ is the mass of our sun. T_P is the Planck time.

Measurements of the actual present density are rather crude and are typically quoted as the ratio of actual density to critical density:

$$\rho \equiv \Omega_0 \, \rho_c; \qquad \Omega_0 \in (0.05, 4)$$
$$= \Omega_0 \, h_0^2 \, (3.64290 \times 10^{-123} m_P \ell_P^{-3}). \qquad (8.24)$$

These observational data constrain the cosmological constant. The canonical estimate is

$$|\Lambda| \quad < \quad 3 \times 10^{-52} \text{ m}^{-2} \approx 8 \times 10^{-122} \ell_P^{-2}, \qquad (8.25)$$

$$|\lambda| \quad \equiv \quad \frac{|\Lambda|\ell_P^2}{8\pi} < 3 \times 10^{-123}, \qquad (8.26)$$

$$|\rho_\Lambda| \quad < \quad 3 \times 10^{-123} \, m_P \ell_P^{-3}. \qquad (8.27)$$

Comparing this with the expected $\lambda_{\text{ZPE}} \approx 1$ we see that we are dealing with one of the absolute worst predictions of modern physics. Many ways of dealing with this problem have been advocated. Suggestions range from ad hoc renormalization to zero, through supersymmetry (SUSY), soaking up the cosmological constant in the extra dimensions of a Kaluza–Klein theory, to superstring theory, and Euclidean wormholes. (See the discussion on pp. 92 and 129.) All of these known attempts are diseased at one level or another.

Exercise: (Literature search) Review the various proposed methods of dealing with the cosmological constant problem.

The observational limit on the cosmological constant can be interpreted as an experimental upper bound on the net vacuum energy density. [Think of the net vacuum energy density as the grand total of a bare vacuum energy density (from the bare cosmological constant), plus the zero point energy density, plus higher-order quantum effects, plus renormalization effects.]

Exercise: Using the observational limit on the cosmological constant, calculate an upper bound for ρ_{vacuum}. Express this upper bound in Joules per cubic metre. Compare this with the energy released when one burns a cubic metre of coal. In light of this experimental upper bound on the net vacuum energy density, critically evaluate the possibility of "mining" the vacuum energy density.

There are tantalizing hints from observational cosmology that Λ might actually be nonzero, though exceedingly small. The data hint at

$$\rho_\Lambda \sim \rho_c, \qquad \lambda \sim +10^{-123} \neq 0. \qquad (8.28)$$

Trying to explain why the cosmological constant might be exactly zero is extremely difficult. Trying to explain why the cosmological constant might be nonzero but exceedingly tiny is even more difficult. For a general discussion, see Peebles [205].

If one is lucky, this problem might prove to be pivotal and central to any true understanding of quantum gravity. With luck this problem may prove as critical and seminal as the ultraviolet catastrophe of classical thermodynamics [157]. The ultraviolet catastrophe was finally laid to rest only with the advent of quantum physics. One might hope that the cosmological constant problem is hiding something equally as fundamental. We will not know whether or not this is the case until the problem is actually solved to the satisfaction of the physics community at large.

Chapter 9

Wormhole taxonomy

Having by now seen some of the many varieties of wormholes that have been considered by the physics community it is useful to temporarily step outside the historical perspective and consider a broad overview of the taxonomy of wormholes.

Wormholes come in a variety of phyla and species. The major phyletic division is the distinction between Lorentzian and Euclidean wormholes. This merely reflects whether or not the manifold in which the wormhole resides is a Lorentzian (pseudo–Riemannian) manifold or a true Riemannian manifold (with Euclidean signature metric). Experimentally, real physics seems to take place in Lorentzian signature. The ontological status of Euclidean signature manifolds is considerably more murky. As a working definition of a wormhole I adopt

Definition 10 *A wormhole is any compact region of spacetime with a topologically simple boundary but a topologically nontrivial interior.*

This working definition will soon be seen to require refinement.

9.1 Lorentzian wormholes

Within the phylum of Lorentzian wormholes speciation occurs into "permanent" ("quasipermanent") and "transient" varieties, each of which has "intra-universe" and "inter-universe" subspecies depending on whether or not the wormhole connects distant regions of one universe to itself or connects different universes in the multiverse. Each subspecies arises in "macroscopic" and "microscopic" varieties [264].

9.1.1   Permanent and quasipermanent wormholes

If one slices a region of spacetime into spacelike hypersurfaces, and each slice of space, thought of as a three-dimensional Riemannian manifold, contains a wormhole, then the wormhole can be thought of as existing throughout a certain duration of time. Such a wormhole will be called a quasipermanent wormhole or simply a permanent wormhole for short. (Quasi-)permanent wormholes are essentially three-dimensional objects that exist for a finite nonzero length of time.

The existence of truly permanent wormholes (as opposed to quasipermanent wormholes) does not violate any of the classical topology change theorems. The creation and destruction of quasipermanent wormholes violates the classical topology change theorems and requires either some mutilation of classical general relativity or retreat into one of the more promiscuous schema for introducing quantum gravity (for example, "almost-everywhere Lorentzian" manifolds or causality violations).

A more technical definition is

Definition 11 *If a Lorentzian spacetime contains a compact region Ω, and if the topology of Ω is of the form $\Omega \sim \Re \times \Sigma$, where Σ is a three-manifold of nontrivial topology, whose boundary has topology of the form $\partial\Sigma \sim S^2$, and if furthermore the hypersurfaces Σ are all spacelike, then the region Ω contains a quasipermanent intra-universe wormhole.*

Quasipermanent inter-universe wormholes are more difficult to characterize. One cannot rely purely on topological (as opposed to geometrical) information as the topology is not enough to uniquely characterize an inter-universe connection. For example, consider the spacetime geometry the spatial slices of which are represented schematically in figure 9.1. The topology of this geometry is trivial, $\Omega \sim \Re^4$. Nevertheless one would wish to interpret the "neck" as a quasipermanent inter-universe wormhole connecting the "baby" universe to its parent.

Note that the "active" region of this topologically trivial wormhole has topology of the form

$$\Omega \sim \Re \times \Re \times S^2 \sim (\text{time}) \times (\text{radius}) \times (\text{throat}). \tag{9.1}$$

The boundary of this "active" region is

$$\partial\Omega \sim (\Re \times S^2) \cup (\Re \times S^2) \cup (\Re \times S^2) \cup (\Re \times S^2), \tag{9.2}$$

that is, two copies of (time) × (throat) plus two copies of (radius) × (throat). Unfortunately this is too weak a topological constraint to uniquely specify objects that we would really want to think of as wormholes. We will just have to live with this limitation and realize that the recognition and definition of inter-universe wormholes requires at least some geometrical information.

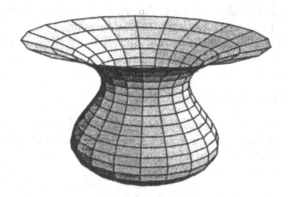

Figure 9.1: Inter-universe wormhole with trivial topology. A "neck" or "throat" connects the "baby" universe to its parent.

9.1.2 Transient wormholes

Transient wormholes are wormholes that pop into and out of existence without (even locally) having a topological structure of the form $\Omega \sim \Re \times \Sigma$. Transient wormholes are intrinsically four-dimensional objects.

Any compact region Ω with boundary $\partial\Omega \sim S^3$ that contains nontrivial topology is certainly an (intra-universe) transient wormhole. By hypothesis the topology of Ω is certainly not of the form $\Re \times \Sigma$.

A transient inter-universe wormhole typically has an "active" region with topology $\Omega \sim \Re \times S^3$. The boundary is then $\partial\Omega \sim S^3 \cup S^3$.

Transient wormholes, because they certainly do not satisfy the classical topology change theorems, require mutilation of classical general relativity to permit classical topology changing processes.

9.1.3 Macroscopic versus microscopic wormholes

Wormholes can be crudely divided into macroscopic and microscopic sub-species depending on the size of the active region relative to the Planck length. (This definition is deliberately a little vague).

9.1.4   Wheeler wormholes

Wheeler wormholes, deriving their existence from the assumed vacuum fluctuations taking place in the spacetime foam, are definitely microscopic in nature. They are typically transient, though by sheer luck might arise with the topology suitable to be considered quasipermanent. The taxonomy of Wheeler wormholes depends critically on one's input assumptions as to what it means to quantize gravity.

9.1.5   Traversable wormholes

We shall give much more detail subsequently, but for now note only that a Lorentzian wormhole that is (quasi-)permanent and macroscopic might prove to be suitable for traversal by human beings.

9.2   Euclidean wormholes

The phylum of Euclidean wormholes is not of immediate interest for the purposes of this monograph. While there is nothing in principle to prevent Euclidean wormholes being quasipermanent and having the topology $\Omega \sim \Re \times \Sigma$, (with $\Re$ an "imaginary time" coordinate and Σ a nontrivial spatial topology), objects of this type are rarely considered.

Having gone to Euclidean signature one either has to restrict the class of manifolds under consideration by assuming compatibility with the existence of a Lorentzian metric or one has to face loss of the various topology change theorems. (There are some topological constraints that survive passage to unrestricted Euclidean signature [113, 110], but these constraints are rather weak.)

Euclidean wormholes are typically "transient". They are commonly thought of as "instantons" in the gravitational field. Many of the specific models considered in the literature in fact have O(4) symmetry. In fact, typically $\mathcal{M} \sim \Re \times S^3$, with $\Re$ a Euclidean radial variable. If one adopts canonical quantization techniques, transient Euclidean wormholes arise only by mutilating the classical phase space of the underlying Lorentzian theory. If one adopts functional integral techniques transient Euclidean wormholes only arise if one takes a rather promiscuous attitude to the class of manifolds contributing to the functional integral. Most practitioners of Euclidean quantum gravity are willing to do exactly this.

Euclidean wormholes have been of considerable interest to the particle physics and relativity communities. They underlie, for instance, Coleman's attempt at explaining the cosmological constant problem [44]. Some feel for the subtleties and ambiguities involved in this proposal may be gleaned from the following comment by Coleman [44, p. 647]:

Although I find this theory to be in many ways very attractive, I must in honesty stress its speculative character. It rests on (Euclidean) wormhole dynamics and the Euclidean formulation of quantum gravity. Thus it is doubly a house built on sand. (Euclidean) wormholes may not exist, or, if they do exist, their effects may be overwhelmed by those of some more exotic configurations. Likewise, the Euclidean formulation of gravity is not a subject with firm foundations and clear rules of procedure; indeed, it is more like a trackless swamp. I think I have threaded my way through it safely, but it is always possible that unknown to myself I am up to my neck in quicksand and sinking fast.

Coleman's ideas have been further explored by a number of authors. Of particular note are the papers: "A wormhole catastrophe" [83], "Escape from the menace of the giant wormholes" [46], and "Return of the giant wormholes" [213]. Additional references include, among many others, [43, 47, 45, 101, 112, 165].

In reference [82] the authors estimate the total number of universes in the multiverse to be greater than $10^{10^{120}}$. (This is the origin of their term googolplexus.) Fortunately, they show that most of these universes are cold, dark, and empty, and so need not concern us. (Clearly, we do not live in one of the cold, dark, empty universes.) This is perhaps the most extreme use of the weak anthropic principle ever invoked in physics [12].

There has been a certain amount of tension within the community over questions of the internal consistency of these ideas. See, for instance, Unruh [256]. That paper presents a nice overview of all of the diseases infesting Euclidean quantum gravity with particular attention to the role of Euclidean wormholes. Quote [256, p. 1053]:

> I ... argue that the presence of (Euclidean) wormhole solutions ... completely destroys the Euclidean quantum theory by producing a highly nonlocal effective Euclidean action which is violently unbounded from below.

I will not explore the issue of Euclidean wormholes in any further depth. For those who wish to pursue the matter further, the quoted references give a good handle on the issues involved.

Extremely bold and determined readers may wish to attack the following little problem:

Research Problem 2 *Estimate the number of baby universes that can dance in a Planck volume. Be prepared to justify your answer to the community at large. Bonus points if you can satisfy both the relativists and the particle physicists.*

Chapter 10

Interregnum

There is a thirty year gap between Wheeler's 1957 work and the 1988 renaissance due to Morris and Thorne. Much was accomplished during this period, but relatively little effort was devoted to Lorentzian wormholes themselves. Considerably more effort was put into such topics as attempting a deeper understanding of the "spacetime foam" picture, and to various (largely unsuccessful) attempts at quantizing gravity. Other topics of interest to the community included the further development of classical cosmology (the standard cosmological model—the big bang), exact solutions of classical general relativity, and development of the standard model of particle physics [SU(3) × SU(2) × U(1)]. More exotic work in particle physics included the "grand unified theories" (GUTs), supersymmetry (SUSY) and supergravity (SUGRA), the Kaluza–Klein theories with their extra spacetime dimensions, and most recently the various superstring theories with their associated "theories of everything" (TOE). During this period Lorentzian wormholes seem to have been considered a curiosity. They were very much relegated to the back burner.

Early books and papers dealing with attempts at quantizing gravity include the Battelle lecture notes of 1967 [73], the DeWitt trilogy [61, 62, 63], and the stunning work of Wyler [292]. Useful surveys may also be gleaned from the Les Houches lectures of 1963 and 1983 [59, 66], and DeWitt's 1975 survey article [64]. Various other more recent sources include [65, 109, 128, 206, 107]. While total confusion reigned (and still reigns) with respect to what it actually means to quantize gravity, good collections of overviews can be found in [134] and [135]. Useful reprint volumes are [133] and [131].

The classical singularity theorems are well summarized in Hawking and Ellis [132]. A more recent elementary text is that of Naber [192]. A number of good modern textbooks covering classical general relativity have also appeared [186, 275, 279]. Certain advanced topics are addressed in [246]. An extensive compilation of exact solutions of the classical field equations is provided in [170], while the cosmological implications of classical general relativity are elucidated in the recent book by Peebles [205].

These references are by no means complete, but should serve as a good starting point for readers interested in further investigation of these issues.

Further afield, the particle physics standard model is exhaustively discussed in reference [74], while the more exotic GUTs are explicated in [231]. Cosmological implications, including inflationary aspects, are expounded upon in [168]. Supersymmetry (SUSY) and supergravity (SUGRA) are covered by a number of books [102, 280, 281]. Higher-dimensional Kaluza–Klein theories are discussed in [9, 291]. (An application to wormhole physics is given in [42].) The recent interest in superstring theory has led to the production of a number of tomes [238, 123, 127].

These issues continue to be of great interest to the communities involved, but will not need to be more deeply addressed in this monograph. For the purposes of this monograph, these issues serve as the backdrop to the wormhole renaissance.

Part III

Renaissance

Chapter 11

Traversable wormholes

11.1 Basic ideas

The varieties of wormhole we have encountered so far have all been diseased in one manner or another, these diseases being sufficiently disastrous so as to preclude even the most optimistic of sentients from seriously considering them as a practical means for interstellar travel.

- Would-be wormholes that contain event horizons (for example, the Einstein–Rosen bridge or nontraversable Schwarzschild wormhole; the Kerr black hole) are bad news for a number of reasons.

  - First, even though the trip down to the event horizon takes a finite amount of time as seen by the traveler, this same trip takes an infinite amount of time as seen by an outside observer watching the traveler ever more slowly approach the event horizon. The person left behind outside the wormhole will be waiting an awfully long time for news from the universe next door.

  - Worse, there are generally nasty and fatal objects hiding behind event horizons. For instance, inner horizons are typically unstable, and attempts at crossing an inner horizon generally lead to a cooked traveler. Curvature singularities will crush you if you hit them. Even if you manage to miss the curvature singularity itself (for instance by hopping through the doughnut hole in the middle of the Kerr black hole) the tidal forces are typically large enough to rip any intrepid explorer to bloody shreds. With just a little bit of mismanagement, tidal forces can be made sufficiently high to disrupt individual nuclei.

- Wheeler wormholes are simply too small. Even if you are willing to grant their existence in the first place, typical sizes are of the order of the Planck length ($\approx 10^{-35}$ m).

- Wormholes based on naked singularities (such as the naked Kerr wormhole) are badly diseased for all of the standard reasons associated with naked singularities themselves. Apart from these issues and related cosmic censorship concerns, tidal forces in the vicinity of the naked singularity are still potentially lethal.

The great 1988 sea change in physicists' willingness to consider wormhole systems came with the realization by Morris and Thorne that it was possible—at least in principle—to cook up suitable pleasingly behaved "traversable" wormhole spacetimes. The word "traversable" is used here to indicate that a human (or alien of similar size and construction) could safely travel through the wormhole in a reasonable amount of time and return to spread the good news. From the above discussion the minimum requirement is that there be no event horizons in the system. (At the very least, event horizons, if present, should be out of the way so that the traveler does not have to cross them.) Avoiding the presence of naked singularities is also a good bet. Combining these two suggested requirements indicates that it is most profitable to consider wormhole spacetimes containing no curvature singularities whatsoever.

Restricting attention to solutions of the Einstein equations that contain no curvature singularities is a difficult task at best. The traditional way of proceeding has been to start by picking one's favorite Lagrangian for the matter fields that one supposes support the wormhole spacetime. One should then calculate the relevant stress-energy tensor and solve the Einstein field equations. Finally, one checks for the presence of curvature singularities in the solution so obtained. All attempts at following this prescription have met with abject failure—the resulting wormholes being diseased in one or more of the manners discussed above.

Morris and Thorne realized that the analysis is radically simplified if one adopts a more engineering oriented approach. Assume the existence of a suitably well-behaved interesting geometry. Then calculate the Riemann tensor associated with this geometry and use the Einstein field equations to deduce what the distribution of stress-energy must be. Finally one asks: Is the deduced distribution of stress-energy physically reasonable? Does the deduced distribution of stress-energy violate any deeply held physical principle?

The answer is that the stress-energy distribution near the throat of a wormhole is certainly peculiar, but that it does not seem to be incompatible with known physics. More technically, we shall soon see that matter near the throat of the wormhole violates the so-called null energy condition (NEC). Traversable wormholes also will be shown to violate the weak, strong, and dominant energy conditions. Precise definitions will be provided later. Very roughly speaking, somewhere near the throat of the wormhole, someone must be able to encounter some negative energy density.

From a classical perspective, violation of the null energy condition is a dubious proposal at best. However, we do know that some quantum effects do lead to (small) measurable experimentally verified violations of the null energy condition. If it were not for the fact that experiment tells us that some quantum effects violate the null energy condition, the work of Morris and Thorne would have been interpreted as the beginning stage of a no-wormhole theorem. See the subsequent discussion of the topological censorship theorem; p. 195ff. See also the discussion of the averaged null energy condition on p. 290ff.

On the other hand, it must be said in all honesty that the observed violations of the null energy condition are small (indeed, miniscule). It is far from clear whether or not one can ever get a large enough violation of the null energy condition to actually support a traversable wormhole.

11.2 Metric

To keep the analysis tractable, Morris and Thorne assumed that their traversable wormholes were time independent, nonrotating, and spherically symmetric bridges between two universes. The manifold of interest is thus a static spherically symmetric spacetime possessing two asymptotically flat regions. Let l denote the proper radial distance. Then without loss of generality the spacetime metric can be put into the form

$$ds^2 = -e^{2\phi(l)}dt^2 + dl^2 + r^2(l)\left[d\theta^2 + \sin^2\theta\,d\varphi^2\right]. \qquad (11.1)$$

- The coordinate l covers the entire range $(-\infty, +\infty)$.

- The assumed absence of event horizons implies that $\phi(l)$ must be everywhere finite.

- The two asymptotically flat regions are assumed to occur at $l \approx \pm\infty$.

- In order for the spatial geometry to tend to an appropriate asymptotically flat limit one must impose

$$\lim_{l\to\pm\infty}\left\{r(l)/|l|\right\} = 1. \qquad (11.2)$$

That is, $r(l) = |l| + O(1)$.

- In order for the spacetime geometry to tend to an appropriate asymptotically flat limit, both the limits

$$\lim_{l\to\pm\infty}\phi(l) = \phi_{\pm} \qquad (11.3)$$

must be finite.

- The radius of the wormhole throat is defined by

$$r_0 = \min\{r(l)\}. \tag{11.4}$$

For simplicity one may assume that there is only one such minimum and that it is an isolated minimum. Generalizing this point is straightforward.

- Without loss of generality, we can take this throat to occur at $l = 0$.

- The metric components should be at least twice differentiable as functions of l.

- These are merely the minimal requirements to obtain a wormhole that is "traversable in principle". For realistic models, "traversable in practice", one should address additional engineering issues such as tidal effects.

- For simplicity one might sometimes wish to assume symmetry under interchange of the two asymptotically flat regions, $l \leftrightarrow -l$, that is, $r(l) = r(-l)$ and $\phi(l) = \phi(-l)$. This requirement is not essential to the definition of a traversable wormhole.

One could perfectly well calculate the Riemann, Ricci, and Einstein tensors using this coordinate system. The results are somewhat unwieldy. (See the exercises.) It turns out to be more efficient to adopt Schwarzschild coordinates, and to reparameterize the functional dependence of the metric, in order to obtain somewhat simpler expressions.

In Schwarzschild (t, r, θ, φ) coordinates

$$ds^2 = -e^{2\phi_\pm(r)}dt^2 + \frac{dr^2}{1 - b_\pm(r)/r} + r^2 \left[d\theta^2 + \sin^2\theta \, d\varphi^2\right]. \tag{11.5}$$

- Two coordinate patches are now required, each one covering the range $[r_0, +\infty)$. Each patch covers one universe, and the two patches join at r_0, the throat of the wormhole.

- For convenience, I shall demand that the t coordinate be continuous across the throat, so that $\phi_+(r_0) = \phi_-(r_0)$.

- The two arbitrary functions $\phi(l)$ and $r(l)$ have been traded in for the four arbitrary functions $\phi_\pm(r)$ and $b_\pm(r)$. This is not an increase in generality since the domains of these functions have been halved.

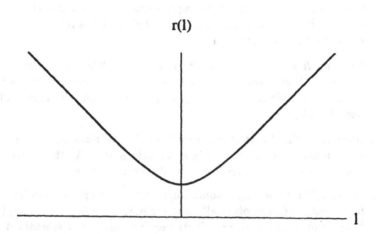

Figure 11.1: Morris–Thorne wormhole: the radius of spherical shells surrounding a traversable wormhole *versus* proper radial distance from the throat. The throat of the wormhole is at proper radial distance zero, and the radius of the wormhole is nonzero at the throat.

- $\phi(r)$ is called the redshift function, while $b(r)$ is called the shape function.

- Proper radial distance is related to the r coordinate by

$$l(r) = \pm \int_{r_0}^{r} \frac{dr'}{\sqrt{1 - b_\pm(r')/r'}}. \tag{11.6}$$

- In order for the spatial geometry to tend to an appropriate asymptotically flat limit one must require both the limits

$$\lim_{r \to \infty} b_\pm(r) = b_\pm \tag{11.7}$$

to be finite. By comparison with the Schwarzschild metric this implies that the mass of the wormhole, as seen from spatial infinity, is given by $b_\pm = 2GM_\pm$. Note that the mass of the wormhole is allowed to be different depending on which universe the observer resides in. Stated differently, the two mouths of the same wormhole can in general have different masses.

- In order for the spacetime geometry to tend to the appropriate flat limit both the limits

$$\lim_{r \to \infty} \phi(r) = \phi_\pm \tag{11.8}$$

must exist and be finite. Note that there is no *a priori* requirement that $\phi_+(\infty) = \phi_-(\infty)$. This implies that time can run at different rates in the two universes. Students of elven folklore will be greatly amused by this observation.

- Since $dr/dl = 0$ at the throat of the wormhole [the throat was defined by the location of the minimum of $r(l)$], we must have $dl/dr \to \infty$ at the throat. This implies $b_\pm(r_0) = r_0$. Away from the throat of the wormhole $b_\pm(r) < r$.

- Away from the throat of the wormhole, the metric components should be at least twice differentiable as functions of r. At the throat itself some delicacy is required—see the discussion below.

- For simplicity one might sometimes wish to assume symmetry under interchange of asymptotically flat regions, $\pm \leftrightarrow \mp$. That is $b_+(r) = b_-(r)$ and $\phi_+(r) = \phi_-(r)$. This requirement is not essential to the definition of a traversable wormhole.

The behavior of the shape and redshift functions at the throat itself is particularly informative. Note that

$$\frac{dr}{dl} = \pm\sqrt{1 - \frac{b}{r}}. \tag{11.9}$$

Now

$$\frac{d^2r}{dl^2} = \frac{d}{dl}\left(\frac{dr}{dl}\right) = \frac{dr}{dl}\frac{d}{dr}\left(\frac{dr}{dl}\right) = \frac{1}{2}\frac{d}{dr}\left[\left(\frac{dr}{dl}\right)^2\right]. \tag{11.10}$$

This yields

$$\frac{d^2r}{dl^2} = \frac{1}{2r}\left(\frac{b}{r} - b'\right). \tag{11.11}$$

Since $r(l)$ is a minimum at the throat and grows as one moves away from the throat, one deduces

$$\exists r_* \Big| \forall r \in (r_0, r_*), \qquad \frac{d^2r}{dl^2} > 0. \tag{11.12}$$

Note that r_* could in principle be as large as $+\infty$. However, typically the point r_* will be "near" the throat r_0. One deduces

$$\exists r_* \Big| \forall r \in (r_0, r_*), \qquad b'(r) < \frac{b(r)}{r}. \tag{11.13}$$

At the throat itself, $b(r_0) = r_0$, so that equation (11.10) simplifies to

$$\left.\frac{d^2r}{dl^2}\right|_{r_0} = \frac{1}{2r_0}\left[1 - b'_+(r_0)\right] = \frac{1}{2r_0}\left[1 - b'_-(r_0)\right]. \tag{11.14}$$

This implies that at the throat

$$b'_+(r_0) = b'_-(r_0). \tag{11.15}$$

Moreover, since $r(l)$ is a minimum at the throat, one also knows that

$$\left.\frac{d^2r}{dl^2}\right|_{r_0} \geq 0, \tag{11.16}$$

whence one obtains the weak inequality

$$b'_\pm(r_0) \leq 1. \tag{11.17}$$

This last result could also be deduced by graphical means. We know that $b_\pm(r) \to 2GM_\pm$ as $r \to \infty$, and that $b_\pm(r) < r$ away from the throat. Recall that the throat is the first and only place where $b_+(r_0) = r_0 = b_-(r_0)$.

b(r)

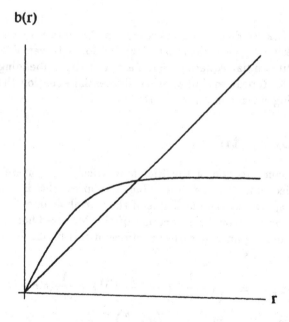

r

Figure 11.2: Morris–Thorne wormhole: the shape function. Since $b(r)$ goes to a finite limit as $r \to \infty$, the outermost crossing of the line $b = r$ must occur at a slope less than or equal to 1.

Similar considerations apply to the redshift function. Note that

$$\frac{d\phi}{dl} = \frac{dr}{dl}\left(\frac{d\phi}{dr}\right) = \sqrt{1 - \frac{b}{r}}\ \phi'. \qquad (11.18)$$

So that

$$
\begin{aligned}
\frac{d^2\phi}{dl^2} &= \sqrt{1 - \frac{b}{r}}\ \frac{d}{dr}\left(\sqrt{1 - \frac{b}{r}}\ \phi'\right) \\
&= \left(1 - \frac{b}{r}\right)\phi'' + \frac{1}{2r}\left(\frac{b}{r} - b'\right)\phi'. \qquad (11.19)
\end{aligned}
$$

Evaluating this at the throat

$$\left.\frac{d^2\phi}{dl^2}\right|_{r_0} = \frac{1}{2r_0}\left[1 - b'_+(r_0)\right]\phi'_+(r_0) = \frac{1}{2r_0}\left[1 - b'_-(r_0)\right]\phi'_-(r_0), \qquad (11.20)$$

which now implies

$$\phi'_+(r_0) = \phi'_-(r_0). \qquad (11.21)$$

Exercise: (Easy) Find some references to the notion of an embedding diagram. Sketch some embedding diagrams for a traversable wormhole. Develop a differential equation that describes the embedding surface in terms of $b(r)$. Repeat to find another differential equation that describes the embedding surface in terms of $r(l)$.

11.3   Curvature

Brute force computation of the Riemann, Ricci, and Einstein tensors is readily carried out. For the time being, I suppress the $\pm$ subscripts that are used for specifying on which side of the wormhole one is residing. The results may be read off by inspection of the Morris–Thorne paper [190, p. 401], or from a glance at earlier portions of this book. For convenience, I restate the results:

$$R^{\hat{i}}{}_{\hat{r}\hat{i}\hat{r}} = \left(1 - \frac{b}{r}\right)\left\{-\phi'' - (\phi')^2\right\} + \frac{1}{2r^2}(b'r - b)\phi', \qquad (11.22)$$

$$R^{\hat{i}}{}_{\hat{\theta}\hat{i}\hat{\theta}} = R^{\hat{i}}{}_{\hat{\varphi}\hat{i}\hat{\varphi}} = -\left(1 - \frac{b}{r}\right)\frac{\phi'}{r}, \qquad (11.23)$$

$$R^{\hat{r}}{}_{\hat{\theta}\hat{r}\hat{\theta}} = R^{\hat{r}}{}_{\hat{\varphi}\hat{r}\hat{\varphi}} = \frac{1}{2r^3}(b'r - b), \qquad (11.24)$$

$$R^{\hat{\theta}}{}_{\hat{\varphi}\hat{\theta}\hat{\varphi}} = \frac{b}{r^3}. \qquad (11.25)$$

All other components of the Riemann tensor, apart from those related to the above by symmetry, vanish.

The nonzero components of the Einstein tensor are

$$G_{\hat{t}\hat{t}} = \frac{b'}{r^2}, \tag{11.26}$$

$$G_{\hat{r}\hat{r}} = -\frac{b}{r^3} + 2\left\{1 - \frac{b}{r}\right\}\frac{\phi'}{r}, \tag{11.27}$$

$$G_{\hat{\theta}\hat{\theta}} = G_{\hat{\varphi}\hat{\varphi}} = \left\{1 - \frac{b}{r}\right\}\left[\phi'' + \phi'\left(\phi' + \frac{1}{r}\right)\right]$$
$$- \frac{1}{2r^2}[b'r - b]\left(\phi' + \frac{1}{r}\right). \tag{11.28}$$

Evaluating the Riemann tensor at the throat itself,

$$R^{\hat{t}}{}_{\hat{r}\hat{t}\hat{r}}\Big|_{r_0} = -\frac{1}{2r_0}[1 - b'(r_0)]\phi'(r_0), \tag{11.29}$$

$$R^{\hat{r}}{}_{\hat{\theta}\hat{r}\hat{\theta}}\Big|_{r_0} = R^{\hat{r}}{}_{\hat{\varphi}\hat{r}\hat{\varphi}}\Big|_{r_0} = -\frac{1}{2r_0^2}[1 - b'(r_0)], \tag{11.30}$$

$$R^{\hat{\theta}}{}_{\hat{\varphi}\hat{\theta}\hat{\varphi}}\Big|_{r_0} = \frac{1}{r_0^2}. \tag{11.31}$$

All other components of the Riemann tensor, apart from those related to the above by symmetry, vanish. Note that because of the continuity conditions applying to $b'_{\pm}(r)$ and $\phi'_{\pm}(r)$ at the throat, we do not have to distinguish the $b_{\pm}$ and $\phi_{\pm}$ in the above formulas. The components of the Riemann tensor are seen to be continuous across the throat, as required. The analogous result for the Einstein tensor is

$$G_{\hat{t}\hat{t}}\Big|_{r_0} = \frac{b'(r_0)}{r_0^2}, \tag{11.32}$$

$$G_{\hat{r}\hat{r}}\Big|_{r_0} = -\frac{1}{r_0^2}, \tag{11.33}$$

$$G_{\hat{\theta}\hat{\theta}}\Big|_{r_0} = G_{\hat{\varphi}\hat{\varphi}}\Big|_{r_0} = \frac{1 - b'(r_0)}{2r_0}\left(\phi' + \frac{1}{r_0}\right). \tag{11.34}$$

11.4 Einstein equations

The nonzero components of the stress-energy tensor are given special symbols:

$$T_{\hat{t}\hat{t}} = \rho; \qquad T_{\hat{r}\hat{r}} = -\tau; \qquad T_{\hat{\theta}\hat{\theta}} = T_{\hat{\varphi}\hat{\varphi}} = p. \tag{11.35}$$

Reminder: ρ denotes the energy density, τ denotes the radial tension (minus the radial pressure), and p denotes the transverse pressure. All other components of the stress-energy tensor vanish by the assumed spherical symmetry. The Einstein equations, $G_{\mu\nu} = 8\pi G\, T_{\mu\nu}$, now yield

$$\rho = \frac{b'}{8\pi G r^2}, \tag{11.36}$$

$$\tau = \frac{1}{8\pi G}\left[\frac{b}{r^3} - 2\left\{1 - \frac{b}{r}\right\}\frac{\phi'}{r}\right], \tag{11.37}$$

$$p = \frac{1}{8\pi G}\left\{\left(1 - \frac{b}{r}\right)\left[\phi'' + \phi'\left(\phi' + \frac{1}{r}\right)\right]\right.$$
$$\left. - \frac{1}{2r^2}(b'r - b)\left(\phi' + \frac{1}{r}\right)\right\}. \tag{11.38}$$

A convenient rearrangement is [190, p. 401]

$$b' = 8\pi G\, \rho\, r^2, \tag{11.39}$$

$$\phi' = \frac{b - 8\pi G\tau r^3}{2r^2(1 - b/r)}, \tag{11.40}$$

$$\tau' = (\rho - \tau)\phi' - 2(p + \tau)/r. \tag{11.41}$$

The first Einstein equation is easily integrated to yield

$$b(r) = b(r_0) + \int_{r_0}^{r} 8\pi G\rho(r')r'^2 dr' = 2G\, m(r), \tag{11.42}$$

where we have defined

$$m(r) \equiv (r_0/2G) + \int_{r_0}^{r} 4\pi\rho r^2 dr \tag{11.43}$$

as the effective mass inside the radius r. The shape function thus has a very direct interpretation in terms of the distribution of mass inside the wormhole. Note that as one moves to spatial infinity

$$\lim_{r\to\infty} m(r) = \frac{r_0}{2G} + \int_{r_0}^{\infty} 4\pi\rho(r)r^2 dr = M. \tag{11.44}$$

At the throat itself

$$\rho|_{r_0} = \frac{b'(r_0)}{8\pi G r_0^2}, \tag{11.45}$$

$$\tau|_{r_0} = \frac{1}{8\pi G r_0^2}, \tag{11.46}$$

$$p|_{r_0} = \frac{1 - b'(r_0)}{16\pi G r_0}\left(\phi' + \frac{1}{r_0}\right). \tag{11.47}$$

In view of the inequality (11.13), the mass function satisfies

$$\exists r_* \Big| \forall r \in (r_0, r_*), \qquad \rho(r) < \frac{m(r)}{4\pi r^3}. \qquad (11.48)$$

At the throat itself

$$\rho(r_0) \leq \frac{1}{8\pi G r_0^2}. \qquad (11.49)$$

The first two Einstein equations can be combined to give

$$
\begin{aligned}
8\pi G(\rho - \tau) &= \frac{1}{r^2}\left(b' - \frac{b}{r}\right) + 2\left(1 - \frac{b}{r}\right)\frac{\phi'}{r}, \\
&= -\frac{1}{r}\left[\left(1 - \frac{b}{r}\right)' - \frac{2\phi'}{r}\left(1 - \frac{b}{r}\right)\right], \\
&= -\frac{e^{2\phi}}{r}\left[e^{-2\phi}\left(1 - \frac{b}{r}\right)\right]'.
\end{aligned}
\qquad (11.50)
$$

But we know that

$$e^{-2\phi}\left(1 - \frac{b}{r}\right)\Big|_{r_0} = 0, \qquad (11.51)$$

while

$$\forall r > r_0, \qquad e^{-2\phi}\left(1 - \frac{b}{r}\right) > 0. \qquad (11.52)$$

Therefore

$$\exists \tilde{r}_* \Big| \forall r \in (r_0, \tilde{r}_*), \qquad \left[e^{-2\phi}\left(1 - \frac{b}{r}\right)\right]' > 0. \qquad (11.53)$$

In principle $\tilde{r}_*$ could be as large as $+\infty$. However, typically the point $\tilde{r}_*$ will be "near" the throat r_0. One deduces that

$$\exists \tilde{r}_* \Big| \forall r \in (r_0, \tilde{r}_*), \qquad (\rho - \tau) < 0. \qquad (11.54)$$

Working at the throat itself results in a weak inequality. From

$$\left[e^{-2\phi}\left(1 - \frac{b}{r}\right)\right]'\Big|_{r_0} \geq 0, \qquad (11.55)$$

one deduces

$$[\rho(r_0) - \tau(r_0)] \leq 0. \qquad (11.56)$$

These inequalities will be central to our subsequent discussion of the energy conditions.

For simplicity the entire discussion has been phrased in terms of an inter-universe wormhole. However, since both mouths of the wormhole reside in asymptotically flat regions of spacetime one can, with minimal distortion of the mouths, rearrange the two mouths to reside in one asymptotically flat region. This would then produce an intra-universe traversable wormhole. Intra-universe wormholes lead to additional levels of complication:

1. If time runs at different rates in the two asymptotically flat regions of the inter-universe wormhole, the associated intra-universe wormhole then has a nonconservative gravitational field. See p. 239ff.

2. When joining the two asymptotically flat regions, one could always add a twist, and thereby generate a non-orientable spacetime. Non-orientable spacetimes are perfectly acceptable at the classical level (classical particles following geodesics), but are problematic at the quantum level. See p. 285ff.

11.5　Exercises: Curvature

Exercise 1:　Use of the mass function, together with the field equations, leads to an interesting result: It allows one to rewrite the formulas for the Riemann tensor as a function of ρ, τ, p, $m(r)$, and r. Show that

$$R^{\hat{i}}{}_{\hat{r}\hat{i}\hat{r}} = +G\left(\frac{2m(r)}{r^3} - 4\pi\left(\rho + \tau + 2p\right)\right), \qquad (11.57)$$

$$R^{\hat{i}}{}_{\hat{\theta}\hat{i}\hat{\theta}} = R^{\hat{i}}{}_{\hat{\varphi}\hat{i}\hat{\varphi}} = +G\left(4\pi\tau - \frac{m(r)}{r^3}\right), \qquad (11.58)$$

$$R^{\hat{r}}{}_{\hat{\theta}\hat{r}\hat{\theta}} = R^{\hat{r}}{}_{\hat{\varphi}\hat{r}\hat{\varphi}} = +G\left(4\pi\rho - \frac{m(r)}{r^3}\right), \qquad (11.59)$$

$$R^{\hat{\theta}}{}_{\hat{\varphi}\hat{\theta}\hat{\varphi}} = +2G\frac{m(r)}{r^3}. \qquad (11.60)$$

This rearrangement of the Riemann tensor has eliminated the redshift function $\phi(r)$ in terms of locally measured quantities, and one piece of global information, $m(r)$.

Exercise 2:　Express the Riemann tensor at the throat in terms of the stress-energy tensor. Show that

$$R^{\hat{i}}{}_{\hat{r}\hat{i}\hat{r}}\bigg|_{r_0} = -4\pi G\left(\rho - \tau + 2p\right), \qquad (11.61)$$

$$R^{\hat{r}}{}_{\hat{\theta}\hat{r}\hat{\theta}}\bigg|_{r_0} = R^{\hat{r}}{}_{\hat{\varphi}\hat{r}\hat{\varphi}}\bigg|_{r_0} = +8\pi G\left(\rho - \tau\right), \qquad (11.62)$$

$$R^{\hat{\theta}}{}_{\hat{\varphi}\hat{\theta}\hat{\varphi}}\bigg|_{r_0} = +8\pi G\,\tau. \qquad (11.63)$$

11.6 The imprint at infinity

In any asymptotically flat spacetime it is possible to define the so-called "imprint at infinity". This observation is the basis for the derivation of asymptotic conservation laws. Since traversable wormholes, by definition, reside in asymptotically flat spacetimes, one can deduce interesting conservation laws that limit the manner in which traversable wormholes interact with each other and with infalling matter.

ADM mass: For the total mass–energy the concept we wish to invoke is the ADM mass. Sufficiently far into the asymptotically flat region the metric is almost that of flat space [*cf.* equation (2.32)]. Indeed, adopting quasi-Minkowski coordinates, in which the active region is assumed to be confined to small r,

$$g_{\mu\nu} = \eta_{\mu\nu} + (\eta_{\mu\nu} + 2V_\mu V_\nu)\frac{2GM}{r} + O(r^{-2}) \qquad (11.64)$$

The total ADM mass M is *defined* by suitably normalizing the coefficient of the $1/r$ term on this expansion.

The ADM mass can also be more formally defined in terms of a suitable limit of surface integrals at spatial infinity. Let Ω be a spacelike two-surface enclosing the active region, then integrating over this surface

$$M \equiv \frac{1}{16\pi G} \lim_{\Omega \to \infty} \int_\Omega \sum_{i,j=1}^{3} (\partial_j g_{ij} - \partial_i g_{jj})\, n^i\, dS. \qquad (11.65)$$

(See, for instance, Wald [275, pp. 293ff.]; here n^i denotes the unit outward normal.) The most important feature of the ADM mass is that it is conserved. Similar incantations can be used to define the total electric charge.

Now consider the case of a traversable wormhole that connects two otherwise separate universes. There are now two asymptotically flat regions, and thus two ADM masses (which can be unequal in general, see p. 103). The conservation law normally deduced for the ADM mass carries through *mutatis mutandis* to a pair of conservation laws. Each ADM mass in each asymptotically flat universe is conserved separately.

Consider the effect of a mass m_i that is initially far away from the wormhole mouth in the "+" universe. Suppose now that this object traverses the wormhole and eventually settles down far from the wormhole in the "−" universe. Then the total ADM masses on both sides of the wormhole satisfy

$$M_{\text{total}}^{+} = M_i^{+} + m_i = M_f^{+}; \qquad (11.66)$$

$$M_{\text{total}}^{-} = M_i^{-} = M_f^{-} + m_f. \qquad (11.67)$$

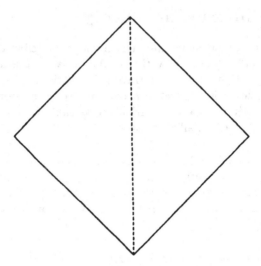

Figure 11.3: Morris–Thorne wormhole: the Penrose diagram of an inter-universe Morris–Thorne wormhole contains two asymptotically flat regions. The dashed line denotes the location of the wormhole throat. Flux lines that thread the throat of the wormhole and penetrate to spatial infinity will be trapped there.

Notation: Here $M_i^{\pm}$ denote the initial masses and $M_f^{\pm}$ denote the final masses of the two wormhole mouths in the "+" and "−" universes, respectively.

That is, when massive objects traverse a wormhole they alter the mass of the wormhole mouths they pass through. The mouth that "absorbs" the object gains mass $(M_f^+ > M_i^+)$, while the mouth that "emits" the object loses mass $(M_f^- < M_i^-)$. For generality I have permitted $m_i \neq m_f$; this allows for the possibility that the object traversing the wormhole may gain or loose some (kinetic) energy in the process.

So far, the argument has been phrased for a wormhole that connects two separate universes but it can also be applied to *intra*-universe wormholes provided

1. the two mouths of the wormhole are sufficiently far apart that their mutual gravitational interaction is negligible, and

2. the initial and final positions of the object traversing the wormhole are sufficiently far away from both mouths of the wormhole.

Then the total ADM mass for the (one!) universe is

$$M_{total} = M_i^+ + M_i^- + m_i = M_f^+ + M_f^- + m_f. \qquad (11.68)$$

Because the two wormhole mouths are assumed to have negligible gravitational interaction we can still assert

$$M_f^+ = M_i^+ + m_i; \qquad M_f^- = M_i^- - m_f. \qquad (11.69)$$

This suggests (but emphatically does not prove) the possibility of a fundamental limit on the total mass that can traverse a wormhole. For a sufficiently large net transfer of mass the final mass of the "emitting" wormhole mouth becomes negative. Under normal circumstances this would be considered a complete disaster. However, we shall soon see that violations of the average null energy condition are required just to hold the wormhole open in the first place. Because of this, the hypotheses used to derive the usual versions of the positive mass theorem do not apply to traversable wormhole spacetimes. There does not seem to be any guarantee that would prevent the total mass of a wormhole mouth from going negative, though such behavior would certainly make one feel very queasy [194].

If the mass of the "emitting" mouth of the wormhole does become negative one has the possibility of a runaway reaction: its mass now being negative the "emitting" mouth will now gravitationally repel the ambient medium. The "absorbing" mouth will continue to accrete matter, becoming ever more massive and attracting more of the surrounding material to itself. On the other hand, the "emitting" mouth will continue to lose mass, and its mass will become more and more negative, thereby reinforcing its gravitational repulsion of the ambient medium. For an example of the peculiarities this leads to, see [49].

Technical point: For simplicity I have concentrated on the ADM mass, since the ADM mass is insensitive to energy loss to null infinity via gravitational or electromagnetic radiation. To investigate the effects of energy radiated to null infinity one should generalize the present discussion by considering the Bondi mass rather than the ADM mass. In this case, one will also need to keep careful track of the total energy radiated.

Electric charge: Similar conservation arguments can be applied to electric charge (in fact to any conserved charge). One can phrase the argument either in terms of the imprint at infinity or in terms of a more physical picture using flux lines. Consider a charged particle (charge $+q$) and an uncharged wormhole. Before the charged particle traverses the wormhole assume that all the flux lines that emanate from it tend to spatial infinity in the "+" universe. These flux lines are frozen in and cannot break as the particle traverses the wormhole. After the particle has traversed the wormhole all of these flux lines must lead back from spatial infinity in the "+"

universe to the wormhole mouth. The wormhole mouth has thus acquired a charge $+q$. In the "$-$" universe, flux lines emerge from the charge and enter the wormhole mouth. Thus the wormhole mouth in the "$-$" universe has acquired a charge $-q$ and the total electric charge in the "$-$" universe is $Q_{\text{total}}^- = -q + q = 0$, which is what it was before the charge transited the wormhole.

If both wormhole mouths are initially charged the argument generalizes to give

$$Q_f^+ = Q_i^+ + q; \qquad Q_f^- = Q_i^- - q. \tag{11.70}$$

Thus a charged particle that traverses a wormhole will alter the charge on both wormhole mouths. Note that there is no need for the charge on the two mouths to be equal. As charge flow continues it will become harder and harder to push additional charge through the wormhole. This is due to the electrostatic repulsion between the incoming charge and the accumulated charge on the wormhole mouth. In the case of electric charge (as opposed to mass) one does not find the runaway modes of the previous section. Rather, left to their own devices, a pair of charged wormhole mouths will tend to minimize their absolute electric charges by preferential charge exchange.

Exercise: Define the "total" charge on the wormhole by adding up the charges in each universe:

$$Q_{\text{total}} = Q_i^+ + Q_i^- = Q_f^+ + Q_f^-. \tag{11.71}$$

Check that this total charge is conserved. Estimate the electrostatic energy in each universe:

$$E^\pm \approx \frac{1}{4\pi\epsilon_0} \frac{(Q^\pm)^2}{r_0}. \tag{11.72}$$

If we could simply add energies in the two universes we would deduce

$$E_{\text{total}} \propto (Q^+)^2 + (Q^-)^2 = (Q_{\text{total}})^2[z^2 + (1-z)^2], \tag{11.73}$$

where z is defined as $z = Q^+/Q_{\text{total}}$. Minimizing total electrostatic energy, subject to fixed total charge, gives $z = 1/2$. This argument assumes that energies in the two universes should be given equal weight. This is true provided $\phi^+(\infty) = \phi^-(\infty)$, but is false if time "runs at different rates" in the two universes. Find the appropriate generalized result.

Chapter 12

Energy conditions

12.1 Definitions

There are at least seven types of energy conditions normally invoked in discussing classical general relativity. These are: The null, weak, strong, and dominant energy conditions (NEC, WEC, SEC, DEC), and the averaged null, weak, and strong energy conditions (ANEC, AWEC, ASEC). To keep the discussion focussed, assume that the energy momentum tensor is of Hawking–Ellis type one (type I) [132, p. 89]. This is quite sufficient for current purposes. In a suitable orthonormal frame, the components of a type one stress-energy tensor are given by

$$T^{\hat{\mu}\hat{\nu}} = \begin{bmatrix} \rho & 0 & 0 & 0 \\ 0 & p_1 & 0 & 0 \\ 0 & 0 & p_2 & 0 \\ 0 & 0 & 0 & p_3 \end{bmatrix}. \tag{12.1}$$

These components are the energy density and the three principal pressures.

12.1.1 Null energy condition

The null energy condition is the assertion that for any null vector

$$\text{NEC} \iff T_{\mu\nu} \, k^\mu k^\nu \geq 0. \tag{12.2}$$

In terms of the principal pressures

$$\text{NEC} \iff \forall j, \quad \rho + p_j \geq 0. \tag{12.3}$$

115

12.1.2 Weak energy condition

The weak energy condition is the assertion that for any timelike vector

$$\text{WEC} \iff T_{\mu\nu} V^{\mu} V^{\nu} \geq 0. \qquad (12.4)$$

If this is true for any timelike vector, it will also (by continuity) imply the null energy condition. The physical significance of this condition is that it forces the local energy density as measured by *any* timelike observer to be positive. In terms of the principal pressures

$$\text{WEC} \iff \rho \geq 0, \quad \text{and} \quad \forall j, \quad \rho + p_j \geq 0. \qquad (12.5)$$

12.1.3 Strong energy condition

The strong energy condition is the assertion that for any timelike vector

$$\text{SEC} \iff \left(T_{\mu\nu} - \frac{T}{2} g_{\mu\nu} \right) V^{\mu} V^{\nu} \geq 0. \qquad (12.6)$$

Here T is the trace of the stress-energy tensor, $T = T_{\mu\nu} \, g^{\mu\nu}$. By continuity, the strong energy condition implies the null energy condition (but it does not imply, in general, the weak energy condition). In terms of the principal pressures

$$T = -\rho + \sum_{j} p_j \qquad (12.7)$$

and

$$\text{SEC} \iff \forall j, \quad \rho + p_j \geq 0, \quad \text{and} \quad \rho + \sum_{j} p_j \geq 0. \qquad (12.8)$$

12.1.4 Dominant energy condition

The dominant energy condition is the assertion that for any timelike vector

$$\text{DEC} \iff T_{\mu\nu} V^{\mu} V^{\nu} \geq 0, \quad \text{and} \quad T_{\mu\nu} V^{\nu} \text{ is not spacelike.} \qquad (12.9)$$

This says that the locally measured energy density is always positive, *and* that the energy flux is timelike or null. The dominant energy condition implies the weak energy condition, and thus also the null energy condition, but does not necessarily imply the strong energy condition. In terms of the principal pressures

$$\text{DEC} \iff \rho \geq 0; \quad \text{and} \quad \forall j, \quad p_j \in [-\rho, +\rho]. \qquad (12.10)$$

12.1.5 Averaged null energy condition

The averaged null energy condition is said to hold on a null curve Γ if

$$\text{ANEC}[\Gamma] \iff \int_\Gamma T_{\mu\nu} \, k^\mu k^\nu \, d\lambda \geq 0. \tag{12.11}$$

Here λ is a generalized affine parameterization of the null curve, the corresponding tangent vector being denoted by k^μ.

Technical points:

1. If Γ is a null geodesic then the generalized affine parameter specializes to the ordinary affine parameter. See, for example, [132, p. 259].

2. Permitting arbitrary parameterizations would not be useful. If arbitrary parameterizations were to be allowed, the ANEC would be equivalent to the ordinary NEC.

3. Because of the multiplicative arbitrariness of the generalized affine parameter and ordinary affine parameter, it is only meaningful to define the ANEC integral up to an overall positive multiplicative constant.

To analyze this condition in terms of the principal pressures, note that one can define a normalization function function ξ, and direction cosines $\cos\psi_i$, by

$$k^\mu \equiv \xi \, (1; \cos\psi_i). \tag{12.12}$$

Then

$$\text{ANEC}[\Gamma] \iff \int_\Gamma \left(\rho + \sum_j \cos^2\psi_j \, p_j \right) \xi^2 \, d\lambda \geq 0. \tag{12.13}$$

In applications one typically requires the ANEC condition to hold on some suitable class $\{\Gamma\}$ of inextendible null geodesics.

12.1.6 Averaged weak energy condition

The averaged weak energy condition is said to hold on a timelike curve Γ if

$$\text{AWEC}[\Gamma] \iff \int_\Gamma T_{\mu\nu} \, V^\mu V^\nu \, ds \geq 0. \tag{12.14}$$

Here s denotes the proper time parameterization of the timelike curve Γ, the corresponding tangent vector being denoted by V^μ. Note that with the choice of s as proper time one has

$$V^\mu = \gamma(1; \beta\cos\psi_i). \tag{12.15}$$

Thus

$$\text{AWEC}[\Gamma] \iff \int_\Gamma \gamma^2 \left(\rho + \beta^2 \sum_j \cos^2 \psi_j \, p_j \right) \, ds \geq 0. \qquad (12.16)$$

In applications one typically requires the AWEC condition to hold on some suitable class $\{\Gamma\}$ of inextendible timelike geodesics.

Technical points:

1. If this class of timelike geodesics is suitably large, its boundary $\partial\{\Gamma\}$ may contain limit points (limit curves) consisting of null geodesics. In this sense, AWEC (on the set $\{\Gamma\}$) can be said to imply ANEC (on the null geodesics in the set $\partial\{\Gamma\}$).

2. However, in a general spacetime there may be null geodesics that are not obtained as the limit of any sequence of timelike geodesics. In general, AWEC and ANEC are independent conditions.

12.1.7   Averaged strong energy condition

The averaged strong energy condition is said to hold on a timelike curve Γ if

$$\text{ASEC}[\Gamma] \iff \int_\Gamma \left\{ T_{\mu\nu} \, V^\mu V^\nu + \frac{1}{2} T \right\} \, ds \geq 0. \qquad (12.17)$$

In terms of the principal pressures this is the rather messy constraint

$$\text{ASEC}[\Gamma] \iff$$
$$\int_\Gamma \left\{ \gamma^2 \left(\rho + \beta^2 \sum_j \cos^2 \psi_j \, p_j \right) - \frac{1}{2}\rho + \frac{1}{2}\sum_j p_j \right\} \, ds \geq 0.$$
$$(12.18)$$

If $\beta \to 1$, then $\gamma \to \infty$, while $\gamma ds \to d\lambda$ and $ds \to 0$. In this limit the ASEC reduces to the ANEC—up to an irrelevant infinite multiplicative factor.

12.2   Applications

The energy conditions, in one form or another, are used in the various classical singularity theorems and theorems of classical black hole thermodynamics. For example:

- The Penrose singularity theorem invokes the weak energy condition. See, for example, [132, §8.2, p. 263].

- The Hawking–Penrose singularity theorem invokes the strong energy condition [132, §8.2, pp. 266–267]. See also Wald [275, pp. 237–241].

- Sundry other singularity theorems discussed by Hawking and Ellis invoke the strong energy condition [132, §8.2, pp. 271–275].

- Tipler's version of the Hawking–Penrose singularity theorem uses the weak energy condition together with an averaged version of the strong energy condition [254].

- Borde's generalizations of Tipler's convergence theorems are phrased in terms of variations on the theme of the averaged null energy condition [23]. (The ANEC integral is only required to be periodically non-negative.)

- Roman's improvement of the Penrose singularity theorem uses the averaged null energy condition along a certain class of future inextendible null geodesics [225, 226].

- The proof of the zeroth law of black hole thermodynamics (the constancy of the surface gravity over the event horizon) relies on the dominant energy condition [275, pp. 333–334].

- The proof of the second law of black hole thermodynamics (the area increase theorem) uses the null energy condition [275, p. 312].

- The Friedman–Schleich–Witt topological censorship theorem invokes the averaged null energy condition [92]. A second, somewhat different, topological censorship theorem proved by these authors uses the dominant energy condition.

- The positive mass theorem of Schoen and Yau addresses the positivity of the ADM mass and hypothesizes the dominant energy condition [236, 235].

- Penrose, Sorkin, and Woolgar have recently provided a different proof of the positivity of the ADM mass which uses Borde's convergence theorems [244, 210].

- Theorems regarding the positivity of the Bondi mass also utilize the dominant energy condition [237, 144].

- The cosmic censorship conjecture (not a theorem) is usually formulated assuming the dominant energy condition [275, p. 305].

There is certainly room for additional work to be done in tidying up these issues.

Research Problem 3 *Work systematically through all of these theorems weakening the energy conditions as far as possible.*

12.3 Known violations

Many physical systems, both theoretical and experimental, are known to violate one or more of the energy conditions in interesting ways.

12.3.1 Classical scalar fields

The most trivial of the violations of the various energy conditions arises when considering the classical free minimally-coupled massive scalar field described by the Lagrangian density [132, pp. 95–96]

$$\mathcal{L} = \frac{1}{2}(\nabla\phi)^2 + \frac{1}{2}m^2\phi^2. \tag{12.19}$$

The stress-energy tensor is

$$T^{\mu\nu} = \nabla^\mu\phi\,\nabla^\nu\phi - \frac{1}{2}g^{\mu\nu}\left([\nabla\phi]^2 + m^2\phi^2\right). \tag{12.20}$$

So for any unit timelike vector

$$\left(T_{\mu\nu} - \frac{T}{2}g_{\mu\nu}\right)V^\mu V^\nu = [V\cdot\nabla\phi]^2 - \frac{1}{2}m^2\phi^2. \tag{12.21}$$

It is thus very easy to violate the strong energy condition. For instance: Take ϕ to be time independent in the frame where $V^\mu = (1,0,0,0)$.

For a null vector, consideration of the quantity

$$T_{\mu\nu}\,k^\mu k^\nu = [k\cdot\nabla\phi]^2, \tag{12.22}$$

shows that the null energy condition is not violated. Indeed, for any unit timelike vector

$$\begin{aligned}
T_{\mu\nu}\,V^\mu V^\nu &= [V\cdot\nabla\phi]^2 + \frac{1}{2}\left([\nabla\phi]^2 + m^2\phi^2\right), \\
&= \frac{1}{2}\left([\eta^{\mu\nu} + 2V^\mu V^\nu]\,\nabla_\mu\phi\,\nabla_\nu\phi + m^2\phi^2\right), \quad (12.23)
\end{aligned}$$

so the weak energy condition is satisfied as well. Finally, with V a unit timelike vector, consider the quantity

$$T_{\mu\nu}V^\nu = \nabla_\mu\phi\,[V\cdot\nabla\phi] - \frac{1}{2}V_\mu\left([\nabla\phi]^2 + m^2\phi^2\right), \tag{12.24}$$

Taking the norm of this vector

$$\|T_{\mu\nu}V^\nu\|^2 = -[V\cdot\nabla\phi]^2 m^2\phi^2 - \frac{1}{4}\left([\nabla\phi]^2 + m^2\phi^2\right)^2. \tag{12.25}$$

Thus the dominant energy condition is satisfied as well.

So all the energy conditions apart from the strong energy condition are satisfied by this simple classical system. Physical examples of systems described by such scalar Lagrangians are the pion field and the Higgs field. The pion is not an elementary particle,[1] while the Higgs scalar has not yet been seen experimentally.[2]

Perhaps the most profitable point of view to take in this matter is to note that in the past, it has proved possible to modify many of the theorems that originally required the strong energy condition in such a manner as to dispense with this particular assumption, replacing it with one of the other energy conditions.

12.3.2 Casimir effect

The Casimir effect is induced when the presence of electrical conductors distorts the zero point energy of the quantum electrodynamics vacuum. The experimental verification of the theoretical calculations of the size and sign of the Casimir effect is direct evidence indicating the reality of the zero point energy.

Casimir's original calculation concerned two parallel conducting plates a small distance a apart. The fact that the plates are conducting implies that one has effectively imposed boundary conditions on the electrodynamic field. If the plates lie in the x-y plane, then the z component of the wave vector is constrained by

$$k_z = \frac{n\pi}{a}. \tag{12.26}$$

This selection rule imposes an a-dependent distortion on the zero point fluctuations. Consider now the stress-energy tensor associated with this distortion in the zero point fluctuations. By symmetry, it can depend only on the spacetime metric $\eta^{\mu\nu}$, the normal vector $\hat{z}^{\mu}$, the distance between the plates a, and possibly the z coordinate. By simple dimensional analysis, this implies that for some pair of dimensionless functions $f_1(z/a)$ and $f_2(z/a)$

$$T^{\mu\nu}_{\text{Casimir}} \equiv \frac{\hbar}{a^4} \left[f_1(z/a)\, \eta^{\mu\nu} + f_2(z/a)\hat{z}^{\mu}\hat{z}^{\nu} \right]. \tag{12.27}$$

Conservation of stress-energy implies that $f_1(z/a)$ and $f_2(z/a)$ are in fact constants.

Exercise: Calculate $\nabla_\mu T^{\mu\nu}$. Because the geometry is flat Minkowski space it suffices to consider $\partial_\mu T^{\mu\nu}$.

---

[1] The pion is a (dressed) quark–antiquark bound state.

[2] Nevertheless, the splendid success of the standard model of particle physics strongly suggests that some Higgs-like object should be there to be found.

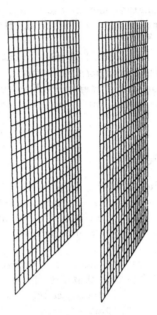

Figure 12.1: Casimir effect: parallel plate geometry.

Because the electromagnetic field is conformally invariant,

$$T \equiv T^{\mu\nu} \, \eta_{\mu\nu} = 0. \tag{12.28}$$

This then allows one to deduce a relationship between f_1 and f_2. Indeed $f_2 = -4f_1$. [Technical point: The electromagnetic field is classically conformally invariant. To apply this argument in the quantum realm you will have to convince yourself that quantization does not introduce any trace (conformal) anomaly. In this case we are lucky, the trace anomaly vanishes.] Thus

$$T^{\mu\nu}_{\text{Casimir}} \propto \frac{\hbar}{a^4} \left(\eta^{\mu\nu} - 4\hat{z}^\mu \hat{z}^\nu \right) = \frac{\hbar}{a^4} \begin{bmatrix} -1 & 0 & 0 & 0 \\ 0 & +1 & 0 & 0 \\ 0 & 0 & +1 & 0 \\ 0 & 0 & 0 & -3 \end{bmatrix}. \tag{12.29}$$

The overall proportionality constant can be calculated in any of a number of ways [37, 211, 19, 17, 99].

$$T^{\mu\nu}_{\text{Casimir}} = \frac{\pi^2}{720} \frac{\hbar}{a^4} \left(\eta^{\mu\nu} - 4\hat{z}^\mu \hat{z}^\nu \right).$$
(12.30)

Since the energy density is negative, $\rho = -(\pi^2 \hbar)/(720 a^4)$, this automatically violates the weak and dominant energy conditions. A quick check shows that the null energy condition is violated ($\rho + p_z < 0$). This implies that the strong energy condition is also violated. (Notation: $\rho \equiv T^{\hat{t}\hat{t}}$; $p_x \equiv T^{\hat{x}\hat{x}} = p_y \equiv T^{\hat{y}\hat{y}}$; $p_z \equiv T^{\hat{z}\hat{z}}$.)

The fact that the Casimir effect is real is experimental evidence that the null, weak, strong, and dominant energy conditions are sometimes violated by quantum effects. Granted, the effect is tiny. For realistic plates, the mass of the plates themselves is always much larger than the Casimir energy. Thus a realistic pair of conducting plates will not, in general, violate the averaged energy conditions. This observation should give one some pause, and suggest that blind reliance on the Casimir energy for traversable wormhole maintenance might not be as profitable a scheme as hoped.

To see this, model the physical plates as sheets of surface mass density σ. Since the plates are normal matter (metal) the internal stresses can be safely ignored. Denoting by $\hat{t}$ a unit vector in the time direction, a more realistic model for the total stress-energy is

$$
\begin{aligned}
T^{\mu\nu}_{\text{Casimir}} = \ & \sigma \, \hat{t}^\mu \hat{t}^\nu \left[\delta(z) + \delta(z - a) \right] \\
& + \Theta(z)\Theta(a - z) \frac{\pi^2}{720} \frac{\hbar}{a^4} \left[\eta^{\mu\nu} - 4\hat{z}^\mu \hat{z}^\nu \right].
\end{aligned}
$$
(12.31)

For a null geodesic that impacts the plate at incidence angle θ, one has $k^\mu = (1, \sin\theta, 0, \cos\theta)$. Then provided $\cos\theta \neq 0$

$$\int_\Gamma T_{\mu\nu} k^\mu k^\nu \, dt = \frac{1}{\cos\theta} \left(2\sigma - 4\cos^2\theta |\rho| a \right).$$
(12.32)

On the other hand, if $\cos\theta = 0$ one is dealing with a null geodesic that is parallel to the plates (fixed z) and never crosses the plates themselves. For a null geodesic parallel to and lying between the plates

$$\int_\Gamma T_{\mu\nu} k^\mu k^\nu \, dt = 0.$$
(12.33)

Thus a necessary condition for the violation of the averaged null energy condition is that

$$\sigma < \frac{\pi^2 \hbar}{360 a^3}.$$
(12.34)

This is physically unreasonable. Let the plates be modeled by a metallic lattice consisting of atoms of atomic mass m separated by a lattice spacing d. Let the thickness of the plate be L. Then

$$\sigma = \frac{mL}{d^3}. \tag{12.35}$$

With regard to violating the ANEC, the Casimir effect can overwhelm the effect of the mass of the plates themselves only if

$$a < d \sqrt[3]{\frac{\pi^2 \hbar}{360 m L}} \ll d. \tag{12.36}$$

This requires the plate separation to be much less than the interatomic spacing between the atoms comprising the plates. Once plate separations are this small one has no justification for believing the original derivation in terms of ideal conducting sheets.

A modification of this argument can be applied to the averaged weak and averaged strong energy conditions. [For electromagnetism, in the absence of any trace (conformal) anomaly, the tracelessness of the stress-energy tensor implies that the averaged weak and averaged strong energy conditions are equivalent.] Consider the timelike geodesic specified by the four-velocity $V^\mu = \gamma(1, \beta \sin\theta, 0, \beta \cos\theta)$. Then for $\cos\theta \neq 0$

$$\int_\Gamma T_{\mu\nu} V^\mu V^\nu \, ds = \frac{\gamma}{\beta \cos\theta} \left\{ 2\sigma - [(1 - \beta^2) + 4\beta^2 \cos^2\theta] \, |\rho| \, a \right\}. \tag{12.37}$$

The averaged weak (averaged strong) energy condition cannot be violated unless σ is unphysically small,

$$\sigma < [(1 - \beta^2) + 4\beta^2] \, \frac{\pi^2 \hbar}{1440 a^3}. \tag{12.38}$$

An exceptional class of geodesics arises for $\cos\theta = 0$. For a geodesic that is parallel to and lies between the plates,

$$\int_\Gamma T^{\mu\nu} V^\mu V^\nu \, ds = -|\rho| \int ds, \tag{12.39}$$

which is clearly negative. This admittedly restricted class of timelike geodesics violates both the averaged weak and averaged strong energy conditions. (One cannot and should not infer a violation of the averaged null energy condition.) This is an explicit counterexample to the suggestion that quantum field theory might always enforce the AWEC in (3+1) dimensions.

Some caution is in order in interpreting the experimental results. Experimentally, one does not measure the stress-energy tensor itself. Instead, one measures the attractive force between the plates. For a pair of large plates of surface area S this force is

$$F = p_z S = -\frac{\pi^2}{240} \frac{\hbar S}{a^4}. \tag{12.40}$$

The total Casimir energy can be calculated in two ways:

$$E = \int_a^\infty F(a') da' = -\frac{\pi^2}{720} \frac{\hbar S}{a^3}, \tag{12.41}$$

and

$$E = \rho \times (\text{volume}) = \rho \, S \, a = -\frac{\pi^2}{720} \frac{\hbar S}{a^3}. \tag{12.42}$$

The fact that the Casimir force is attractive implies that the Casimir energy is negative. The negativity of the Casimir energy is simply a statement that the system with conducting plates present is more tightly bound than the vacuum—it has less energy than the vacuum. The Casimir energy is the observable part of the zero point energy. It is the finite difference between the otherwise formally infinite zero point energies calculated in the presence and the absence of the conducting plates. ("Formally infinite" presumably means "cut off at the Planck scale", that is, of order m_P/ℓ_P^3.)

A final warning: For other geometries (spheres, ellipsoids, random surfaces) calculations of the Casimir energy are much more difficult. Casimir energies are not always negative, they can be (and sometimes are) positive. See, for example, [19].

12.3.3 Topological Casimir effect

A variation on the theme of the original Casimir effect occurs when one uses topology, rather than physical metal plates, to distort the vacuum. This occurs, for instance, if one imposes periodic boundary conditions on the universe. For example, take the universe to be periodic in the z direction, with periodicity a. This is equivalent to assuming that the universe is rolled up into a cylinder of circumference a. Then for every quantum field, the z component of momentum is constrained by

$$k_z = n\frac{2\pi}{a}. \tag{12.43}$$

This is slightly different from the constraint obtained for conducting plates.

The symmetries of the problem constrain the stress-energy tensor to be of the same form as for the parallel plate geometry, but with a different multiplicative prefactor [99]

$$T^{\mu\nu}_{\text{Casimir}} = \frac{\pi^2}{45} \frac{\hbar}{a^4} \left(\eta^{\mu\nu} - 4\hat{z}^\mu \hat{z}^\nu \right). \tag{12.44}$$

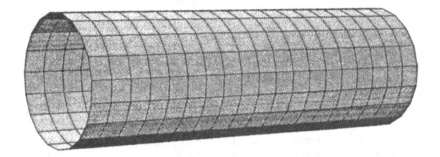

Figure 12.2: Casimir effect: topological Casimir effect in a periodic universe.

As was previously the case, the form of this stress-energy tensor implies violation of the null, weak, strong, and dominant energy conditions.

The major difference is now that there are no plates, and no associated delta-function contributions to the stress-energy. Consequently, taking inextendible null and timelike geodesics (which are in this case infinite straight lines) the averaged null, averaged weak, and averaged strong energy conditions can also be violated. For a null geodesic in the $k^\mu = (1, \sin\theta, 0, \cos\theta)$ direction

$$\int_\Gamma T_{\mu\nu} \, k^\mu k^\nu \, dt = -4\cos^2\theta \, |\rho| \int dt. \qquad (12.45)$$

For a timelike geodesic in the $V^\mu = \gamma(1, \beta\sin\theta, 0, \beta\cos\theta)$ direction

$$\int_\Gamma T_{\mu\nu} \, V^\mu V^\nu \, ds = -[(1 - \beta^2) + 4\beta^2 \cos^2\theta] \, |\rho| \int ds. \qquad (12.46)$$

These observations are of some interest as they show that at least in some $(3 + 1)$-dimensional geometries almost all geodesics violate the averaged energy conditions. In fact, this is an explicit counterexample to the suggestion that the axioms of quantum field theory might universally enforce the averaged null energy condition.

12.3.4 Squeezed vacuum

The squeezed vacuum is an exotic distortion of the quantum-electrodynamic vacuum. Squeezing a quantum state [278] results in some of the quantum mechanical degrees of freedom having an unusually low variance. (Unusually low variance means, less variance than would be expected on the basis of the equipartition theorem.) The canonically conjugate degrees of freedom must, by the Heisenberg uncertainty principle, have an unusually high variance.

This trick can be used to excavate energy from one place in the ordinary vacuum state at the cost of piling up excess energy elsewhere [28, 190]. The regions from which energy has been excavated are lower in energy than an equal volume of ordinary vacuum and so by definition have a locally negative energy density.

The squeezed vacuum state violates the null, weak, strong, and dominant energy conditions, but does not violate the averaged energy conditions [28, 190].

Hochberg and Kephart have argued that vacuum squeezing is directly relevant to the issue of traversable wormhole existence [138].

12.3.5 Hawking evaporation

The existence of Hawking radiation from evaporating black holes has not (yet) been verified experimentally. This is due to a dearth of suitable small black holes in our immediate vicinity—we have not yet been able to acquire one for laboratory investigations.

Nevertheless the Hawking evaporation process is a relatively clean "theoretical laboratory". That several different calculational schemes converge on the same result gives us considerable confidence that we know what we are doing.

The very fact that Hawking evaporation occurs at all violates the area increase theorem (second law) for classical black holes. This implies that the quantum processes underlying the Hawking evaporation process must also induce a violation of one or more of the input assumptions used in proving the classical area increase theorem. The only input assumption that seems vulnerable to quantum violation is the assumed applicability of the null energy condition. Indeed, it is possible to numerically calculate the renormalized vacuum expectation value of the quantum stress-energy tensor in the vicinity of the event horizon and in this way to verify directly that the null energy condition is violated by quantum effects.

For a black hole that is evaporating into vacuum, the relevant vacuum state is the so-called Unruh vacuum. On the other hand, a black hole that is in thermal equilibrium with its surroundings is described by the so-called Hartle–Hawking vacuum state [257, 39, 276].

12.3.6   Hartle–Hawking vacuum

Explicit numerical calculations [147, 32, 146] show that, at the horizon of a Schwarzschild black hole, a conformally-coupled scalar field (in the Hartle–Hawking vacuum state) has stress-energy[3]

$$\langle 0_H | T^{\mu\nu} | 0_H \rangle \approx \hbar \, \frac{\pi^2}{90} \left(\frac{1}{8\pi M} \right)^4 \begin{bmatrix} -37.728 & 0 & 0 & 0 \\ 0 & +37.728 & 0 & 0 \\ 0 & 0 & +10.29 & 0 \\ 0 & 0 & 0 & +10.29 \end{bmatrix}.$$

$$(12.47)$$

Other papers investigating the renormalized stress-energy tensor include [7, 30, 34, 97, 39, 33, 158, 200, 98]. Note that these calculations are all performed in the test-field limit; gravitational back-reaction is ignored.

Exercise 1: (Trivial) Check that at the event horizon the null, weak, strong, and dominant energy conditions are all violated. (Reminder: $\rho \equiv T^{\hat{t}\hat{t}}$; $p \equiv T^{\hat{\theta}\hat{\theta}} = T^{\hat{\phi}\hat{\phi}}$; $\tau \equiv -T^{\hat{r}\hat{r}}$.) Hints: (1) The violation of the weak and dominant energy conditions follows from the negativity of ρ. (2) Checking that $\rho + p < 0$, one also sees that the null and strong energy conditions are violated.

Exercise 2: (Literature search) Repeat the previous exercise for the Unruh vacuum.

Exercise 3: (Literature search) Find out what the Boulware vacuum is. Try to repeat the previous exercise for the Boulware vacuum. To what extent does this exercise even make sense?

The precise status of the averaged energy conditions is considerably more complicated. For simplicity I limit discussion to the Hartle–Hawking vacuum:

- The numerical data of Howard [146] indicate that, outside the horizon, $\rho - \tau > 0$. (At the horizon itself, $\rho = \tau$.) The numerical data cover the range from $r = 2M$ to $r = 6.8M$. For larger radial distances one resorts to Page's analytic approximation, which becomes increasingly good as one moves further away from the horizon [146, 200]. This implies that the averaged null energy condition is satisfied for radial null geodesics. Caveat: The ANEC integral has to be modified to a one–sided integral by cutting it off at the event horizon.

---

[3]Warning: The paper by Fawcett [81] unfortunately contains an error. See [147, 146].

- For the unstable circular photon orbit at $r = 3m$ the ANEC integral is proportional to $\rho + p$. Inspection of Howard's numerical data indicates that $\rho + p > 0$ at $r = 3M$, so the ANEC is satisfied along this particular null geodesic.

- For any circular curve at fixed r (*not* a geodesic except in the case of $r = 3M$) the ANEC integral is still proportional to $\rho + p$. Inspection of Howard's numerical data indicates that $\rho + p < 0$ for $r \lesssim 2.25\,M$, so the ANEC is *not* satisfied for this particular class of nongeodesic null curves. (Page's analytic approximation gives a slightly different result, $\rho + p < 0$ for $r < 2.18994\,M$.)

- Null geodesics that come in from infinity and return to infinity never get closer to the origin than $r = 3\,M$. See, for instance, [186, pp. 672–678]. The ANEC integral may be written as

$$\int_{\Gamma} (\rho - \tau \cos^2 \psi + p \sin^2 \psi) \xi^2 d\lambda. \tag{12.48}$$

Inspection of Howard's numerical data indicates that the integrand is strictly positive for $r \geq 3\,M$, so the ANEC is satisfied along all null geodesics that come from, and return to, infinity.

Exercise: Look at incoming null geodesics that have smaller than critical impact parameter. These geodesics come in from infinity, they may then circle the black hole a large number of times, but are guaranteed to finally plunge into the event horizon. Show that the ANEC is satisfied along these null geodesics.

Research Problem 4 *Generalize these observations. Go beyond the numerics. Develop some analytic arguments. Repeat the analysis for the Reissner–Nordström geometry, and then for arbitrary spherically symmetric black holes.*

12.3.7 Cosmological inflation

It is widely believed that the universe was subject to an inflationary epoch shortly after the big bang [168]. The mechanism driving cosmological inflation is essentially a change in the effective cosmological constant. This change is presumably due to the universe entering a false vacuum state after a GUT scale phase transition. This illustrates what is perhaps the most troubling aspect of the cosmological constant problem. The present-day cosmological constant is zero to fantastic accuracy $|\Lambda_{\text{now}}| < 8 \times 10^{-122} \ell_P^{-2}$. However, during the inflationary epoch, the best estimates are

$$|\Lambda_{\text{inflation}}| \approx +(m_{\text{GUT}}/m_P)^4 \ell_P^{-2} \approx +10^{-12} \ell_P^{-2} \gg 10^{+110} |\Lambda_{\text{now}}|. \tag{12.49}$$

By rewriting the total stress-energy tensor as

$$T^{\mu\nu}_{\text{total}} \equiv T^{\mu\nu}_{\text{ordinary}} - \frac{\Lambda}{8\pi G}\, g^{\mu\nu}, \qquad (12.50)$$

one sees that a positive (effective) cosmological constant corresponds to a positive vacuum energy density (and a negative vacuum pressure).

Considered in isolation, a positive (effective) cosmological constant satisfies the NEC, WEC, and DEC, but violates the SEC. Similarly, considered in isolation, a negative (effective) cosmological constant satisfies the NEC and SEC, but violates the WEC and DEC. In particular, the inflationary epoch (with its positive effective cosmological constant) violates the SEC.

12.3.8   Comments

Many other violations of the energy conditions are known:

- The quantum stress-energy tensor of a free massive scalar field violates the weak and dominant energy conditions. A proof using formal axiomatic field theory was provided in [79].[4] A more direct analysis in terms of interference effects between states of different particle occupation number is given in [52]. See also [294, 166].

- Moving mirrors lead to violations of the local energy conditions [100, 51, 85, 86].

- Certain interacting field theories are known to violate the SEC without violating the WEC [229, 230].

- Massive Dirac particles in a Kerr background are known to violate the WEC [274].

While the classical validity of the energy conditions are perfectly reasonable assumptions it is clear that (semiclassical) [5] quantum effects are capable of violating the null, weak, strong, and dominant energy conditions. The precise conditions under which quantum effects in $(3 + 1)$-dimensional spacetime are capable of violating the averaged energy conditions is unknown. Further discussion of this point will be delayed until we address the question of topological censorship.

The quantum-induced violations of the null, weak, strong, and dominant energy conditions are typically very small. After all, by definition these are

---

[4]Exercise: (Not trivial) Generalize the argument of that paper to show that the null and strong energy conditions are also violated. This argument can be viewed as a precursor to the squeezed vacuum analysis.

[5]Semiclassical in the sense that quantum field theory is invoked on a fixed (possibly curved) background spacetime manifold—no attempt being made to quantize geometry itself.

order $\hbar$ effects! It is far from clear whether or not it is possible to get the large violation of the energy conditions that we shall soon see is required to support a macroscopic traversable wormhole.

12.4 Wormhole spacetimes

12.4.1 NEC, WEC, DEC, and SEC

The analysis of the traversable wormhole system has already shown that

$$\exists \tilde{r}_* \Big| \forall r \in (r_0, \tilde{r}_*), \qquad (\rho - \tau) < 0. \tag{12.51}$$

At the throat itself a weaker inequality holds:

$$[\rho(r_0) - \tau(r_0)] \leq 0. \tag{12.52}$$

From inspection of the preceding definitions, this shows that the null energy condition is violated over the finite range $(r_0, \tilde{r}_*)$ in the vicinity of the wormhole throat. This automatically implies violations of the weak, strong, and dominant energy conditions in this same range. At the throat itself, the null energy condition is either violated or on the verge of being violated.

Thus that the wormhole throat must be threaded by "exotic matter", where I shall define "exotic matter" to be matter that violates the null energy condition. Morris and Thorne define exotic matter as matter that violates the weak energy condition. This minor definitional difference is of no great import.

12.4.2 ANEC

To begin to answer the question "how much exotic matter is needed?", we consider the averaged null energy condition.

There is a general argument, based on the Raychaudhuri equation [132, pp. 86–88,95], that implies that the ANEC must be violated along radial null geodesics in a spherically symmetric traversable wormhole spacetime. The argument is essentially due to Page, and a version is described in the original Morris–Thorne paper [190]. Generalizing the argument away from spherical symmetry ultimately leads to the Friedman–Schleich–Witt topological censorship theorem [92]. For the time being, in the interest of keeping the discussion explicit, I shall focus on spherical symmetry.

Consider a bundle of radial null geodesics (light rays) that enters a spherical traversable wormhole at one mouth and exits at the other. The cross-sectional area of this bundle must first be decreasing and then, after passing through the throat, must increase.

The Raychaudhuri equation for null geodesics states [132, pp. 86–88,95]

$$\frac{d\hat{\theta}}{d\lambda} = -R_{\mu\nu} \, k^{\mu}k^{\nu} - 2\hat{\sigma}^2 - \frac{1}{2}\hat{\theta}^2 + 2\hat{\omega}^2. \qquad (12.53)$$

I adopt the notation of Hawking and Ellis: here $\hat{\theta}$ is the expansion of the bundle of null geodesics, the fractional rate of change of cross-sectional area. The symbol $\hat{\sigma}$ denotes the shear of the bundle. The symbol $\hat{\omega}$ denotes the vorticity of the bundle.

Exercise:

- Show that $\hat{\theta} = \pm(1/r^2)(dr^2/d\lambda) = \pm(2/r)(dr/d\lambda)$.

- Show that far away from the throat, in the asymptotically flat region, $\hat{\theta} \to \pm 2/r$.

- Show that the shear $\hat{\sigma}$ is identically zero. Hint: The shear tensor for radial null geodesics is by definition orthogonal to both $\hat{r}$ and $\hat{t}$, and is by definition traceless in the remaining θ-φ subspace. Apply spherical symmetry.

- Show that the vorticity $\hat{\omega}$ is identically zero. Hint: Everything is moving radially.

The results of this exercise imply that the Raychaudhuri equation reduces to

$$\frac{d\hat{\theta}}{d\lambda} = -R_{\mu\nu} \, k^{\mu}k^{\nu} - \frac{1}{2}\hat{\theta}^2. \qquad (12.54)$$

Now, by definition, the moment of minimal cross-sectional area occurs at the throat. So at the throat $\hat{\theta} = 0$, but $d\hat{\theta}/d\lambda \geq 0$, which implies

$$R_{\mu\nu} \, k^{\mu}k^{\nu} \leq 0. \qquad (12.55)$$

This is a borderline violation of the null convergence condition. Using the Einstein field equations this translates into a borderline violation of the null energy condition

$$T_{\mu\nu} \, k^{\mu}k^{\nu} \leq 0. \qquad (12.56)$$

A stronger result follows from integrating the Raychaudhuri equation along a radial null geodesic:

$$\int_{\lambda_1}^{\lambda_2} R_{\mu\nu} \, k^{\mu}k^{\nu} d\lambda = -\frac{1}{2}\int_{\lambda_1}^{\lambda_2} \hat{\theta}^2 d\lambda \; - \; \hat{\theta}\Big|_{\lambda_1}^{\lambda_2}. \qquad (12.57)$$

Now let $\lambda_1 \to -\infty$; while $\lambda_2 \to +\infty$. Since $\hat{\theta}^2$ is positive semidefinite, not identically zero, and since the boundary terms are now zero, one deduces

$$\int_{-\infty}^{+\infty} R_{\mu\nu} k^\mu k^\nu d\lambda < 0. \qquad (12.58)$$

Equivalently, using the Einstein field equations

$$\int_{-\infty}^{+\infty} T_{\mu\nu} k^\mu k^\nu d\lambda < 0. \qquad (12.59)$$

So the ANEC must be violated along inextendible radial null geodesics. This result can be made more explicit by direct evaluation of the ANEC integral along a radial null geodesic

$$I_\Gamma = \int_\Gamma (\rho - \tau) \, \xi^2 \, d\lambda. \qquad (12.60)$$

To calculate this explicitly, we will need to express the affine parameter λ in terms of the radial variable r. Because we are dealing with a radial null curve we know

$$\sqrt{g_{tt}} \, dt = \sqrt{g_{rr}} \, dr \quad \Rightarrow \quad e^\phi dt = \frac{dr}{\sqrt{1 - b/r}} = dl. \qquad (12.61)$$

For an affine parameterization, the inner product between the Killing vector K and the tangent vector k is a constant. Without loss of generality we can set this constant to 1. (Remember: The ANEC integral is only defined up to an irrelevant overall factor.) Thus

$$K^\mu \, g_{\mu\nu} \, k^\nu = g_{tt} \, k^t = g_{tt} \, \frac{dt}{d\lambda} = 1 \quad \Rightarrow \quad d\lambda = e^{+2\phi} dt = e^\phi dl. \qquad (12.62)$$

Finally, $k^t = e^{-2\phi}$ implies $\xi \equiv k^l = e^{-\phi}$. Pulling this all together

$$I_\Gamma = \int_{l=-\infty}^{l=+\infty} (\rho - \tau) \, e^{-\phi} \, dl. \qquad (12.63)$$

The use of the field equations allows one to rewrite this integral in terms of geometrical quantities. Splitting the integral into two pieces (from spatial infinity down to the throat, and then out to the other spatial infinity) one has

$$
\begin{aligned}
8\pi G \, I_\Gamma^\pm &= -\int_{r_0}^\infty \frac{1}{r} \left[\left(1 - \frac{b_\pm}{r}\right)' - 2\phi_\pm' \left(1 - \frac{b_\pm}{r}\right) \right] \frac{e^{-\phi_\pm}}{\sqrt{1 - b_\pm/r}} dr, \\
&= -\int_{r_0}^\infty \frac{2}{r} \left[e^{-\phi_\pm} \sqrt{1 - \frac{b_\pm}{r}} \right]' dr. \qquad (12.64)
\end{aligned}
$$

Integrate by parts, noting that the boundary contributions from the throat and from spatial infinity vanish. Then

$$4\pi G\, I_\Gamma^\pm = -\int_{r_0}^\infty \frac{1}{r^2} e^{-\phi_\pm} \sqrt{1 - \frac{b_\pm}{r}}\; dr. \qquad (12.65)$$

Since the integral is now strictly negative, we confirm that spherically symmetric wormhole spacetimes violate the averaged null energy condition.

Exercise: Obtain this result directly from the integrated Raychaudhuri equation. (This allows one to delay use of the Einstein field equations until the very last step.)

1. Trivial: Evaluate $\hat\theta(\infty)$ and $\hat\theta(r_0)$.

2. Easy: Evaluate $\int_{r_0}^\infty \hat\theta^2 d\lambda$.

Note that this result gives slightly more information than just saying that the ANEC is violated along *inextendible* radial null geodesics. It also guarantees that the ANEC is violated along certain *partial* radial null geodesics. In particular the ANEC is violated along partial radial null geodesics that start at the throat and go to spatial infinity. In the case of a multiple throat wormhole [multiple local minima of $R(l)$], the ANEC is violated along partial radial null geodesics that proceed from one throat to another.

The total ANEC integral, including contributions from both asymptotically flat regions, is

$$I_\Gamma = -\frac{1}{4\pi G} \int_{r_0}^\infty \frac{1}{r^2} \left[e^{-\phi_+} \sqrt{1 - \frac{b_+}{r}} + e^{-\phi_-} \sqrt{1 - \frac{b_-}{r}} \right] dr. \qquad (12.66)$$

We can place some bounds on this integral. Suppose that the surface of the wormhole occurs at $R^\pm > r_0$. That is, suppose the stress energy (exotic or otherwise) to be confined to radii less than or equal to $R_\pm$, so that the spacetime outside $R_\pm$ is given by a piece of the Schwarzschild solution. Then for $r > R_\pm$

$$b(r) = 2Gm_\pm; \qquad e^{\phi_\pm(r)} = e^{\phi_\pm(\infty)} \sqrt{1 - \frac{2Gm_\pm}{r}}. \qquad (12.67)$$

And so

$$4\pi G\, I_\Gamma^\pm = -\frac{e^{-\phi_\pm(\infty)}}{R_\pm} - \int_{r_0}^{R_\pm} \frac{1}{r^2} e^{-\phi_\pm} \sqrt{1 - \frac{b_\pm}{r}}\; dr. \qquad (12.68)$$

Whence

$$I_\Gamma < -\frac{1}{4\pi G} \left[\frac{e^{-\phi_+(\infty)}}{R_+} + \frac{e^{-\phi_-(\infty)}}{R_-} \right]. \qquad (12.69)$$

For a symmetric wormhole we can, without loss of generality, set $\phi(\infty) = 0$. Thus

$$I_\Gamma < -\frac{1}{2\pi G}\frac{1}{R}. \tag{12.70}$$

The amount by which a spherically symmetric traversable wormhole violates the averaged null energy condition is bounded above by the radius at which the wormhole first deviates from the nontraversable Schwarzschild wormhole.

To obtain a lower bound on the ANEC integral, note that by construction $\phi_\pm(r)$ remains finite and bounded throughout the interval $[r_0, \infty)$. Thus $\phi_\pm^{min} \equiv \min\{\phi_\pm(r)\}$ exists and is finite. From equation (12.66) one has

$$I_\Gamma \;>\; -\frac{1}{4\pi G}\int_{r_0}^\infty \frac{1}{r^2}\left[e^{-\phi_+} + e^{-\phi_-}\right]\, dr, \tag{12.71}$$

$$\;>\; -\frac{1}{4\pi G}\frac{1}{r_0}\left[e^{-\phi_+^{min}} + e^{-\phi_-^{min}}\right]. \tag{12.72}$$

So, modulo technical fiddles concerning the minimum ϕ^{min}, the violations of the averaged null energy condition are bounded below by the radius of the throat. This minimum, corresponding to a minimum of the gravitational potential e^ϕ, will typically occur at the throat of the wormhole, but there is no general requirement that it occurs there.

Various other inequalities along these lines can also be derived. One could also work with nonradial geodesics, but this gets rather messy.

Research Problem 5 *Develop a clean and natural measure to characterize the "total amount" of exotic matter in the wormhole. Derive inequalities for the "total amount" of exotic matter that are variations on the above.*

12.4.3 AWEC

Analyzing the averaged weak energy condition requires consideration of timelike geodesics. Let us again restrict attention to radial geodesics; then the AWEC integral is

$$I_\Gamma[\text{AWEC}] = \int_\Gamma \gamma^2\,(\rho - \beta^2\tau)\, ds. \tag{12.73}$$

To proceed, express the proper time parameter s in terms of the radial variable l by noting $dl = \gamma\beta\, ds$. Using $\gamma \equiv 1/\sqrt{1 - \beta^2}$ gives

$$I_\Gamma[\text{AWEC}] = \int_\Gamma \left\{\sqrt{\gamma^2 - 1}(\rho - \tau) + \frac{\rho}{\sqrt{\gamma^2 - 1}}\right\}\, dl. \tag{12.74}$$

Reminder: In the orthonormal basis velocity components are defined by $V^{\hat{\mu}} = (\gamma, \gamma\beta, 0, 0)$. In the coordinate basis

$$V^\mu = \left(e^{-\phi}\gamma, \sqrt{1 - b/r}\; \gamma\beta, 0, 0 \right). \tag{12.75}$$

Exercise: (Trivial) Check that $g_{\mu\nu} V^\mu V^\nu = -1$.

For a timelike geodesic parameterized by proper time, the inner product between the Killing vector K and the tangent vector V is a constant, thus

$$K^\mu \; g_{\mu\nu} \; V^\nu = g_{tt} \; V^t = e^{2\phi} \frac{dt}{ds} = e^\phi \gamma = e^{\phi(\infty)} \gamma_\infty. \tag{12.76}$$

Thus

$$\gamma = e^{-\phi} e^{+\phi(\infty)} \gamma_\infty. \tag{12.77}$$

Consider the limit as $\gamma_\infty \to \infty$, that is $\beta_\infty \to 1$. Then, among other things, $\gamma \to \infty$, while $\beta \to 1$. Now evaluate

$$
\begin{aligned}
\lim_{\gamma_\infty \to \infty} \left\{ \frac{I_\Gamma[\text{AWEC}]}{e^{+\phi(\infty)} \gamma_\infty} \right\} &= \lim_{\gamma_\infty \to \infty} \left\{ \frac{\int_\Gamma \gamma^2 \left(\rho - \beta^2\tau \right) ds}{e^{+\phi(\infty)} \gamma_\infty} \right\}, \\
&= \lim_{\gamma_\infty \to \infty} \left\{ \int_\Gamma \gamma e^{-\phi} \left(\rho - \beta^2\tau \right) ds \right\}, \\
&= \lim_{\gamma_\infty \to \infty} \left\{ \int_\Gamma (\beta^{-1}\rho - \beta\tau) \, e^{-\phi} \, dl \right\}, \\
&= \int_\Gamma (\rho - \tau) e^{-\phi} dl \\
&= I_\Gamma[\text{ANEC}]. \tag{12.78}
\end{aligned}
$$

Thus the appropriate limit of the averaged weak energy condition, obtained by taking timelike radial geodesics that traverse the wormhole increasingly rapidly, is in fact the averaged null energy condition, as required. Having already shown that (spherically symmetric) traversable wormhole spacetimes violate the averaged null energy condition, it follows that sufficiently high speed timelike radial geodesics must violate the averaged weak energy condition.

(Remember, the ANEC is defined only up to an arbitrary positive multiplicative constant. This is why we are allowed to divide by a suitable normalizing factor in the preceding calculation.)

While we could certainly explicitly exhibit the AWEC integral in terms of the geometry (the shape and redshift functions) and the parameter γ_∞, no additional insight would be gained by doing so. Explicitly exhibiting the ASEC integral leads to even messier formulas.

Chapter 13

Engineering considerations

13.1 Tidal disruption

If the Riemann curvature is too large our intrepid traveler will be ripped apart by tidal effects. Keeping the tidal forces small is a fundamental engineering consideration in making a wormhole "traversable in practice" as opposed to merely "traversable in principle".

13.1.1 Radial motion

Consider a radially moving traveler with four-velocity V^μ. (There is no requirement that the motion be geodesic [190]. The following calculation applies to arbitrarily accelerated travelers as long as the motion is radial.) Consider two points in his/her/its body that are separated by a vector amount $(\Delta\xi)^\mu$. By going to the traveler's rest frame it is easy to see that $V^\mu(\Delta\xi)_\mu = 0$. These two points are subject to slightly different accelerations, this difference being governed by the inhomogeneity of the gravitational field. This difference in accelerations is the definition of a tide, and these tidal effects are proportional to the Riemann tensor. Explicitly

$$(\Delta a)^\mu = -R^\mu{}_{\alpha\nu\beta}\, V^\alpha(\Delta\xi)^\nu V^\beta. \tag{13.1}$$

Warning: This looks like, but is not quite, the equation for *geodesic* deviation. Thus one might be worried that the above equation can only be applied in cases of geodesic motion. That this equation applies also to accelerated motion can be seen from the argument by Morris and Thorne [190,

137

p. 404] which is itself based on [186, box 37.1, pp. 1007–1009]. The sub-tleties involved in the argument are largely a matter of choosing appropriate definitions.

The symmetries of the Riemann tensor now automatically imply that $(\Delta a)^\mu V_\mu = 0$. Resolve this tidal equation into radial and transverse com-ponents. Let n^μ be the radial vector perpendicular to V^μ. That is,

$$V^{\hat\mu} = (\gamma, \gamma\beta, 0, 0), \qquad \text{while} \qquad n^{\hat\mu} = (\gamma\beta, \gamma, 0, 0). \tag{13.2}$$

The hats on the indices indicate the use of an orthonormal frame. Thus $\beta = v/c$ and $\gamma = 1/\sqrt{1 - \beta^2}$ take on their usual Minkowski space values.

Now decompose

$$(\Delta\xi)^\mu = (\Delta\xi)_{||} \, n^\mu + (\Delta\xi)^\nu_\perp, \tag{13.3}$$

where by definition $(\Delta\xi)^\mu_\perp V_\mu = 0 = (\Delta\xi)^\mu_\perp n_\mu$. Perform the same decom-position on the relative acceleration vector Δa. Then a brief calculation shows

$$(\Delta a)_{||} \equiv -(R_{\mu\alpha\nu\beta} \, n^\mu V^\alpha n^\nu V^\beta) \, (\Delta\xi)_{||} = -R_{\hat t \hat r \hat t \hat r} \, (\Delta\xi)_{||}. \tag{13.4}$$

Introducing the projection operator

$$g^{\mu\nu}_\perp = g^{\mu\nu} + V^\mu V^\nu - n^\mu n^\nu \tag{13.5}$$

which projects tensors onto the transverse θ-φ subspace,

$$
\begin{aligned}
(\Delta a)^\sigma_\perp &\equiv -\frac{1}{2}(R_{\mu\alpha\nu\beta} \, g^{\mu\nu}_\perp V^\alpha V^\beta) \, (\Delta\xi)^\sigma_\perp \\
&= -\gamma^2 \left(R_{\hat\theta\hat t\hat\theta\hat t} + \beta^2 R_{\hat\theta\hat r\hat\theta\hat r} \right) (\Delta\xi)^\sigma_\perp.
\end{aligned} \tag{13.6}
$$

In terms of the redshift and shape function

$$(\Delta a)_{||} = \left[\left(1 - \frac{b}{r}\right) \{-\phi'' - (\phi')^2\} + \frac{1}{2r^2}(b'r - b)\phi' \right] (\Delta\xi)_{||}, \tag{13.7}$$

$$(\Delta a)_\perp = \frac{\gamma^2}{r^2} \left[(r - b)\phi' + \frac{\beta^2}{2}\left(b' - \frac{b}{r}\right) \right] (\Delta\xi)_\perp. \tag{13.8}$$

If an object of size ℓ is designed to withstand a maximum tidal acceleration g and is to traverse the wormhole safely, one must have

$$\left| \left(1 - \frac{b}{r}\right) \{-\phi'' - (\phi')^2\} + \frac{1}{2r^2}(b'r - b)\phi' \right| \leq \frac{g}{\ell}, \tag{13.9}$$

$$\frac{\gamma^2}{r^2} \left| (r - b)\phi' + \frac{\beta^2}{2}\left(b' - \frac{b}{r}\right) \right| \leq \frac{g}{\ell}. \tag{13.10}$$

The first of these inequalities can be interpreted as a constraint on the gradient of the redshift, while the second is effectively a constraint on the speed at which the traveler can safely traverse the wormhole.

To get a feel for the size of these tidal effects, denote the Earth's surface gravity by $g_\oplus = 9.8$ m s^{-2}. Consider a typical length scale of $\ell_0 = 2$ m. Then, in units where $c = 1$,

$$\frac{g}{\ell} \approx \frac{g/g_\oplus}{\ell/\ell_0} \frac{1}{(10^8 \text{ m})^2}. \tag{13.11}$$

The tidal inequalities are particularly simple when evaluated at the throat of the wormhole:

$$|\phi'| \leq \frac{2g \, r_0}{(1-b') \, \ell}, \tag{13.12}$$

$$\gamma^2 \beta^2 \leq \frac{2g \, r_0^2}{(1-b') \, \ell}. \tag{13.13}$$

In many examples the tidal stresses are maximum at the instant of traversing the throat, but there is no general requirement to this effect.

The constraint on transit velocity is actually a side effect of a more general result: If the tidal forces are velocity independent then the wormhole cannot be traversable. To see how this comes about, note that it is only the transverse part of tidal accelerations that is ever velocity dependent, and that this velocity dependence cancels out if and only if

$$R_{\hat\theta\hat t\hat\theta\hat t} = -R_{\hat\theta\hat r\hat\theta\hat r}. \tag{13.14}$$

This places a constraint on the geometry

$$b' - \frac{b}{r} = -2r \left(1 - \frac{b}{r}\right) \phi'. \tag{13.15}$$

Integrating this constraint yields

$$e^{2\phi(r)} = e^{2\phi(\infty)} \left(1 - \frac{b}{r}\right). \tag{13.16}$$

This is enough to inform us that our would-be wormhole is not traversable. (Since $g_{tt} = 0$ at $r = r_0$, we know there is a horizon where we wanted the throat to be.)

Finally, for comparison purposes it is useful to see what happens in the Schwarzschild metric:

$$(\Delta a)_{\|} = +2\frac{GM}{r^3}(\Delta\xi)_{\|}, \tag{13.17}$$

$$(\Delta a)_\perp = -\frac{GM}{r^3}(\Delta\xi)_\perp. \tag{13.18}$$

Indeed, a "small" Schwarzschild black hole will kill an infalling (human) observer with tidal effects long before the central singularity is reached ("small" meaning $2GM < 10^8$ metres, $M < 10^5 M_\odot$.)

Further consideration of these tidal inequalities will be deferred until a subsequent section where we shall discuss a few specific examples.

13.1.2   Nonradial motion

The extra technical fiddles required to deal with tidal deformations under nonradial motion are sufficiently messy to warrant a separate treatment. Let the traveler be moving with four-velocity V^μ. Again, there is no requirement that the motion be geodesic. We have already seen that any two points separated by a vector ξ^μ will be subject to slightly different accelerations:

$$(\Delta a)^\mu = -R^\mu{}_{\alpha\nu\beta} V^\alpha (\Delta \xi)^\nu V^\beta. \tag{13.19}$$

Without loss of generality, we can take any nonradial motion to be equatorial. That is, again adopting an orthonormal frame,

$$V^{\hat\mu} \equiv (V^{\hat t}, V^{\hat r}, V^{\hat\theta}, V^{\hat\phi}) = (\gamma, \gamma\beta\cos\psi, 0, \gamma\beta\sin\psi). \tag{13.20}$$

It is useful to introduce an orthonormal triad, $(n_I; I \in \{1,2,3\})$, that is perpendicular to the four-velocity V. Specifically take

$$
\begin{aligned}
(n_1)^{\hat\mu} &\equiv (\gamma\beta, \gamma\cos\psi, 0, \gamma\sin\psi), & (13.21) \\
(n_2)^{\hat\mu} &\equiv (0, -\sin\psi, 0, \cos\psi), & (13.22) \\
(n_3)^{\hat\mu} &\equiv (0, 0, 1, 0). & (13.23)
\end{aligned}
$$

Then

$$g_{\mu\nu} (n_I)^\mu (n_J)^\nu = \delta_{IJ}, \qquad \text{and} \qquad g_{\mu\nu} V^\mu (n_I)^\nu = 0. \tag{13.24}$$

Exercise: (Trivial) Check that these vectors form an orthonormal tetrad.

Now recall that, by definition, $V^\mu (\Delta\xi)_\mu = 0$. Furthermore $(\Delta a)^\mu V_\mu = 0$. This allows one to decompose both $\Delta\xi$ and Δa as follows:

$$
\begin{aligned}
(\Delta\xi)^\mu &= (\Delta\xi)^I (n_I)^\mu, & (13.25) \\
(\Delta a)^\mu &= (\Delta a)^I (n_I)^\mu. & (13.26)
\end{aligned}
$$

Thus the equation governing tidal accelerations can be reduced to a 3×3 matrix equation

$$(\Delta a)^I = Q^I{}_J (\Delta\xi)^J, \tag{13.27}$$

where the matrix $Q^I{}_J = Q_{IJ}$ is defined by

$$Q_{IJ} \equiv -R_{\mu\alpha\nu\beta} \, (n_I)^\mu V^\alpha (n_J)^\nu V^\beta. \tag{13.28}$$

Tidal forces are fully described by this relatively simple 3×3 matrix. A brief agony of algebra yields

$$Q_{11} = -R_{\hat{t}\hat{r}\hat{t}\hat{r}} \cos^2 \psi - R_{\hat{t}\hat{\varphi}\hat{t}\hat{\varphi}} \sin^2 \psi, \tag{13.29}$$

$$Q_{22} = -\gamma^2 \left(R_{\hat{t}\hat{r}\hat{t}\hat{r}} \sin^2 \psi + R_{\hat{t}\hat{\varphi}\hat{t}\hat{\varphi}} \cos^2 \psi + \beta^2 R_{\hat{r}\hat{\varphi}\hat{r}\hat{\varphi}}\right), \tag{13.30}$$

$$Q_{33} = -\gamma^2 \left(R_{\hat{t}\hat{\varphi}\hat{t}\hat{\varphi}} + \beta^2 R_{\hat{r}\hat{\varphi}\hat{r}\hat{\varphi}} \cos^2 \psi + \beta^2 R_{\hat{\theta}\hat{\varphi}\hat{\theta}\hat{\varphi}} \sin^2 \psi\right), \tag{13.31}$$

$$Q_{12} = -\gamma \left(R_{\hat{t}\hat{r}\hat{t}\hat{r}} - R_{\hat{t}\hat{\varphi}\hat{t}\hat{\varphi}}\right) \sin \psi \cos \psi, \tag{13.32}$$

$$Q_{13} = Q_{23} = 0. \tag{13.33}$$

These final expressions are actually somewhat simpler than intermediate stages of the calculation. The components of the Riemann tensor have been calculated in previous sections of the book, and the present form of the matrix Q_{IJ} makes clear the velocity and angular dependence of the tidal forces. Note that $Q_{12} \neq 0$, this implies the presence of tidally induced shear.

13.1.3 Exercises

Exercise 1: (Tedious) Check these formulas for Q_{IJ}. You may find it easier to do the calculation in stages. Define

$$X_{\mu\nu} \equiv R_{\mu\alpha\nu\beta} \, V^\alpha V^\beta. \tag{13.34}$$

Show that, in the natural orthonormal frame,

$$X_{\hat{t}\hat{t}} = \gamma^2 \beta^2 \left(R_{\hat{t}\hat{r}\hat{t}\hat{r}} \cos^2 \psi + R_{\hat{t}\hat{\varphi}\hat{t}\hat{\varphi}} \sin^2 \psi\right), \tag{13.35}$$

$$X_{\hat{r}\hat{r}} = \gamma^2 \left(R_{\hat{t}\hat{r}\hat{t}\hat{r}} + \beta^2 R_{\hat{r}\hat{\varphi}\hat{r}\hat{\varphi}} \sin^2 \psi\right), \tag{13.36}$$

$$X_{\hat{\theta}\hat{\theta}} = \gamma^2 \left(R_{\hat{t}\hat{\varphi}\hat{t}\hat{\varphi}} + \beta^2 R_{\hat{r}\hat{\varphi}\hat{r}\hat{\varphi}} \cos^2 \psi + \beta^2 R_{\hat{\theta}\hat{\varphi}\hat{\theta}\hat{\varphi}} \sin^2 \psi\right), \tag{13.37}$$

$$X_{\hat{\varphi}\hat{\varphi}} = \gamma^2 \left(R_{\hat{t}\hat{\varphi}\hat{t}\hat{\varphi}} + \beta^2 R_{\hat{r}\hat{\varphi}\hat{r}\hat{\varphi}} \cos^2 \psi\right), \tag{13.38}$$

$$X_{\hat{t}\hat{r}} = -\gamma^2 \beta R_{\hat{t}\hat{r}\hat{t}\hat{r}} \cos \psi, \tag{13.39}$$

$$X_{\hat{t}\hat{\varphi}} = -\gamma^2 \beta R_{\hat{t}\hat{\varphi}\hat{t}\hat{\varphi}} \sin \psi, \tag{13.40}$$

$$X_{\hat{r}\hat{\varphi}} = -\gamma^2 \beta^2 R_{\hat{r}\hat{\varphi}\hat{r}\hat{\varphi}} \sin \psi \cos \psi, \tag{13.41}$$

$$X_{\hat{t}\hat{\theta}} = X_{\hat{r}\hat{\theta}} = X_{\hat{\theta}\hat{\varphi}} = 0. \tag{13.42}$$

Once you have done this, calculate

$$Q_{IJ} = -X_{\mu\nu} \, (n_I)^\mu \, (n_J)^\nu = -X_{\hat{\mu}\hat{\nu}} \, (n_I)^{\hat{\mu}} \, (n_J)^{\hat{\nu}}. \tag{13.43}$$

Exercise 2: (High tech) Learn how to use any one of the standard symbolic manipulation packages available for modern workstations. Write an appropriate code fragment to have a computer check the algebra.

Exercise 3: (Easy) By setting $\psi = 0$, check that these formulas reduce to the results previously given for radial motion. In the present notation:

$$Q_{11} \;\; = \;\; -R_{\hat{t}\hat{r}\hat{t}\hat{r}}, \tag{13.44}$$

$$Q_{22} \;\; = \;\; Q_{33} = -\gamma^2 \left(R_{\hat{t}\hat{\varphi}\hat{t}\hat{\varphi}} + \beta^2 R_{\hat{r}\hat{\varphi}\hat{r}\hat{\varphi}} \right), \tag{13.45}$$

$$Q_{12} \;\; = \;\; Q_{13} = Q_{23} = 0. \tag{13.46}$$

Note that the shear now vanishes.

Exercise 4: (Easy) By setting $\psi = \pi/2$, deduce the tidal forces exerted on an object in circular motion around the wormhole mouth. (Remember, this does not have to be geodesic motion.) You should find

$$Q_{11} \;\; = \;\; -R_{\hat{t}\hat{\varphi}\hat{t}\hat{\varphi}}, \tag{13.47}$$

$$Q_{22} \;\; = \;\; -\gamma^2 \left(R_{\hat{t}\hat{r}\hat{t}\hat{r}} + \beta^2 R_{\hat{r}\hat{\varphi}\hat{r}\hat{\varphi}} \right), \tag{13.48}$$

$$Q_{33} \;\; = \;\; -\gamma^2 \left(R_{\hat{t}\hat{\varphi}\hat{t}\hat{\varphi}} + \beta^2 R_{\hat{\theta}\hat{\varphi}\hat{\theta}\hat{\varphi}} \right), \tag{13.49}$$

$$Q_{12} \;\; = \;\; Q_{13} = Q_{23} = 0. \tag{13.50}$$

Again, note that the shear vanishes.

Exercise 5: Evaluate, for arbitrary impact angle, the tidal forces acting just as the object passes the throat. Use the fact that the Riemann simplifies considerably at the throat to deduce

$$Q_{11} \;\; = \;\; -\frac{1}{2r_0}[1 - b'(r_0)]\phi'(r_0)\cos^2\psi, \tag{13.51}$$

$$Q_{22} \;\; = \;\; -\frac{\gamma^2}{2r_0}[1 - b'(r_0)] \left(\phi'(r_0)\sin^2\psi - \frac{\beta^2}{r_0} \right), \tag{13.52}$$

$$Q_{33} \;\; = \;\; +\frac{\gamma^2\beta^2}{r_0^2} \left(\frac{1}{2}[1 - b'(r_0)]\cos^2\psi - \sin^2\psi \right), \tag{13.53}$$

$$Q_{12} \;\; = \;\; -\frac{\gamma}{2r_0}[1 - b'(r_0)]\phi'(r_0)\sin\psi\cos\psi, \tag{13.54}$$

$$Q_{13} \;\; = \;\; Q_{23} = 0. \tag{13.55}$$

Exercise 6: Consider an object that executes circular motion while remaining on the throat of the wormhole. (The throat of the wormhole is typically a minimum of the gravitational potential). Check that the tidal forces are described by

$$Q_{11} = Q_{12} = Q_{13} = Q_{23} = 0, \tag{13.56}$$

$$Q_{22} = -\frac{\gamma^2}{2r_0}[1 - b'(r_0)]\left(\phi'(r_0) - \frac{\beta^2}{r_0}\right), \tag{13.57}$$

$$Q_{33} = -\frac{\gamma^2\beta^2}{r_0^2}. \tag{13.58}$$

Exercise 7: (Open ended) Use these formulas to fully investigate tidal forces in all of the various model wormholes considered in this monograph. Cook up a few toy models of your own and repeat the analysis.

Exercise 8: (Easy) In addition to being ripped apart by differential accelerations (tidal effects), the observer is subject to overall "g forces" due to the acceleration of his/her center of mass. Show that for radial motion the acceleration of the center of mass is given by

$$a = \pm\sqrt{1 - \frac{b}{r}}\, e^{-\phi}(\gamma e^{\phi})' = e^{-\phi}\frac{d}{dl}(\gamma e^{\phi}). \tag{13.59}$$

What happens in the case of geodesic radial motion? What can be said about the quantity γe^{ϕ}?

Exercise 9: (Open ended) Analyze nonradial accelerated motion.

13.2 Nongeometrical disruption?

The traversable wormhole geometries considered in this chapter have been carefully constructed to minimize and control the geometrical difficulties associated with wormhole travel—tidal forces and event horizons have been kept under careful control. The cost of abolishing the event horizon has been the introduction of exotic matter.

Unfortunately, the combination of spherical symmetry with the deduced existence of exotic matter somewhere in the wormhole implies that any and all travelers passing through this class of wormhole must pass through a region containing exotic matter. It is far from clear what the direct, nongravitational, effect of exotic matter would be on any potential visitor to the exotic region.

For instance, at the throat itself there is a generic prediction for the tension carried by the exotic matter threading the throat:

$$\tau\Big|_{r_0} = \frac{1}{8\pi G r_0^2} \approx 5 \times 10^{+42} \text{ J m}^{-3} \times \left(\frac{1 \text{ metre}}{r_0}\right)^2 \tag{13.60}$$

$$\approx \; 5 \times 10^{+10} \; \text{J m}^{-3} \times \left(\frac{1 \; \text{light-year}}{r_0} \right)^2 . \quad (13.61)$$

On the other hand, consider a sample of ordinary bulk matter. Since interatomic binding energies are of order an electron volt, and interatomic spacings are of order angstroms, the maximum conceivable tension supportable by bulk matter is of order

$$\tau_{\text{bulk}} \approx \frac{1 \, \text{eV}}{(1 \, \text{Å})^3} \approx 1.6 \times 10^{+11} \; \text{J m}^{-3}. \quad (13.62)$$

Thus the tension known to be present at the throat is larger than the maximum conceivable tension supportable by bulk matter unless the wormhole mouth has a truly macroscopic size of order a light-year. (For related comments along these lines see [240].)

Two important points should be made:

1. It is far from clear whether or not exotic matter need couple to ordinary matter. It is conceivable, though perhaps unlikely, that the truly enormous tension resident at the throat of the wormhole could be of a sufficiently bizarre character such that ordinary matter (you/me/whatever) did not feel its presence directly. As long as exotic matter can be detected only through its gravitational effects, nongeometrical problems of this type can be safely ignored. (Hypothetical forms of matter that can *only* be seen via their gravitational effects are sometimes referred to as "shadow matter".)

2. If one generalizes the analysis to avoid spherical symmetry, one might hope to hide all the exotic matter "in the woodwork". This point will be addressed in subsequent chapters.

13.3   Construction *ex nihilo*?

In view of the previous discussion of topology-changing processes it is far from clear whether or not a traversable wormhole might ever be constructed *ex nihilo*. Even an arbitrarily advanced civilization might find topological constraints on wormhole production to be insurmountable.

Classically (or even semiclassically) the situation is reasonably clear—topology change is a diseased process. One's best bet might be to find a wormhole that was built into the fabric of spacetime back in the big bang, and by suitable means to modify it and make it suitable for traversal. Of course, this would greatly limit the utility of such traversable wormholes since there would be little control over where the other side of the wormhole might lead. On the other hand, considerations of this type do not inhibit the production of pocket universes. Building a baby universe and enlarging it to suitable size need not require any change in topology. The neck connecting the pocket universe to its parent would locally appear to be a traversable wormhole even though the topology is trivial. (See figure 9.1, p. 91.)

If one appeals to quantum gravity to permit topology change the situation might be somewhat improved. The issue is now (presumably): What is the precise set of metrics one should functionally integrate over? Depending on taste, topology change may be permitted. The crude picture one might now envisage is that one might grab a transient wormhole out of the Wheeler spacetime foam, stabilize it, and then blow it up to usable size. (For instance, Roman has proposed using cosmological inflation to extract a wormhole from the spacetime foam [227, 228]. See also [140].) A certain lack of control as to the whereabouts (or whenabouts) of the other end of such a hypothetical entity is again manifest.

Traversable wormhole construction *ex nihilo* is a proposition much more dubious than the endeavor of wormhole mining.

13.4 Examples

13.4.1 Zero radial tides

The class of traversable wormholes considered so far has depended on two freely specifiable functions—the shape function $b(r)$ and the redshift $\phi(r)$. It is useful to consider several more restrictive classes of traversable wormholes with somewhat simpler structure. For instance, to keep the radial tides zero it is sufficient to set the redshift $\phi(r)$ equal to zero, resulting in an ultrastatic wormhole. The wormhole metric still possesses one arbitrarily specifiable function—the shape function $b(r)$—which is quite enough to keep life interesting. The metric now simplifies to

$$ds^2 = -dt^2 + \frac{dr^2}{1 - b_\pm(r)/r} + r^2 \left[d\theta^2 + \sin^2 \theta \, d\varphi^2\right]. \tag{13.63}$$

The nonzero components of the Riemann tensor are

$$R^{\hat{r}}_{\;\hat{\theta}\hat{r}\hat{\theta}} = R^{\hat{r}}_{\;\hat{\varphi}\hat{r}\hat{\varphi}} = \frac{1}{2r^3}(b'r - b), \tag{13.64}$$

$$R^{\hat{\theta}}_{\;\hat{\varphi}\hat{\theta}\hat{\varphi}} = \frac{b}{r^3}. \tag{13.65}$$

The nonzero components of the Einstein tensor are

$$G_{\hat{t}\hat{t}} = \frac{b'}{r^2}, \tag{13.66}$$

$$G_{\hat{r}\hat{r}} = -\frac{b}{r^3}, \tag{13.67}$$

$$G_{\hat{\theta}\hat{\theta}} \;=\; G_{\hat{\phi}\hat{\phi}} = -\frac{1}{2r^3}(b'r - b). \tag{13.68}$$

This implies an algebraic relationship between the components of the stress-energy tensor

$$2p = \tau - \rho. \tag{13.69}$$

The ANEC integral, governing the averaged null energy condition, reduces to

$$I_\Gamma = \frac{-1}{4\pi G} \int \frac{1}{r^2}\sqrt{1 - \frac{b}{r}}\; dr. \tag{13.70}$$

Finally, the tidal accelerations are

$$(\Delta a)^{\mu}_{\|} \;=\; 0, \tag{13.71}$$

$$(\Delta a)^{\mu}_{\perp} \;=\; \frac{\gamma^2 \beta^2}{2r^2}\left(b' - \frac{b}{r}\right)(\Delta\xi)_{\perp}. \tag{13.72}$$

13.4.2   Zero density

A particularly simple class of traversable wormhole is specified by demanding that the intrinsic geometry of the constant time spatial slices is the same as that for the spatial slices of the nontraversable Einstein–Rosen bridge (the nontraversable Schwarzschild wormhole, or Schwarzschild black hole). This is accomplished by setting $b(r) = r_0 = 2GM$, but leaving the redshift function freely specifiable.

The metric is

$$ds^2 = -e^{2\phi_{\pm}(r)}\,dt^2 + \frac{dr^2}{1 - r_0/r} + r^2\left[d\theta^2 + \sin^2\theta\,d\phi^2\right]. \tag{13.73}$$

The nonzero components of the Riemann tensor are

$$R^{\hat{t}}{}_{\hat{r}\hat{t}\hat{r}} \;=\; \left(1 - \frac{r_0}{r}\right)\{-\phi'' - (\phi')^2\}\frac{1}{2r^2}r_0\phi', \tag{13.74}$$

$$R^{\hat{t}}{}_{\hat{\theta}\hat{t}\hat{\theta}} \;=\; R^{\hat{t}}{}_{\hat{\phi}\hat{t}\hat{\phi}} = -\left(1 - \frac{r_0}{r}\right)\frac{\phi'}{r}, \tag{13.75}$$

$$R^{\hat{r}}{}_{\hat{\theta}\hat{r}\hat{\theta}} \;=\; R^{\hat{r}}{}_{\hat{\phi}\hat{r}\hat{\phi}} = -\frac{r_0}{2r^3}, \tag{13.76}$$

$$R^{\hat{\theta}}{}_{\hat{\phi}\hat{\theta}\hat{\phi}} \;=\; \frac{r_0}{r^3}. \tag{13.77}$$

The nonzero components of the Einstein tensor are

$$G_{\hat{t}\hat{t}} \;=\; 0, \tag{13.78}$$

$$G_{\hat{r}\hat{r}} \;=\; -\frac{r_0}{r^3} + 2\left\{1 - \frac{r_0}{r}\right\}\frac{\phi'}{r}, \tag{13.79}$$

$$G_{\hat{\theta}\hat{\theta}} = G_{\hat{\varphi}\hat{\varphi}} = \left\{1 - \frac{r_0}{r}\right\}\left[\phi'' + \phi'\left(\phi' + \frac{1}{r}\right)\right]$$
$$+ \frac{r_0}{2r^2}\left(\phi' + \frac{1}{r}\right). \tag{13.80}$$

The ANEC integral, governing the averaged null energy condition, reduces to

$$I_\Gamma = \frac{-1}{4\pi G}\int \frac{1}{r^2}e^{-\phi}\sqrt{1 - \frac{r_0}{r}}\,dr. \tag{13.81}$$

Finally, the tides are described by

$$(\Delta a)_{||} = \left[\left(1 - \frac{r_0}{r}\right)\left\{-\phi'' - (\phi')^2\right\} - \frac{r_0}{2r^2}\phi'\right](\Delta\xi)_{||}, \tag{13.82}$$

$$(\Delta a)_\perp = \frac{\gamma^2}{r^2}\left[(r - r_0)\phi' - \beta^2\frac{r_0}{2r}\right](\Delta\xi)_\perp. \tag{13.83}$$

At the throat this simplifies to

$$(\Delta a)_{||} = -\frac{1}{2r_0}\phi'\,(\Delta\xi)_{||}, \tag{13.84}$$

$$(\Delta a)_\perp = -\frac{\beta^2\gamma^2}{2r_0^2}\,(\Delta\xi)_\perp. \tag{13.85}$$

The key issue making this wormhole traversable is the absence of a horizon—the finiteness of $\phi(r_0)$. Keeping the tidal forces manageable is at some level a refinement on the notion of "traversable in principle" to the notion of "traversable in practice".

13.4.3 Proximal Schwarzschild

Consider the metric

$$ds^2 = -(1 - r_0/r + \epsilon/r^2)dt^2 + \frac{dr^2}{1 - r_0/r} + r^2\left[d\theta^2 + \sin^2\theta\,d\varphi^2\right]. \tag{13.86}$$

If $\epsilon = 0$ this is just the ordinary Schwarzschild solution, describing the ordinary Schwarzschild black hole (the nontraversable Einstein–Rosen bridge) with $r_0 = 2GM$. Something very interesting happens as soon as $\epsilon > 0$, even infinitesimally. The metric then describes the geometry of a wormhole that is "traversable in principle". Note that this is a special case of the zero density wormholes of the previous subsection. The key to traversability is that $g_{rr} \to \infty$ at $r = r_0 \equiv 2GM$, while g_{tt} remains finite there: $g_{tt}(r_0) = \epsilon/r_0^2$. Thus r_0 is the radius of the wormhole throat, and there is no event horizon there, so the wormhole is at least in principle traversable.

The first two Einstein equations are

$$\rho = 0, \tag{13.87}$$

$$\tau = \frac{\epsilon}{8\pi G r^4} \frac{2 - r_0/r}{1 - r_0/r + \epsilon/r^2}. \tag{13.88}$$

For $\epsilon \ll r_0^2$, the radial tension is very sharply peaked in the vicinity of the throat. Note that at the throat

$$\tau \rightarrow \frac{1}{8\pi G r_0^2}, \tag{13.89}$$

as indeed (based on our previous general analysis) it must. The thickness of this region of non-negligible tension is of order $\sqrt{\epsilon}$.

Tidal accelerations at the throat are

$$(\Delta a)_\| = -\frac{1}{4\epsilon} (\Delta \xi)_\|, \tag{13.90}$$

$$(\Delta a)_\perp = -\frac{\beta^2 \gamma^2}{2r_0^2} (\Delta \xi)_\perp. \tag{13.91}$$

The size of the radial tidal forces is governed by the thickness of the region of significant tension, while the scale of the transverse tidal forces is governed by the width of the wormhole throat.

The ANEC integral is

$$4\pi G \, I_\Gamma = -\int_{r_0}^{\infty} \frac{1}{r^2} \frac{\sqrt{1 - r_0/r}}{\sqrt{1 - r_0/r + \epsilon/r^2}} dr. \tag{13.92}$$

This ANEC integral is bounded by

$$-\frac{1}{4\pi G \, r_0} < I_\Gamma < 0. \tag{13.93}$$

The lesson to be learned is this: "infinitesimal" perturbations of the Schwarzschild metric can easily lead to wormholes that are "traversable in principle". In the example above, if ϵ is sufficiently small ($\epsilon \ll r_0^2$), this wormhole metric is experimentally indistinguishable from the ordinary Schwarzschild metric until one gets very close to where one would have expected the event horizon to be.

Warning: While this perturbation is infinitesimal in the sense that the metric is hardly disturbed for $r > 2GM$, the perturbation of the global geometry of the maximally extended spacetime is by no means small.

Is there any reason to suspect that metrics of this type are relevant to real physics? One has one's suspicions. Considerable work has gone into the calculation of how an externally imposed geometry affects quantum fields—such computations lead, for instance, to calculations of the renormalized expectation value of the stress-energy tensor in a Schwarzschild background [147, 146]. But if the stress-energy tensor is nonzero, it should modify the background geometry. Ultimately one would like to be able to do a self-consistent calculation: Find a geometry such that its influence on the quantum fields produces a stress-energy tensor that is precisely that required to satisfy the Einstein equations. Such a self-consistent calculation has proved elusive. Perturbative solutions obtained in an iterative manner have been discussed by York [293], by Anderson *et al.* [8], by Hochberg and Kephart [139], and by Hochberg *et al.* [141].

13.5 Exercises

13.5.1 Proper radial coordinates

Consider the wormhole metric expressed in terms of the proper radial distance:

$$ds^2 = -e^{2\phi(l)}dt^2 + dl^2 + R^2(l)\left[d\theta^2 + \sin^2\theta\, d\varphi^2\right]. \tag{13.94}$$

Calculate the Riemann and Einstein tensors in this coordinate system. You should find[1]

$$R^{\hat{t}}{}_{\hat{r}\hat{t}\hat{r}} = -\phi'' - (\phi')^2, \tag{13.95}$$

$$R^{\hat{t}}{}_{\hat{\theta}\hat{t}\hat{\theta}} = R^{\hat{t}}{}_{\hat{\varphi}\hat{t}\hat{\varphi}} = -\frac{\phi' R'}{R}, \tag{13.96}$$

$$R^{\hat{r}}{}_{\hat{\theta}\hat{r}\hat{\theta}} = R^{\hat{r}}{}_{\hat{\varphi}\hat{r}\hat{\varphi}} = -\frac{R''}{R}, \tag{13.97}$$

$$R^{\hat{\theta}}{}_{\hat{\varphi}\hat{\theta}\hat{\varphi}} = \frac{1-(R')^2}{R^2}. \tag{13.98}$$

All other components of the Riemann tensor, apart from those related to the above by symmetry, vanish.

The nonzero components of the Einstein tensor are

$$G_{\hat{t}\hat{t}} = -\frac{2R''}{R} + \frac{1-(R')^2}{R^2}, \tag{13.99}$$

$$G_{\hat{r}\hat{r}} = \frac{2\phi' R'}{R} - \frac{1-(R')^2}{R^2}, \tag{13.100}$$

$$G_{\hat{\theta}\hat{\theta}} = G_{\hat{\varphi}\hat{\varphi}} = \phi'' + (\phi')^2 + \frac{\phi' R' + R''}{R}. \tag{13.101}$$

---

[1]Notation: A prime now denotes a derivative with respect to the proper radial distance $X' = dX/dl$.

Check every calculation in the previous chapters by recalculating using this coordinate system. Physically correct results will not depend on the coordinate system.

13.5.2   Isotropic coordinates

The wormhole metric could also be expressed in terms of isotropic coordinates. In that case

$$ds^2 = -e^{2\phi(r)}dt^2 + e^{-2\psi(r)}\left[dr^2 + r^2(d\theta^2 + \sin^2\theta\, d\varphi^2)\right]. \qquad (13.102)$$

This geometry describes a traversable wormhole provided

- $\phi(r)$ and $\psi(r)$ are everywhere finite.

- $C(r) \equiv 2\pi r e^{-\psi(r)}$ has a minimum at $r_0 \neq 0$. Here r_0 is the location of the throat.

- The two asymptotically flat regions are at $r = +\infty$ and $r = 0$. To see that the second asymptotically flat region is "hiding" at $r = 0$ will require a bit of work.

- $\phi(0)$ and $\phi(\infty)$ must both be finite.

- $\psi(\infty)$ must be finite, while $\psi(r) \to 2\ln r$ as $r \to 0$. That is, $e^{-2\psi(r)} \to r^{-4}$ as $r \to 0$.

Calculate the Riemann and Einstein tensors in this coordinate system. You should find

$$R^{\hat{i}}{}_{\hat{r}\hat{i}\hat{r}} = -e^{2\psi}[\phi'' + \phi'(\phi' + \psi')], \qquad (13.103)$$

$$R^{\hat{i}}{}_{\hat{\theta}\hat{i}\hat{\theta}} = R^{\hat{i}}{}_{\hat{\varphi}\hat{i}\hat{\varphi}} = e^{2\psi}\phi'\left(\psi' - \frac{1}{r}\right), \qquad (13.104)$$

$$R^{\hat{r}}{}_{\hat{\theta}\hat{r}\hat{\theta}} = R^{\hat{r}}{}_{\hat{\varphi}\hat{r}\hat{\varphi}} = e^{2\psi}\left(\psi'' + \frac{\psi'}{r}\right), \qquad (13.105)$$

$$R^{\hat{\theta}}{}_{\hat{\varphi}\hat{\theta}\hat{\varphi}} = -e^{2\psi}\psi'\left(\psi' - \frac{2}{r}\right). \qquad (13.106)$$

All other components of the Riemann tensor, apart from those related to the above by symmetry, vanish.

The nonzero components of the Einstein tensor are

$$G_{\hat{t}\hat{t}} = e^{2\psi}\left[2\psi'' - (\psi')^2 + \frac{4\psi'}{r}\right], \qquad (13.107)$$

$$G_{\hat{r}\hat{r}} = e^{2\psi}\left[(\psi')^2 - 2\phi'\psi' + \frac{2(\phi' - \psi')}{r}\right], \qquad (13.108)$$

$$G_{\hat{\theta}\hat{\theta}} = G_{\hat{\varphi}\hat{\varphi}} = e^{2\psi}\left[\phi'' + (\phi')^2 - \psi'' + \frac{\phi' - \psi'}{r}\right]. \qquad (13.109)$$

Check every calculation in the previous chapters by recalculating using this coordinate system. Physically correct results will not depend on the coordinate system.

13.6 Generalizations

To keep the analysis tractable, Morris and Thorne had to assume spherical symmetry and time independence, though the matter distribution in the wormhole throat was permitted to spread over a finite radial extent. It is very useful to relax these constraints. Unfortunately, to do so we will have to introduce other restrictions on the distribution of matter. In the next chapter we shall see that by concentrating all the matter in the wormhole onto thin shells it is possible to analyze other systems such as:

- Static thin-shell wormholes of arbitrary symmetry.

- Dynamic thin-shell wormholes of spherical symmetry.

- Dynamic loop-based wormholes of arbitrary symmetry.

Contrast this with the present Morris–Thorne analysis:

- Static thick-shell wormholes of spherical symmetry.

Further generalizations are considerably more difficult. It would be nice to develop general formalisms capable of handling

- Dynamic thick-shell wormholes of spherical symmetry (probably relatively easy: a variation on the theme of Roman [227] will probably do the job).

- Static thick-shell wormholes of arbitrary symmetry (difficult: a good research problem).

- Dynamic thin-shell wormholes of arbitrary symmetry (probably difficult).

- Dynamic thick-shell wormholes of arbitrary symmetry (extremely difficult: a direct attack is probably hopeless).

Research Problem 6 *After reading the next few chapters for suitable background, attack the problem of developing some of the generalizations sketched above.*

There are at least two very good reasons for wanting to have a more general formalism available:

1. A dynamic analysis is the only way to begin to address stability questions.

2. Spherical symmetry has its own problems: The direct nongravitational effect of exotic matter is always an issue in spherically symmetric systems.

Other possible generalizations involve changing one's views as to the appropriate classical theory of gravity. One can replace Einstein gravity by Einstein–Gauss–Bonnet gravity [16] (this particular extension being relevant only in greater than four dimensions), higher-derivative gravity [108, 137], nonsymmetric gravity [188], or even Brans–Dicke gravity [176]. Numerical studies are also of interest [193].

Finally, I should remind the reader that the introduction of traversable wormholes, even very small traversable wormholes, implies the possibility of drastically changing the global causal structure of spacetime without significantly affecting the local geometry. This suggests the possibility of using traversable wormholes to probe the interior of a black hole [96].

Chapter 14

Thin shells: Formalism

The thin-shell formalism, also known as the junction condition formalism, is the general relativistic technique for analyzing the gravitational field of a thin (delta-function) layer of matter. It is a general technique of wide applicability. For the purposes of this book, it is the major tool suited to analyzing traversable wormholes without spherical symmetry.

Consider the case of electrostatics. As every undergraduate knows a delta function layer of electric charge of volume density $\rho(\vec{x}) = \sigma\delta(z)$ and surface density σ generates a discontinuity in the electric field

$$\vec{E}(\vec{x}) = \vec{E}_0 + \text{sign}(z)\ \hat{z}\ \frac{\sigma}{\epsilon_0}, \qquad (14.1)$$

and a kink in the electrostatic potential

$$\phi(\vec{x}) = \vec{E}_0 \cdot \vec{x} + |z|\ \frac{\sigma}{\epsilon_0}. \qquad (14.2)$$

The gravitational analog of this situation is that a thin shell of matter generates a gravitational field such that:

- The metric is continuous, but with a kink.

- The connexion (Christoffel symbol) has a step function discontinuity.

- The Riemann tensor has a delta-function contribution.

Historically, development of the thin-shell formalism was first initiated by Sen [239] and Lanczos [172, 173] in the early part of this century, culminating in the work of Israel [150], of Taub [248], of Barrabès [10], and of Barrabès and Israel [11]. The Lanczos–Sen–Israel version of the formalism relates the Ricci tensor generated by the gravitational influence of a thin shell of stress-energy to the discontinuity of the second fundamental forms

153

defined by considering the thin shell to be an immersed submanifold of
the full spacetime. For the purposes of this book it is useful to derive, in
addition, an expression for the full Riemann tensor.

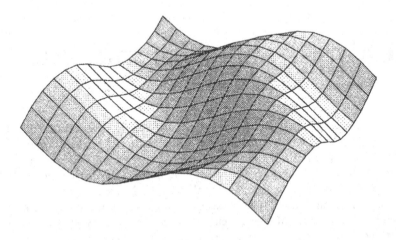

Figure 14.1: Thin shell: This chapter exhibits a formalism for calculating
the gravitational field induced by a thin shell of matter.

The formalism is presented for a timelike shell (spacelike normal). The
modifications required to deal with a spacelike shell (timelike normal) are
trivial. The general analysis for the case of a shell of null stress-energy is
considerably more tedious, and will not be needed in this book. The reader
is referred to the work of Redmount [220] where an analysis in terms of the
Newman–Penrose formalism is presented, and to the more recent analysis
of Barrabès and Israel [11].

14.1 Second fundamental form

Consider a thin shell of stress-energy situated in some smooth ambient
space. Define a function $\eta(x)$ such that $\eta = 0$ is the location of the shell;
$\eta(x)$ is at this stage otherwise arbitrary. Without loss of generality the
metric can, in the vicinity of the shell, be cast in the form

$$g_{\alpha\beta}(x) = \Theta(\eta(x))\, g^{+}_{\alpha\beta}(x) + \Theta(-\eta(x))\, g^{-}_{\alpha\beta}(x). \qquad (14.3)$$

The metric is continuous at the shell so that (independent of the choice of coordinate system)

$$g^+_{\alpha\beta}(x)\Big|_{\eta=0} = g^-_{\alpha\beta}(x)\Big|_{\eta=0}. \tag{14.4}$$

In old-fashioned language, the first fundamental forms (first groundforms, induced metrics) of the two sides of the shell are equal.

The unit normal to the shell is defined by

$$n_\alpha \propto \partial_\alpha\eta = \nabla^\pm_\alpha\eta; \qquad n^\alpha n_\alpha = +1. \tag{14.5}$$

(Note that this unit normal vector field is defined not only on the shell itself, but also on a four-dimensional region surrounding the shell. The notation $\nabla^\pm$ refers to the two covariant derivatives constructed from the two metrics $g^\pm$.) The second fundamental forms associated with the two sides of the shell $\eta = 0$ are defined to be

$$K^\pm_{\mu\nu} \equiv \nabla^\pm{}_{(\mu}n_{\nu)} \equiv \frac{1}{2}\left(\nabla_\mu{}^\pm n_\nu + \nabla_\nu{}^\pm n_\mu\right). \tag{14.6}$$

If one adopts Gaussian normal coordinates near the shell, things simplify somewhat. To do this, first place three-dimensional coordinates, denoted $x^i_\perp$, on the surface $\eta = 0$. Extend this to a four-dimensional coordinate patch by considering geodesics that intersect the shell at normal incidence. Let the fourth coordinate be the distance from the shell, as measured along such a normal geodesic. Call this fourth coordinate $\eta(x)$—this construction is equivalent to making a specific choice for the function $\eta(x)$ previously considered. This construction for $\eta(x)$ may break down if one gets too far from the shell. (This happens if the normal geodesics ever intersect.) Typically, Gaussian normal coordinates can be set up only in the immediate vicinity of the shell. They do not, typically, cover the whole manifold.

With this specific choice of $\eta(x)$, the unit normal to the thin shell now satisfies

$$n_\alpha = \partial_\alpha\eta = \nabla^\pm_\alpha\eta; \qquad n^\alpha n_\alpha = +1. \tag{14.7}$$

Using these Gaussian normal coordinates, the metric can be cast into the form

$$g_{\alpha\beta}(\eta, x_\perp) = \Theta(\eta)\, g^+_{\alpha\beta}(\eta, x_\perp) + \Theta(-\eta)\, g^-_{\alpha\beta}(\eta, x_\perp). \tag{14.8}$$

This is a specialization of equation (14.3) above. Equivalently

$$g^\pm_{\alpha\beta}(\eta, x_\perp)\, dx^\alpha dx^\beta = g^\pm_{ij}(\eta, x_\perp)\, dx^i_\perp dx^j_\perp + d\eta^2, \tag{14.9}$$

or, in matrix form

$$g^\pm_{\mu\nu}(\eta, x_\perp) = \begin{bmatrix} g^\pm_{ij} & 0 \\ 0 & +1 \end{bmatrix}. \tag{14.10}$$

With this choice of η, the second fundamental forms simplify

$$K^{\pm}_{\mu\nu} \equiv \nabla^{\pm}_{(\mu}n_{\nu)} = \nabla^{\pm}_{\mu}\nabla^{\pm}_{\nu}\eta, \qquad (14.11)$$

which is now manifestly symmetric. This implies

$$n^{\alpha}\, K^{\pm}_{\alpha\beta} = 0. \qquad (14.12)$$

In Gaussian normal coordinates, the second fundamental forms can also be interpreted as the normal derivatives of the metric

$$K^{\pm}_{\alpha\beta} = \frac{1}{2}\frac{\partial g_{\alpha\beta}}{\partial\eta}\bigg|_{0^{\pm}} = \frac{1}{2}\frac{\partial g^{\pm}_{\alpha\beta}}{\partial\eta}. \qquad (14.13)$$

Exercise: By explicitly calculating the connexion (Christoffel symbols), show that this expression for the second fundamental form as a normal derivative agrees with the formal definition.

14.2 Connexion

Now consider the metric derivatives $g_{\alpha\beta,\gamma}$. In view of the fact that

$$\frac{d}{dx}\Theta(\pm\eta) = \pm\delta(\eta), \qquad (14.14)$$

and of the continuity of the metric at $\eta = 0$, one derives

$$g_{\alpha\beta,\gamma} = \Theta(\eta)\, g^{+}_{\alpha\beta,\gamma} + \Theta(-\eta)\, g^{-}_{\alpha\beta,\gamma}. \qquad (14.15)$$

It is now a simple application of Gaussian normal coordinates to show [173]

$$g^{+}_{\alpha\beta,\gamma} - g^{-}_{\alpha\beta,\gamma} = \left(\frac{g^{+}_{\alpha\beta}}{\partial\eta} - \frac{g^{-}_{\alpha\beta}}{\partial\eta}\right) n_{\gamma}. \qquad (14.16)$$

For future convenience define the discontinuity in the second fundamental form to be

$$\kappa_{\alpha\beta} \equiv K^{+}_{\alpha\beta} - K^{-}_{\alpha\beta}, \qquad (14.17)$$

so that

$$g^{+}_{\alpha\beta,\gamma} - g^{-}_{\alpha\beta,\gamma} = 2\kappa_{\alpha\beta}\, n_{\gamma}. \qquad (14.18)$$

Though most easily derived using Gaussian normal coordinates this result is general.

The second derivatives of the metric are easily evaluated:

$$
\begin{aligned}
g_{\alpha\beta,\gamma\delta} &= \Theta(\eta)\, g^{+}_{\alpha\beta,\gamma\delta} + \Theta(-\eta)\, g^{-}_{\alpha\beta,\gamma\delta} \\
&\quad + \delta(\eta)\left\{ g^{+}_{\alpha\beta,\gamma} - g^{-}_{\alpha\beta,\gamma} \right\} n_\delta \\
&= \Theta(\eta)\, g^{+}_{\alpha\beta,\gamma\delta} + \Theta(-\eta)\, g^{-}_{\alpha\beta,\gamma\delta} \\
&\quad + 2\delta(\eta)\, \kappa_{\alpha\beta}\, n_\gamma n_\delta.
\end{aligned}
\tag{14.19}
$$

In a very similar vein, one may consider the connexion (Christoffel symbol)

$$
\Gamma_{\alpha\beta\gamma} \equiv \frac{1}{2}\left[g_{\alpha\beta,\gamma} + g_{\alpha\gamma,\beta} - g_{\beta\gamma,\alpha} \right] = \Theta(\eta)\, \Gamma^{+}_{\alpha\beta\gamma} + \Theta(-\eta)\, \Gamma^{-}_{\alpha\beta\gamma}.
\tag{14.20}
$$

Exercise: (Trivial) Show that at the shell ($\eta = 0$), one has

$$
\Gamma^{+}_{\alpha\beta\gamma} - \Gamma^{-}_{\alpha\beta\gamma} = \frac{1}{2}\left(\kappa_{\alpha\beta} n_\gamma + \kappa_{\alpha\gamma} n_\beta - \kappa_{\beta\gamma} n_\alpha \right).
\tag{14.21}
$$

14.3 Riemann tensor

With most of the preparatory work out of the way, it is now a simple matter of invoking the standard definition of the Riemann tensor to show that

$$
\begin{aligned}
R_{\alpha\beta\gamma\delta} &= -\frac{1}{2}\left(g_{\alpha\gamma,\beta\delta} + g_{\beta\delta,\alpha\gamma} - g_{\alpha\delta,\beta\gamma} - g_{\beta\gamma,\alpha\delta} \right) \\
&\quad - g^{\sigma\rho}\left(\Gamma_{\alpha\gamma\sigma}\Gamma_{\beta\delta\rho} - \Gamma_{\alpha\delta\sigma}\Gamma_{\beta\gamma\rho} \right).
\end{aligned}
\tag{14.22}
$$

Thus the Riemann tensor in the vicinity of the thin shell is

$$
\begin{aligned}
R_{\alpha\beta\gamma\delta} &= -\delta(\eta)\left[\kappa_{\alpha\gamma} n_\beta n_\delta + \kappa_{\beta\delta} n_\alpha n_\gamma - \kappa_{\alpha\delta} n_\beta n_\gamma - \kappa_{\beta\gamma} n_\alpha n_\delta \right] \\
&\quad + \Theta(\eta)\, R^{+}_{\alpha\beta\gamma\delta} + \Theta(-\eta)\, R^{-}_{\alpha\beta\gamma\delta}.
\end{aligned}
\tag{14.23}
$$

This is what was expected—smooth away from the shell with a delta-function contribution at the location of the shell.

It is easy to see that the known symmetries of the Riemann tensor are correctly reflected in this expression. A little more work establishes that the full (uncontracted) Bianchi identities are satisfied.

Exercise 1: Check the Bianchi identities. Note that $\partial_\alpha \delta(\eta) = \delta'(\eta)\partial_\alpha \eta = \delta'(\eta) n_\alpha$, and that $\nabla^{\pm}_\alpha n_\beta = K^{\pm}_{\alpha\beta}$ is symmetric. The proof of the Bianchi identities follows by considering the antisymmetrization properties of the covariant derivative $R_{\alpha\beta[\gamma\delta;\epsilon]}$.

Exercise 2: With hindsight, the above expression for the full Riemann tensor can be easily derived from standard textbook results. For instance, starting from equations (21.82), (21.75), and (21.76) of Misner, Thorne, and Wheeler [186], one might perform a Gaussian pillbox integration similar to that discussed on pp. 551–556. Check that this leads to the above result.

By contraction on appropriate indices one may recover the more usual Lanczos–Sen–Israel version of the junction conditions [172, 173, 239, 150, 248, 10]. It is useful to introduce the projection tensor

$$h_{\mu\nu} = g_{\mu\nu} - n_\mu n_\nu. \tag{14.24}$$

This projection tensor projects general tensors down onto the subspace spanned by the thin shell.

Warning: Do not confuse $h_{\mu\nu}$ in the sense of a projection tensor with $h_{\mu\nu}$ regarded as a small fluctuation in the spacetime metric. Unfortunately, both usages are standard. What is meant should be clear from context.

In terms of the Ricci tensor, the Ricci scalar, and the Einstein tensor:

$$
\begin{aligned}
R_{\mu\nu} &= -\delta(\eta)\left[\kappa_{\mu\nu} + \kappa\, n_\mu n_\nu\right] \\
&\quad +\Theta(\eta)\, R^+_{\mu\nu} + \Theta(-\eta)\, R^-_{\mu\nu}, \tag{14.25} \\
R &= -2\kappa\,\delta(\eta) + \Theta(\eta)\, R^+ + \Theta(-\eta)\, R^-, \tag{14.26} \\
G_{\mu\nu} &= -\delta(\eta)\left[\kappa_{\mu\nu} - \kappa h_{\mu\nu}\right] \\
&\quad +\Theta(\eta)\, G^+_{\mu\nu} + \Theta(-\eta)\, G^-_{\mu\nu}. \tag{14.27}
\end{aligned}
$$

14.4   Stress-energy tensor

The total stress-energy tensor may, without loss of generality, be written in the form

$$T_{\mu\nu} = \delta(\eta)S_{\mu\nu} + \Theta(\eta)\, T^+_{\mu\nu} + \Theta(-\eta)\, T^-_{\mu\nu}. \tag{14.28}$$

Here $S^{\mu\nu}$ denotes the surface stress-energy. Covariant conservation of stress-energy, $\nabla_\nu T^{\mu\nu} = 0$, implies several constraints. First, by isolating the term proportional to $\delta'(\eta)$, it is easy to see that

$$S^{\mu\nu} n_\nu = 0. \tag{14.29}$$

Secondly, isolating the term proportional to $\delta(\eta)$,

$$\bar{\nabla}_\nu S^{\mu\nu} + n^\nu \left(T^+_{\mu\nu} - T^-_{\mu\nu}\right) = 0. \tag{14.30}$$

This relates the change in stress-energy on the thin shell itself to the flux of incoming material. There is a potential subtlety arising from the fact that

∇S has to be evaluated on the thin shell itself and is therefore ambiguous. Which connexion should one use, Γ^+ or Γ^-? The answer is that one should use the average[1]

$$\bar{\nabla} = \frac{1}{2}\{\nabla^+ + \nabla^-\}. \tag{14.31}$$

Evaluating the component of equation (14.30) parallel to the normal vector requires attention to this subtlety. Note

$$n_\mu \bar{\nabla}_\nu S^{\mu\nu} = \bar{\nabla}_\nu(n_\mu S^{\mu\nu}) - (\bar{\nabla}_\nu n_\mu)S^{\mu\nu} = -(\bar{\nabla}_\nu n_\mu)S^{\mu\nu}. \tag{14.32}$$

One can relate $\bar{\nabla} n$ to the average of the two second fundamental forms

$$\bar{\nabla}_\mu n_\nu = \frac{1}{2}\left(K^+_{\mu\nu} + K^-_{\mu\nu}\right). \tag{14.33}$$

Thus, the normal component of the conservation of stress-energy yields a pressure balance equation

$$\frac{1}{2}\left(K^+_{\mu\nu} + K^-_{\mu\nu}\right)S^{\mu\nu} = n^\mu n^\nu \left(T^+_{\mu\nu} - T^-_{\mu\nu}\right). \tag{14.34}$$

Note that the normal component of the stress-energy tensor is by definition the pressure acting on the shell: $p \equiv T_{\mu\nu} n^\mu n^\nu$. Thus

$$\left(T^+_{\mu\nu} - T^-_{\mu\nu}\right)n^\mu n^\nu = p^+ - p^-. \tag{14.35}$$

Finally

$$\frac{1}{2}\left(K^+_{\mu\nu} + K^-_{\mu\nu}\right)S^{\mu\nu} = p^+ - p^-. \tag{14.36}$$

This relates the pressure difference across the shell to a combination of geometrical quantities (the $K^\pm$) defined at the shell itself and the stress-energy of the shell.

14.5   Einstein field equations

From the Einstein field equations, $G_{\mu\nu} = 8\pi G\, T_{\mu\nu}$, one now deduces

$$S_{\mu\nu} = -\frac{1}{8\pi G}\left[\kappa_{\mu\nu} - \kappa h_{\mu\nu}\right]. \tag{14.37}$$

Inverting this relationship

$$\kappa_{\mu\nu} = -8\pi G\left[S_{\mu\nu} - \frac{1}{2}S h_{\mu\nu}\right]. \tag{14.38}$$

---

[1]Exercise: Check this assertion.

Since the shell's stress-energy is now directly related to geometrical quantities the pressure balance equation may be rewritten as

$$[\text{tr}(K_+^2) - \text{tr}(K_+)^2] - [\text{tr}(K_-^2) - \text{tr}(K_-)^2] = -16\pi G\, [p_+ - p_-]. \quad (14.39)$$

This is now a pure relationship between geometry and pressure differences. The cognoscenti will recognize a deep relationship with the ADM $(3+1)$ formalism.

By adopting an orthonormal frame specially adapted to the surface one may put the normal in the form $n_{\hat\mu} = (0,0,0,1)$, and the metric and projection tensors in the form

$$g_{\hat\mu\hat\nu} = \begin{bmatrix} -1 & 0 & 0 & 0 \\ 0 & +1 & 0 & 0 \\ 0 & 0 & +1 & 0 \\ 0 & 0 & 0 & +1 \end{bmatrix}, \qquad h_{\hat\mu\hat\nu} = \begin{bmatrix} -1 & 0 & 0 & 0 \\ 0 & +1 & 0 & 0 \\ 0 & 0 & +1 & 0 \\ 0 & 0 & 0 & 0 \end{bmatrix}. \quad (14.40)$$

More importantly, one can diagonalize S and κ:

$$S_{\hat\mu\hat\nu} = \begin{bmatrix} \sigma & 0 & 0 & 0 \\ 0 & -\vartheta_1 & 0 & 0 \\ 0 & 0 & -\vartheta_2 & 0 \\ 0 & 0 & 0 & 0 \end{bmatrix}, \qquad \kappa_{\hat\mu\hat\nu} = \begin{bmatrix} \kappa_0 & 0 & 0 & 0 \\ 0 & \kappa_1 & 0 & 0 \\ 0 & 0 & \kappa_2 & 0 \\ 0 & 0 & 0 & 0 \end{bmatrix}. \quad (14.41)$$

Interpretation:

- σ is the surface energy density.

- ϑ_1 and ϑ_2 are the principal surface tensions (minus the principal pressures).

- κ_0, κ_1, and κ_2 measure the extrinsic curvature at the shell. They are related to the principal radii of curvature of the thin shell. (See below.)

In terms of these quantities the Einstein field equations are

$$\sigma = -\frac{1}{8\pi G}\,[\kappa_1 + \kappa_2], \qquad (14.42)$$

$$\vartheta_1 = -\frac{1}{8\pi G}\,[\kappa_2 - \kappa_0], \qquad (14.43)$$

$$\vartheta_2 = -\frac{1}{8\pi G}\,[\kappa_1 - \kappa_0]. \qquad (14.44)$$

14.6 Exercises

14.6.1 Principal radii of curvature

The significance of the second fundamental form can be understood in yet another way—in terms of the principal radii of curvature. Consider a curve γ (either timelike or spacelike) that is confined to lie on the surface Σ. Let the unit tangent to the curve γ be t^μ and consider the quantity

$$K_{\mu\nu} t^\mu t^\nu = (\nabla_\mu n_\nu) t^\mu t^\nu = -n_\nu [t^\mu \nabla_\mu] t^\nu. \tag{14.45}$$

The last step uses the fact that, by definition, $n^\mu t_\mu = 0$. But the quantity

$$a^\nu \equiv [t^\mu \nabla_\mu] t^\nu \tag{14.46}$$

is just the rate of change of the tangent vector t along the curve γ. If the curve γ is timelike the vector a^ν is called the four-acceleration. If the curve γ is spacelike then one defines the radius of curvature, R, and unit normal, $\tilde{n}$, of the curve γ by

$$a^\nu \equiv [t^\mu \nabla_\mu] t^\nu = \frac{\tilde{n}^\nu}{R}. \tag{14.47}$$

Then

$$K^{\pm}_{\mu\nu} t^\mu t^\nu = \frac{n^\mu \tilde{n}_\mu}{R} = \frac{\cos \psi}{R}. \tag{14.48}$$

The quantity $\cos \psi / R$ is called the normal curvature of the surface. The principal directions of the surface are defined to be the eigendirections of the second fundamental form. That is,

$$K^{\pm}_{\mu\nu} t^\nu = \lambda t_\mu. \tag{14.49}$$

(Note that for a timelike surface one of these principal directions is timelike; the other two are spacelike.) The principal radii of curvature of the surface are defined to be the eigenvalues of the second fundamental form.

Show that for a timelike surface, diagonalizing the second fundamental form yields

$$K^{\pm}_{\hat{\mu}\hat{\nu}} = \begin{bmatrix} 1/R_0 & 0 & 0 & 0 \\ 0 & 1/R_1 & 0 & 0 \\ 0 & 0 & 1/R_2 & 0 \\ 0 & 0 & 0 & 0 \end{bmatrix}. \tag{14.50}$$

Here $|1/R_0|$ is the magnitude of the four-acceleration of the timelike principal direction. On a generic shell, these principal radii of curvature vary from place to place.

14.6.2   Jacobi tensor

Tidal effects are sometimes reformulated in terms of the Jacobi tensor [186]

$$J_{\alpha\beta\gamma\delta} \equiv \frac{1}{2}(R_{\alpha\gamma\beta\delta} + R_{\alpha\delta\beta\gamma}), \qquad (14.51)$$

$$R_{\alpha\beta\gamma\delta} \equiv \frac{2}{3}(J_{\alpha\gamma\beta\delta} + J_{\alpha\delta\beta\gamma}). \qquad (14.52)$$

Show that for a thin shell of matter

$$
\begin{aligned}
J_{\alpha\beta\gamma\delta} = &-\delta(\eta)\Big[\kappa_{\alpha\beta}\, n_\gamma\, n_\delta + \kappa_{\gamma\delta}\, n_\alpha\, n_\beta \\
&-\frac{1}{2}\kappa_{\alpha\delta}\, n_\beta\, n_\gamma - \frac{1}{2}\kappa_{\beta\gamma}\, n_\alpha\, n_\delta \\
&-\frac{1}{2}\kappa_{\alpha\gamma}\, n_\beta\, n_\delta - \frac{1}{2}\kappa_{\beta\delta}\, n_\alpha\, n_\gamma\Big] \\
&+\Theta(\eta)\, J_{\alpha\beta\gamma\delta}^{+} + \Theta(-\eta)\, J_{\alpha\beta\gamma\delta}^{-}. \qquad (14.53)
\end{aligned}
$$

This result is, of course, completely equivalent to the result for the Riemann tensor.

14.6.3   Weyl tensor

The Weyl tensor describes the part of the curvature that is invariant under conformal rescalings of the metric. Start from the definition [186, 132]

$$
\begin{aligned}
C_{\alpha\beta\gamma\delta} \equiv &+ R_{\alpha\beta\gamma\delta} \\
&- \frac{1}{2}[g_{\alpha\gamma}\, R_{\beta\delta} + g_{\beta\delta}\, R_{\alpha\gamma} - g_{\alpha\delta}\, R_{\beta\gamma} - g_{\beta\gamma}\, R_{\alpha\delta}] \\
&+ \frac{1}{12}R\,[g_{\alpha\gamma}\, g_{\beta\delta} + g_{\beta\delta}\, g_{\alpha\gamma} - g_{\alpha\delta}\, g_{\beta\gamma} - g_{\beta\gamma}\, g_{\alpha\delta}]. \quad (14.54)
\end{aligned}
$$

To simplify life, recall that the projection tensor is $h_{\alpha\beta} \equiv g_{\alpha\beta} - n_\alpha n_\beta$. Define the transverse traceless part of the discontinuity in the second fundamental form to be

$$\tilde{\kappa}_{\alpha\beta} \equiv \kappa_{\alpha\beta} - \frac{1}{3}\kappa h_{\alpha\beta}. \qquad (14.55)$$

Show that

$$
\begin{aligned}
C_{\alpha\beta\gamma\delta} = &-\delta(\eta)\Big[\tilde{\kappa}_{\alpha\gamma}\Big(n_\beta\, n_\delta - \frac{1}{2}g_{\beta\delta}\Big) + \tilde{\kappa}_{\beta\delta}\Big(n_\alpha\, n_\gamma - \frac{1}{2}g_{\alpha\gamma}\Big) \\
&-\tilde{\kappa}_{\alpha\delta}\Big(n_\beta\, n_\gamma - \frac{1}{2}g_{\beta\gamma}\Big) - \tilde{\kappa}_{\beta\gamma}\Big(n_\alpha\, n_\delta - \frac{1}{2}g_{\alpha\delta}\Big)\Big] \\
&+\Theta(\eta)\, C_{\alpha\beta\gamma\delta}^{+} + \Theta(-\eta)\, C_{\alpha\beta\gamma\delta}^{-}. \qquad (14.56)
\end{aligned}
$$

Verify that this satisfies the relevant trace identities. Show that the tensors $\tilde{\kappa}^\alpha{}_\beta \equiv \kappa^\alpha{}_\beta - \frac{1}{3}\kappa h^\alpha{}_\beta$, and $n^\alpha n_\beta$ are separately invariant under conformal transformations of the metric.

14.6.4 Balloons and soap bubbles

There is a nice application of part of this rather complicated differential geometric machinery to a pair of very simple classical physical systems— balloons and soap bubbles—that serves to illustrate the basic physical principles.

An inflated balloon is a simple thin sheet of rubber that resides in ordinary Minkowski space. (Ignore the back-reaction—I am not interested in the gravitational field of the balloon.) Because the spacetime metric inside the balloon is the same as that outside the balloon (both geometries are pieces of flat Minkowski space), the second fundamental forms defined on either side of the rubber are equal. That is,

$$K^+_{\mu\nu} = K^-_{\mu\nu}. \tag{14.57}$$

Show that in terms of the principal radii of curvature

$$K^\pm_{\mu\hat{\nu}} = \begin{bmatrix} 0 & 0 & 0 & 0 \\ 0 & 1/R_1 & 0 & 0 \\ 0 & 0 & 1/R_2 & 0 \\ 0 & 0 & 0 & 0 \end{bmatrix}. \tag{14.58}$$

(This is of course a special and simpler case of one of the earlier exercises.)

The rubber sheet making up the surface of the balloon can still be described by a stress-energy tensor, which still depends on the surface mass density and surface tensions. The stress-energy tensor is still covariantly conserved—after all the conservation of energy and momentum works perfectly well in special relativity or even in Newtonian mechanics. Therefore the pressure balance equation still holds. [Note, I did not use Einstein's gravitational field equations in the derivation of equation (14.36).]

For every point on the surface of an arbitrarily shaped balloon, show that

$$\frac{\vartheta_{11}}{R_1} + \frac{\vartheta_{22}}{R_2} = \Delta p. \tag{14.59}$$

Here Δp is the pressure difference between the inside and outside of the balloon, while ϑ_{ij} is the surface tension tensor.

Repeat this analysis for a soap bubble. Show that

$$\vartheta\left(\frac{1}{R_1} + \frac{1}{R_2}\right) = \Delta p. \tag{14.60}$$

(The symbol ϑ now denotes the isotropic surface tension.) The net result of this exercise is that we have derived the well-known Laplace–Young equation from the covariant conservation of stress-energy. (For more traditional ways of analyzing soap bubbles, see, for instance, Isenberg [148].)

Finally, consider a soap film supported by some arbitrarily complicated wire frame. (For instance, see some of the delightful illustrations in Isenberg [148]). In this case $\Delta p = 0$; the pressure difference across the film is zero. Show that the trace of the extrinsic curvature vanishes:

$$K \equiv \left(\frac{1}{R_1} + \frac{1}{R_2} \right) = 0. \tag{14.61}$$

So this relatively abstract machinery reproduces the well-known result that soap films are maximally embedded surfaces [148]; this is equivalent to the statement that soap films are surfaces of minimum area.

Chapter 15

Thin shells: Wormholes

By applying the thin-shell formalism, some particularly simple examples of traversable wormholes can be constructed [261, 260].

15.1 Static aspheric wormholes

15.1.1 Minkowski surgery

The models of interest in this section can be very easily described. The basic idea is to take two ordinary flat universes and use surgical cut and paste techniques to join them together in a suitably simple way. These models are notable both because the analysis is not limited to spherical symmetry, and because it is possible to in some sense minimize the use of exotic matter. In particular, it is possible for a traveler to traverse such a wormhole without passing through a region of exotic matter [261].

Mathematically: Consider the following construction. Take *two* copies of flat Minkowski space, and remove from each identical regions of the form $\Re \times \Omega$, where Ω is a three-dimensional compact spacelike hypersurface and $\Re$ is a timelike straight line (e.g., the time axis). Then identify these two incomplete spacetimes along the timelike boundaries $\Re \times \partial\Omega$. The resulting spacetime is well behaved (it is in fact geodesically complete) and possesses two asymptotically flat regions (two universes) connected by a wormhole. The throat of the wormhole is just the junction $\partial\Omega$ at which the two original Minkowski spaces are identified. By construction it is clear that the resulting spacetime is everywhere Riemann flat except possibly at the throat. Consequently we know that the stress-energy tensor in this spacetime is concentrated at the throat, with a delta-function singularity there. Note that this is a mathematical construction—not a physical prescription for building a wormhole. One should not envisage taking a meat cleaver to spacetime and sewing up the resulting mess with needle and thread.

Technical point: "Geodesic completeness" is the statement that all geodesics in the spacetime are well behaved and do *not* suddenly terminate at "peculiar" points.

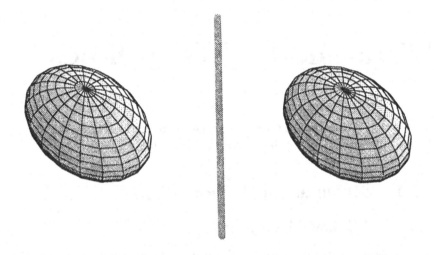

Figure 15.1: Thin-shell wormhole: generic construction. Take two copies of flat spacetime. Remove identical hypervolumes from each spacetime. Identify at the boundaries.

Applying the results of the preceding chapter one obtains, for this class of wormhole spacetimes, the exact result

$$R_{\alpha\beta\gamma\delta} = -\delta(\eta) \left[\kappa_{\alpha\gamma} \, n_\beta \, n_\delta + \kappa_{\beta\delta} \, n_\alpha \, n_\gamma - \kappa_{\alpha\delta} \, n_\beta \, n_\gamma - \kappa_{\beta\gamma} \, n_\alpha \, n_\delta \right]. \quad (15.1)$$

Suitable contractions yield

$$R_{\mu\nu} = -\delta(\eta) \left[\kappa_{\mu\nu} + \kappa \, n_\mu n_\nu \right]; \quad (15.2)$$

$$R = -2\kappa \, \delta(\eta); \quad (15.3)$$

$$G_{\mu\nu} = -\delta(\eta) \left[\kappa_{\mu\nu} - \kappa \, h_{\mu\nu} \right]. \quad (15.4)$$

The Einstein field equations give

$$T_{\mu\nu} = -\frac{1}{8\pi G} \delta(\eta) \left[\kappa_{\mu\nu} - \kappa \, h_{\mu\nu} \right]. \quad (15.5)$$

Because of the particularly simple form of the geometry, the second fundamental form at the throat is easily calculated. (Either see the preceding exercises; or note that because everything is static the time coordinate just comes along for the ride. So one can just copy results from any of the standard elementary texts that discuss two-surfaces embedded in three-space. See, for example, [182].) Suitable coordinates can be constructed by first going to the rest frames of the individual throats, and second, choosing the second fundamental forms to be diagonal. Then

$$K_{\hat{\mu}\hat{\nu}}^{\pm} = \pm \begin{bmatrix} 0 & 0 & 0 & 0 \\ 0 & 1/R_1 & 0 & 0 \\ 0 & 0 & 1/R_2 & 0 \\ 0 & 0 & 0 & 0 \end{bmatrix}. \tag{15.6}$$

Here R_1 and R_2 are the principal radii of curvature of the two-dimensional surface $\partial\Omega$. In general R_1 and R_2 will depend on position. (A convex surface has positive radii of curvature, a concave surface has negative radii of curvature.) The overall $\pm$ sign comes from the fact that η is an outward pointing coordinate in one universe, but an inward pointing coordinate in the other universe. In terms of the surface energy density σ and principal surface tensions $\vartheta_{1,2}$, Einstein's field equations now yield

$$\sigma = -\frac{1}{4\pi G} \cdot \left(\frac{1}{R_1} + \frac{1}{R_2} \right); \tag{15.7}$$

$$\vartheta_1 = -\frac{1}{4\pi G} \cdot \frac{1}{R_2}; \tag{15.8}$$

$$\vartheta_2 = -\frac{1}{4\pi G} \cdot \frac{1}{R_1}. \tag{15.9}$$

This implies that in general ($\partial\Omega$ convex) we will be dealing with *negative* surface energy density and *negative* surface tensions.

It is now easy to see how to build a wormhole such that a traveler encounters no exotic matter. Simply choose Ω to have at least one flat face. On that face the two principal radii of curvature are infinite, so $\kappa_{\mu\nu} = 0$, and the Riemann tensor and stress-energy tensors are both zero. A traveler encountering such a flat face will feel no tidal forces and see no matter, exotic or otherwise. Such a traveler will simply be shunted into the other universe.

We have just seen that in order for the throat of the wormhole to be convex, the surface energy density must be negative. This behavior may be rephrased as a violation of the weak and dominant energy conditions at the throat of the wormhole. By looking along the principal directions and considering the null trajectories $k_{\hat{\mu}}^{(1)} = (1,1,0,0)$ and $k_{\hat{\mu}}^{(2)} = (1,0,1,0)$, one finds

$$T^{\mu\nu} k_{\mu}^{(1)} k_{\nu}^{(1)} = (\sigma - \vartheta_1)\,\delta(\eta) = -\frac{1}{4\pi G}\,\frac{1}{R_1}\,\delta(\eta), \tag{15.10}$$

$$T^{\mu\nu} \, k_\mu^{(2)} k_\nu^{(2)} = (\sigma - \vartheta_2) \, \delta(\eta) = -\frac{1}{4\pi G} \, \frac{1}{R_2} \, \delta(\eta). \qquad (15.11)$$

This implies that at a convex section of the throat, the null and strong energy conditions are also violated. Since these violations of the energy conditions are all proportional to a delta function, it automatically follows that the averaged energy conditions are also violated.

The violation of the various energy conditions has been discussed previously for the spherically symmetric case. Very general arguments (the topological censorship theorem) for this behavior will be discussed in subsequent chapters. For the traversable wormholes discussed in this chapter, a very simple and relatively general argument can be given: Consider a bundle of light rays impinging on the throat of the wormhole. A little thought (following some geodesics through the wormhole), will convince one that the throat of the wormhole acts as a "perfect mirror", except that the "reflected" light is shunted into the other universe. This is enough to imply that a convex portion of $\partial\Omega$ will defocus the bundle, whereas a concave portion of $\partial\Omega$ will focus it. The "focusing theorem" for null geodesics then immediately implies that convex portions of $\partial\Omega$ violate both the null energy condition and the averaged null energy condition. (See, e.g., reference [186, exercise 22.14, p. 582], or [132].)

The discussion so far has been carried out in terms of an inter-universe wormhole. The same construction can be applied to obtain an intra-universe wormhole. For an intra-universe wormhole the surface $\partial\Omega_1$ is to be identified with a translated, rotated, mirror-image copy of itself, $\partial\Omega_2$.

There is an extra degree of freedom for intra-universe wormholes: If $\partial\Omega_1$ and $\partial\Omega_2$ are mirror images of each other, then the resulting space is orientable. If $\partial\Omega_1$ and $\partial\Omega_2$ are not mirror images (that is, they are simple translates) then the space is not orientable. At the classical level, non-orientable spaces are not a problem—they are equally pleasant or unpleasant in their behavior. At the quantum level, there are good reasons to believe that space must be orientable. See the discussion on p. 285.

15.1.2 Cubic wormholes

Now that we have discussed non-symmetric solutions in general, it becomes useful to consider some more special cases. What I am trying to do here is to minimize the use of exotic matter as much as possible. Let the compact set Ω be a cube whose edges and corners have been smoothed by rounding. Then the throat $\partial\Omega$ consists of six flat planes (the faces), twelve quarter cylinders (the edges), and eight octants of a sphere (the corners). Let the edge of the cube be of length L, and let the radius of curvature of the cylinders and spheres be r, with $r \ll L$. The stress-energy tensor on the six faces is zero. On the twelve quarter cylinders comprising the edges the surface stress-energy tensor takes the form:

$$S^{\mu\nu} = \frac{1}{4\pi Gr} \begin{bmatrix} +1 & 0 & 0 & 0 \\ 0 & -1 & 0 & 0 \\ 0 & 0 & 0 & 0 \\ 0 & 0 & 0 & 0 \end{bmatrix}. \tag{15.12}$$

This means that each quarter cylinder supports an energy per unit length of

$$\mu = \sigma \cdot \frac{2\pi r}{4} = -\frac{1}{4\pi Gr} \cdot \frac{\pi r}{2} = -\frac{1}{8G}, \tag{15.13}$$

and a tension

$$\tau = \vartheta_1 \cdot \frac{2\pi r}{4} = -\frac{1}{8G} = \mu. \tag{15.14}$$

Note that μ and τ are independent of r. In particular, they are well behaved and finite (though negative) as $r \to 0$.

The total energy concentrated on each of the eight octants at the corners is

$$E \equiv \sigma \cdot \frac{4\pi r^2}{8} = -\frac{1}{4\pi Gr} \cdot \frac{\pi r^2}{4} = -\frac{r}{8G}, \tag{15.15}$$

which tends to zero as $r \to 0$.

The net material force acting on each of the eight corners is independent of r, and is equal to

$$\vec{F} = \mu \left(\pm \hat{x} \pm \hat{y} \pm \hat{z} \right). \tag{15.16}$$

That is, the material comprising the wormhole throat has to provide exactly this much internal stress in order to support the wormhole against gravitational forces.

Take the $r \to 0$ limit. In this limit Ω becomes an ordinary (sharp cornered) cube. The stress-energy tensor for the wormhole is then concentrated entirely on the edges of the cube where

$$\mu = \tau = -\frac{1}{8G} = -\frac{1}{8} \cdot \frac{m_P}{\ell_P} = -1.52 \times 10^{43} \text{ J m}^{-1}. \tag{15.17}$$

Needless to say, energies and tensions of this magnitude (let alone *sign*) are well beyond current technological capabilities. To get a feel for the fundamental limits of ordinary atomic matter, consider a monomolecular chain of atoms. Since interatomic binding energies are of order electron volts, while interatomic spacings are of order angstroms, the maximum conceivable linear tension supportable by a monomolecular chain is of order

$$\tau_{\max} \approx \frac{1 \text{ eV}}{1 \text{ Å}} \approx 10^{-9} \text{ J m}^{-1} \approx 10^{-53} \frac{m_P}{\ell_P}. \tag{15.18}$$

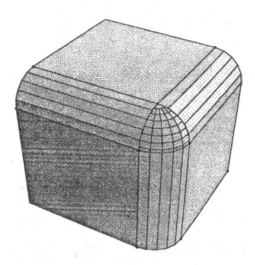

Figure 15.2: Cubic wormhole: rounded edges. The throat of this wormhole can be divided into six faces (planes), twelve edges (quarter cylinders), and eight corners (octants of a sphere).

Caveats: A quick overview of the logic is called for: Given that one finds a particular geometry interesting this type of analysis calculates the stress-energy that must be there in order to satisfy Einstein's equations of general relativity. As yet, no attempt has been made to derive a microscopic model describing the physics of this matter. In particular, this type of analysis does not and cannot uniquely specify the matter action. Nor does it allow one to derive unique equations of motion for the matter. (In particular, one cannot perform a stability analysis.)

It is possible, however, to show that the stress-energy present at the edges of the cube can be mimicked by the stress-energy tensor of a *negative tension* classical cosmic string. (The corners are a littler dicier—whatever is going on at the corners cannot be mimicked by a simple piece of cosmic string.) To see this, note that classical cosmic strings are described by the Nambu–Goto action, and write the Nambu–Goto action [123, 238, 127] in the form:

$$S = \tau \int d^2\xi \, d^4x \, \delta^4(x - X(\xi)) \, \sqrt{-\det(h_{AB})}. \qquad (15.19)$$

Here ξ_A are a pair of coordinates on the world-sheet swept out by the

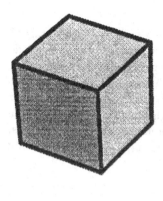

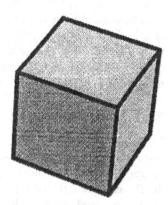

Figure 15.3: Cubic wormhole: sharp edges. This idealized wormhole is the limit of the rounded edge wormhole as the edges become sharp. All the stress-energy is now concentrated on the sharp edges. The flat faces carry no stress-energy.

cosmic string. $X(\xi)$ is the location of the world-sheet in $(3+1)$-dimensional spacetime, and the induced metric on the world-sheet is defined by

$$h_{AB}(\xi) = \partial_A X^\mu(\xi)\, \partial_B X^\nu(\xi)\, g_{\mu\nu}(X(\xi)). \qquad (15.20)$$

Varying with respect to the spacetime metric yields the classical spacetime stress-energy tensor

$$T^{\mu\nu}(x) = -\tau \int d^2\xi\, \delta^4(x - X(\xi))\, h^{AB}\, \partial_A X^\mu\, \partial_B X^\nu. \qquad (15.21)$$

For a classical cosmic string stretched along the x axis we may choose the world-sheet coordinates such that $X^\mu(\xi) \equiv X^\mu(\xi^0, \xi^1) = (\xi^0, \xi^1, 0, 0)$. The stress-energy tensor is then quickly calculated to be

$$T^{\mu\nu}(t,x,y,z) = +\tau\,\delta(y)\,\delta(z) \begin{bmatrix} +1 & 0 & 0 & 0 \\ 0 & -1 & 0 & 0 \\ 0 & 0 & 0 & 0 \\ 0 & 0 & 0 & 0 \end{bmatrix}. \qquad (15.22)$$

This is exactly the algebraic form of the stress-energy tensor just obtained for the edges of a cubical wormhole. Note however, that field theoretic models of cosmic strings lead to positive string tensions. No natural mechanism for generating negative string tension is known. This calculation shows that the stress-energy on the edges is compatible with the stress-energy tensor of a Nambu–Goto string. We shall soon enough see that it is compatible with a lot of other things as well. Also, one should note that the corners, though well behaved in the sense that there is no energy there, are certainly not pieces of string in any way, shape, or form.

15.1.3   Polyhedral wormholes

Having dealt with cubic wormholes, generalizations are immediate. First, note that the length of the edge L nowhere enters into the calculation. This implies that any rectangular prism would do just as well. The generalization to Ω being an arbitrary polyhedron is also straightforward. Consider an arbitrary polyhedron with edges and corners smoothed by rounding. At each edge the geometry is locally that of two planes joined by a fraction $(\phi/2\pi)$ of a cylinder. Here ϕ is the "bending angle" at the edge in question. The local geometry at each corner is that of some fraction of a sphere. As previously, the energy concentrated on the corners tends to zero as $r \to 0$. The energy per unit length concentrated on each edge is now

$$\mu = -\frac{1}{4\pi Gr} \cdot \phi r = -\frac{\phi}{4\pi G}. \qquad (15.23)$$

As previously, the limit $r \to 0$ is well behaved, in which case we obtain

$$\mu = \tau = \frac{-\phi}{4\pi G}. \qquad (15.24)$$

It is instructive to compare this with the known geometry of a single infinite length classical cosmic string. (Some technical issues related to cosmic strings will be discussed on pp. 187–193 and on pp. 220–222.)

Note that each edge is surrounded by a total of $2(\pi + \phi)$ radians ($\pi + \phi$ radians in each universe). Thus the deficit angle ($\Delta\theta$) at each edge is given by $\Delta\theta = -2 \cdot \phi$. In terms of this deficit angle $\mu = T = (\Delta\theta/8\pi G)$, which is the usual relationship for classical cosmic strings. An edge of the polyhedron is said to be convex if the bending angle is positive. If an edge is convex the tension is negative. Conversely, if an edge is concave, the tension at that edge is positive.

15.1.4 Dihedral wormholes

Perhaps the simplest example of the class of polyhedral wormholes is the dihedral wormhole. Take an arbitrary self-avoiding curve that lies in some fixed plane. This self-avoiding curve specifies the edges of a (degenerate) two-sided polytope—a dihedron. Two such dihedra can be identified in the manner described above to produce a traversable wormhole. One could also take a cubic wormhole and collapse two opposite faces on top of each other. The deficit angle is now $\Delta\theta = -2\pi$ so the linear mass density and tension at the edges is

$$\mu = T = -\frac{1}{8G} = -\frac{1}{4} \cdot \frac{m_P}{\ell_P}. \tag{15.25}$$

Figure 15.4: Dihedral wormhole. The dihedral wormhole can be obtained by collapsing a thin-shell wormhole along some axis. This particular example has been obtained by collapsing a sharp edge cubic wormhole along one axis. All the stress-energy in a dihedral wormhole is concentrated at the single edge framing the wormhole.

15.1.5 How much exotic matter?

It is instructive to ask: "How much exotic matter is required to build a traversable wormhole?" The best answer I can give for short-throat wormholes is to consider the integral

$$\text{``}M\text{''} = \int \rho \, dV = \int_{\partial\Omega} \sigma \, dA = \frac{1}{8\pi G} \int_{\partial\Omega} \kappa \, dA. \qquad (15.26)$$

If this is expressed explicitly in terms of the principal radii of curvature, one can easily see

$$\text{``}M\text{''} = -\frac{m_P}{4\pi\ell_P} \int_{\partial\Omega} \left(\frac{1}{R_1} + \frac{1}{R_2} \right) dA = -m_P \, F_{\text{shape}} \, \frac{L}{\ell_P}. \qquad (15.27)$$

Here F_{shape} is a dimensionless number of order one, characteristic of the precise shape of the wormhole throat. L is some suitable measure of the linear dimension of the throat. Inserting some numbers,

$$
\begin{aligned}
\text{``}M\text{''} \;&=\; -m_P \, F_{\text{shape}} \, \frac{L}{\ell_P} \\
&\approx\; -10^{27} \text{ kg} \times \frac{L}{1 \text{ metre}} \\
&\approx\; -10^{-3} M_\odot \times \frac{L}{1 \text{ metre}}.
\end{aligned} \qquad (15.28)
$$

So building a one metre wide thin-throat traversable wormhole requires about one Jupiter mass of exotic negative energy matter. This result is rather disappointing. Note that "M" is *not* the total mass of the wormhole as seen from infinity. The total mass of this class of wormholes is in fact zero. The non-linearity of the Einstein field equations leads to this interesting effect.

To get a quick feel for how this happens—consider a lump of negative energy. Since it is energy, albeit negative energy, it will be surrounded by a gravitational field of some sort. Insofar as it is meaningful to define the energy of the gravitational field in general relativity, the energy of the gravitational field may be positive, negative, or zero. In the Newtonian approximation, positive mass sources typically lead to negative gravitational energy. By extension, negative mass sources typically lead to positive gravitational energy. The net result is that the total net mass, as measured from infinity, may be positive, negative, or zero depending on the precise details of the arrangement of the negative energy that constitutes the core of the system under consideration.

One further formal development is of note. Define $A(\eta)$ to be the area of the two-dimensional surface that everywhere lies a proper distance η above the throat of the wormhole. Then the integrated trace of the second fundamental form satisfies

$$\int_{\partial\Omega} K \, dA = \left.\frac{\partial A}{\partial \eta}\right|_0. \qquad (15.29)$$

Because of the symmetry between the two sides of this type of wormhole one can then assert

$$\text{``}M\text{''} = -\frac{m_P}{4\pi\ell_P}\frac{\partial A}{\partial\eta}\bigg|_0. \tag{15.30}$$

The fact that $A(\eta)$ has to increase as one moves away from the throat then guarantees that "M" is negative, even if the surface $\partial\Omega$ is strongly concave in places.

Summary: In this section I have investigated some rather simple examples of traversable wormholes. I have been able to avoid the use of spherical symmetry. Although these wormholes require the presence of exotic matter, it is possible to exhibit wormholes and geodesics such that the traveler does not have to encounter the exotic matter directly. The big open question, naturally, is whether a significant amount of exotic matter is in fact obtainable in the laboratory. The theoretical problems are daunting, and the technological problems seem completely beyond our reach.

15.1.6 Exercise: Polyhedral universes

A useful exercise is to essentially turn one of these polyhedral wormholes inside out and construct a polyhedral universe. Such objects may be interesting as potential candidates for exotic cosmologies. This is essentially a variation on the theme of Redmount's cellular cosmological models for a void-dominated universe [221].

For definiteness, consider the construction of the cubical wormhole. After cutting a cube out of flat Minkowski space, throw away the exterior region, keeping only the interior cubical region. Repeat. Now identify faces on the two cubes. The result is a spatially closed universe that is almost everywhere flat. The only stress-energy present is at the edges of the cubes, where one now has a positive linear energy density, and a positive tension

$$\mu = T = +\frac{1}{8G} = +\frac{1}{8}\cdot\frac{m_P}{\ell_P}. \tag{15.31}$$

There is now a positive deficit angle $\Delta\theta = +\pi$.

The whole procedure can be repeated with arbitrary polyhedra, in fact with arbitrarily large numbers of polyhedra identified at appropriate faces. Proceeding in this way one can build up a static spatially closed universe that is almost everywhere flat. The only stress-energy that is present is a network of straight segments of cosmic string that join at vertices of the polyhedra. The results are certainly entertaining but perhaps of dubious relevance to cosmology.

Generalize this construction to an expanding universe.

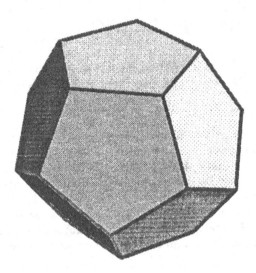

Figure 15.5: Polyhedral universe: reverse the construction used for thin-shell wormholes. Discard the external region and patch two polyhedreal interiors together by identifying appropriate faces. Again, all the stress-energy is concentrated on the edges.

15.2 Dynamic spheric wormholes

In this section I wish to attempt a dynamical analysis of a time-dependent wormhole. To build a suitably tractable class of wormholes, adopt the constraint of spherical symmetry, and assume that all matter is confined to a thin boundary layer between universes. Thus these models are a subset of the "absurdly benign" wormholes of Morris and Thorne [190]. The models are constructed by surgically grafting two Schwarzschild spacetimes together in such a way that no event horizon is permitted to form. This surgery concentrates a nonzero stress-energy on the thin shell (boundary layer) between the two universes. This class of traversable wormholes is sufficiently simple for a (partial) dynamical stability analysis to be carried out. The stability analysis places constraints on the equation of state of the exotic matter that comprises the throat of the wormhole.

It is most illuminating to first construct the class of models to be considered in the static case, ignoring stability questions. Once details of the static case have been spelled out, I shall give more attention to the dynamical analysis of stability. I shall be limited to considering spherically symmetric motions of the wormhole throat.

We have already seen that the static analysis enforces the presence of "exotic stress-energy" (i.e., violation of the null energy condition.) The stability analysis places further constraints on the equation of state of this exotic stress-energy. Suitable candidates for the equation of state of this exotic stress-energy are identified.

15.2.1 Schwarzschild surgery

To construct the wormholes of interest, consider the ordinary Schwarzschild solution to the vacuum Einstein field equations:

$$ds^2 = -(1 - 2M/r)dt^2 + \frac{dr^2}{(1 - 2M/r)} + r^2(d\theta^2 + \sin^2\theta \, d\varphi^2). \quad (15.32)$$

Use the ordinary Schwarzschild coordinates, and do *not* maximally extend the manifold, as that would prove to be unprofitable. Now take *two* copies of this manifold, and remove from them the four-dimensional regions described by

$$\Omega_{1,2} \equiv \{r_{1,2} \leq a \mid a > 2M\}. \quad (15.33)$$

One is left with two geodesically incomplete manifolds with boundaries given by the timelike hypersurfaces

$$\partial\Omega_{1,2} \equiv \{r_{1,2} = a \mid a > 2M\}. \quad (15.34)$$

Now identify these two timelike hypersurfaces (i.e., $\partial\Omega_1 \equiv \partial\Omega_2$). The resulting spacetime $\mathcal{M}$ is geodesically complete and possesses two asymptotically flat regions connected by a wormhole. The throat of the wormhole is at $\partial\Omega$. Because $\mathcal{M}$ is piecewise Schwarzschild, the stress-energy tensor is everywhere zero, except at the throat itself. At $\partial\Omega$ one expects the stress-energy tensor to be proportional to a delta function. This situation is again made to order for an application of the "thin-shell" formalism, also known as the "junction condition" formalism. Note that the condition $a > 2M$ is necessary to prevent the formation of an event horizon.

Exercise: Repeat this analysis for the case $M_1 \neq M_2$. The wormhole throat is now taken to have different masses in the two different asymptotically flat regions. Pay particular attention to the time coordinate. What is the matching condition for t_1 and t_2 across the shell?

Since all the stress-energy is concentrated on the throat, the throat may be viewed as behaving like a domain wall between the two universes. The simplest domain wall one can construct is the classical membrane, but this will be shown to be unstable. More generally one may consider a domain

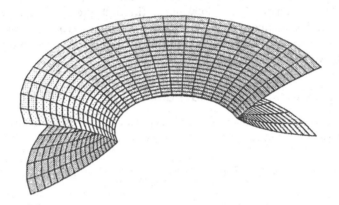

Figure 15.6: Schwarzschild surgery: Two copies of the Schwarzschild geometry are truncated at some finite radius outside the event horizon. The two segments are then pasted together. The seam where the geometries are joined becomes the throat of the wormhole.

wall consisting of a membrane that has some (2+1)-dimensional matter trapped on it. Domain walls of this type can in principle possess essentially arbitrary equations of state. We shall use the stability analysis to constrain the equation of state.

I also wish to mention that the analysis soon to be presented generalizes immediately to traversable wormholes based on surgical modifications of the Reissner–Nordström spacetime. Merely repeat the above discussion using the metric

$$ds^2 = -(1 - 2M/r + Q^2/r^2)dt^2 + \frac{dr^2}{(1 - 2M/r + Q^2/r^2)}$$
$$+ r^2 \left(d\theta^2 + \sin^2\theta \, d\varphi^2 \right). \qquad (15.35)$$

15.2.2   Einstein equations

Applying the thin-shell formalism, the exact result for the Riemann tensor is

$$R_{\alpha\beta\gamma\delta} = -\delta(\eta)\left[\kappa_{\alpha\gamma}\, n_\beta\, n_\delta + \kappa_{\beta\delta}\, n_\alpha\, n_\gamma - \kappa_{\alpha\delta}\, n_\beta\, n_\gamma - \kappa_{\beta\gamma}\, n_\alpha\, n_\delta\right]$$

$$+R^{\text{Schwarzschild}}_{\alpha\beta\gamma\delta}. \tag{15.36}$$

For the spherically symmetric and reflection symmetric case at hand considerable simplifications occur. First, $K^+ = -K^- = \frac{1}{2}\kappa$. Second, $T^\pm_{\mu\nu} \equiv 0$, so the pressure balance constraint is vacuous. Third, the surface stress-energy is covariantly conserved, $\bar{\nabla}_\mu S^{\mu\nu} = 0$. This can be reformulated in terms of the induced metric on the shell itself as $S^{ij}_{|j} = 0$. Fourth, spherical symmetry implies that

$$\kappa^{\hat{\mu}\hat{\nu}} = \begin{bmatrix} \kappa^{\hat{t}\hat{t}} & 0 & 0 & 0 \\ 0 & \kappa^{\hat{\theta}\hat{\theta}} & 0 & 0 \\ 0 & 0 & \kappa^{\hat{\theta}\hat{\theta}} & 0 \\ 0 & 0 & 0 & 0 \end{bmatrix}, \tag{15.37}$$

while the surface stress-energy tensor may be written in terms of the surface energy density σ and surface tension ϑ as

$$S^{\hat{\mu}\hat{\nu}} = \begin{bmatrix} \sigma & 0 & 0 & 0 \\ 0 & -\vartheta & 0 & 0 \\ 0 & 0 & -\vartheta & 0 \\ 0 & 0 & 0 & 0 \end{bmatrix}. \tag{15.38}$$

The Einstein field equations are

$$\sigma = -\frac{1}{4\pi} \cdot \kappa^{\hat{\theta}\hat{\theta}}; \qquad \vartheta = -\frac{1}{8\pi} \cdot \{\kappa^{\hat{t}\hat{t}} + \kappa^{\hat{\theta}\hat{\theta}}\}. \tag{15.39}$$

This has now reduced the computation of the stress-energy tensor to that of computing the two nontrivial components of the second fundamental form. This is very easy to do in the static case, and still quite manageable if the throat is in motion.

15.2.3 Static wormholes

The static case is particularly simple. In the Gaussian normal coordinates appropriate to a spherical shell the Schwarzschild metric is

$$ds^2 = -[1 - 2M/R(\eta)]dt^2 + d\eta^2 + R(\eta)^2 \left(d\theta^2 + \sin^2\theta \, d\varphi^2\right). \tag{15.40}$$

In terms of the proper radial distance η, the function $R(\eta)$ satisfies

$$\frac{dR}{d\eta} = \pm\sqrt{1 - \frac{2M}{R}}. \tag{15.41}$$

The second fundamental form (extrinsic curvature) is, in these Gaussian normal coordinates, simply

$$K^\pm_{\mu\nu} = \frac{1}{2} \left.\frac{\partial g_{\mu\nu}}{\partial \eta}\right|_{R=a} = \frac{1}{2} \left.\frac{dR}{d\eta}\right|_{R=a} \cdot \left.\frac{\partial g_{\mu\nu}}{\partial R}\right|_{R=a}. \tag{15.42}$$

Thus

$$K^{\pm}_{\hat{t}\hat{t}} = \mp \frac{M/a^2}{\sqrt{1 - 2M/a}}, \tag{15.43}$$

$$K^{\pm}_{\hat{\theta}\hat{\theta}} = \pm \frac{\sqrt{1 - 2M/a}}{a}, \tag{15.44}$$

which immediately leads to

$$\sigma = -\frac{1}{2\pi a} \cdot \sqrt{1 - 2M/a}, \tag{15.45}$$

$$\vartheta = -\frac{1}{4\pi a} \cdot \frac{1 - M/a}{\sqrt{1 - 2M/a}}. \tag{15.46}$$

Note that the energy density is negative. This is again just a special case of the defocusing arguments considered in previous chapters. The surface tension is also negative, but this merely implies that we are dealing with a surface *pressure*, not a tension. It should not be too surprising that a positive pressure is needed to prevent collapse of the wormhole throat.

Two special cases are of immediate interest:

The classical membrane: The three-dimensional generalization of the Nambu–Goto action satisfies the equation of state $\sigma = \vartheta$. From the Einstein equations (15.46), the radius of a static classical membrane is related to the total mass of the system by $a = 3M$. Fortunately this is safely outside the event horizon. One sees

$$\sigma = \vartheta = -\frac{1}{2\pi a} \cdot \frac{1}{\sqrt{3}} = -\frac{1}{2\pi M} \cdot \frac{1}{3\sqrt{3}}. \tag{15.47}$$

This should be compared with the discussion in the preceding chapter, wherein negative tension classical cosmic strings were used to construct spherically asymmetric wormholes with polyhedral throats. When we turn to the dynamical analysis, we shall quickly see that this type of wormhole is dynamically unstable.

Traceless stress-energy: The case $S^{\mu}{}_{\mu} = 0$ (*i.e.*, $\sigma + 2\vartheta = 0$) is of interest because it describes massless stress-energy confined to the throat. (Such a stress-energy tensor arises from considering the Casimir effect for massless fields confined to propagate only on the throat, a popular way of obtaining exotic stress-energy.) Unfortunately, in this case no solution to the Einstein field equations exists. (σ, ϑ prove to be imaginary.) This result is rather depressing as it indicates that consideration of the Casimir effect associated with *massless* fields is rather less useful than expected.

Exercise 1: Check the above. Use the Einstein equations (15.46) and the equation of state to infer $a = (3/2)M$. This is inside the event horizon.

Exercise 2: Following the earlier discussion, the analysis immediately generalizes. Show that for a traversable wormhole based on surgical modification of the Reissner–Nordström spacetime,

$$\sigma = -\frac{1}{2\pi a} \cdot \sqrt{1 - 2M/a + Q^2/a^2}; \tag{15.48}$$

$$\vartheta = -\frac{1}{4\pi a} \cdot \frac{1 - M/a}{\sqrt{1 - 2M/a + Q^2/a^2}}. \tag{15.49}$$

Exercise 3: Repeat this analysis for the case $M_1 \neq M_2$. (The wormhole throat is now taken to have different masses in the two different asymptotically flat regions.) Evaluate the surface energy density and surface tension.

15.2.4 Dynamic wormholes

To analyze the dynamics of the wormhole, we permit the radius of the throat to become a function of time $a \mapsto a(\tau)$. The symbol τ is used to denote the proper time as measured by a comoving observer on the wormhole throat. Note that by an application of Birkhoff's theorem we can be confident that at large radius, in fact for any $r > a(\tau)$, the geometry will remain that of a piece of Schwarzschild spacetime (or Reissner–Nordström spacetime). In particular, the assumed spherical symmetry is a sufficient condition for us to conclude that there is no gravitational radiation regardless of the behavior of $a(\tau)$. Let the position of the throat be described by $x^\mu(\tau, \theta, \varphi) \equiv (t(\tau), a(\tau), \theta, \phi)$, so that the four-velocity of a piece of stress-energy at the throat is

$$V^\mu \equiv \left(\frac{dt}{d\tau}, \frac{da}{d\tau}, 0, 0 \right) = \left(\frac{\sqrt{1 - \frac{2M}{a} + \dot{a}^2}}{1 - \frac{2M}{a}}, \ \dot{a}, \ 0, \ 0 \right). \tag{15.50}$$

Note that an overdot is used to denote a derivative with respect to τ. The unit normal to the throat $\partial\Omega$ is

$$n^\mu = \left(\frac{\dot{a}}{1 - \frac{2M}{a}}, \ \sqrt{1 - \frac{2M}{a} + \dot{a}^2}, \ 0, \ 0 \right). \tag{15.51}$$

Exercise: (Trivial) Check that $V^\mu V_\mu = -1$; $n^\mu n_\mu = +1$; and $V^\mu n_\mu = 0$.

In this dynamical geometry $\eta(x)$ is defined as the shortest distance from the point x to the throat of the wormhole—this is equivalent to the statement that $\eta(x)$ is a normal coordinate to the wormhole throat. Note that at the throat itself $n^\mu = \nabla^\mu \eta(x)$.

Calculating the $\hat\theta\hat\theta$ and $\hat\varphi\hat\varphi$ components of the second fundamental form is easy:

$$K_{\hat\theta\hat\theta}^\pm \equiv K_{\hat\varphi\hat\varphi}^\pm = \frac{1}{r} \cdot \left.\frac{\partial r}{\partial \eta}\right|_{r=a} = \pm\frac{1}{a} \cdot \sqrt{1 - \frac{2M}{a} + \dot{a}^2}. \qquad (15.52)$$

Evaluating the $\tau\tau$ component of K is more tedious. One may, naturally, proceed via brute force. It is more instructive to present a short digression. Note that

$$
\begin{aligned}
K_{\hat\tau\hat\tau}^\pm &\equiv K_{\mu\nu}^\pm V^\mu V^\nu \\
&= +\nabla_\mu^\pm n_\nu V^\mu V^\nu \\
&= -V^\mu n_\nu \nabla_\mu^\pm V^\nu \\
&= -n_\nu (V^\mu \nabla_\mu^\pm V^\nu) \\
&= -n_\nu A_\pm^\nu.
\end{aligned}
\qquad (15.53)
$$

Here $A_\pm^\mu$ is the four-acceleration of the throat. (This is a simplified special case of the earlier general analysis.) By the spherical symmetry of the problem, the four-acceleration is proportional to the unit normal,

$$A_\pm^\mu \equiv A_\pm \cdot n^\mu. \qquad (15.54)$$

By the symmetry under interchange of universes, $A_\pm = \pm A$. Thus

$$K_{\hat\tau\hat\tau}^\pm = \mp A \equiv \mp(\text{magnitude of the four-acceleration}). \qquad (15.55)$$

To explicitly evaluate the four-acceleration, utilize the fact that the underlying Schwarzschild geometry possesses a Killing vector

$$k^\mu \equiv \left(\frac{\partial}{\partial t}\right)^\mu \equiv (1,0,0,0). \qquad (15.56)$$

Warning: Notation—I have previously used K to denote a Killing vector and k to denote a null tangent vector. However, here k is the Killing vector, while K is the second fundamental form. Unfortunately there are simply too many standard uses of the symbols K and k, the reader will have to deduce the meaning by context.

Note that, at the throat,

$$k_\mu = (-[1 - 2M/a], 0, 0, 0),$$ (15.57)

so that

$$k_\mu n^\mu = -\dot{a} \quad \text{and} \quad k_\mu V^\mu = -\sqrt{1 - \frac{2M}{a} + \dot{a}^2}.$$ (15.58)

With hindsight, it proves interesting to evaluate

$$
\begin{aligned}
\frac{d}{d\tau}(k_\mu V^\mu) &= k_{\mu;\nu} V^\nu V^\mu + k_\mu \frac{dV^\mu}{d\tau} \\
&= k_\mu A n^\mu \\
&= -A \dot{a}.
\end{aligned}
$$ (15.59)

On the other hand,

$$\frac{d}{d\tau}(k_\mu V^\mu) = -\frac{d}{d\tau}\sqrt{1 - \frac{2M}{a} + \dot{a}^2}$$ (15.60)

$$= -\frac{1}{\sqrt{1 - \frac{2M}{a} + \dot{a}^2}} \cdot \left(\frac{M}{a^2} + \ddot{a}\right) \cdot \dot{a}.$$ (15.61)

Comparing the two calculations, we find that the four-acceleration of the throat is

$$A = \frac{\left(\ddot{a} + \frac{M}{a^2}\right)}{\sqrt{1 - \frac{2M}{a} + \dot{a}^2}}.$$ (15.62)

Note that as $M \to 0$ one recovers the correct special relativistic formula for the four-acceleration $A \to \ddot{a}/\sqrt{1 + \dot{a}^2}$, complete with appropriate time dilation factors.

As $\dot{a} \to 0$ one has $A \to (M/a^2)/\sqrt{1 - 2M/a}$. This is the correct result for the gravitational pseudoforce experienced by an observer held at rest in a Schwarzschild geometry.

The Einstein field equations yield

$$\sigma = -\frac{1}{2\pi a} \cdot \sqrt{1 - \frac{2M}{a} + \dot{a}^2};$$ (15.63)

$$\vartheta = -\frac{1}{4\pi a} \cdot \frac{\left(1 - \frac{M}{a} + \dot{a}^2 + a\ddot{a}\right)}{\sqrt{1 - \frac{2M}{a} + \dot{a}^2}}.$$ (15.64)

It is relatively easy to check that equations (15.64) imply the conservation of stress-energy

$$\dot{\sigma} = -2(\sigma - \vartheta)\frac{\dot{a}}{a} \quad \text{or equivalently,} \quad \frac{d}{d\tau}(\sigma a^2) = \vartheta \cdot \frac{d}{d\tau}(a^2).$$ (15.65)

As is usual, there is a redundancy between the Einstein field equations and the covariant conservation of stress-energy. With the field equations for a moving throat in hand, the dynamical stability analysis will prove simple.

15.2.5   Stability analysis

The Einstein equations obtained in the preceding section may be recast as the pair

$$\dot{a}^2 - \frac{2M}{a} - [2\pi\sigma(a)a]^2 = -1; \qquad (15.66)$$

$$\dot{\sigma} = -2(\sigma - \vartheta)\frac{\dot{a}}{a}. \qquad (15.67)$$

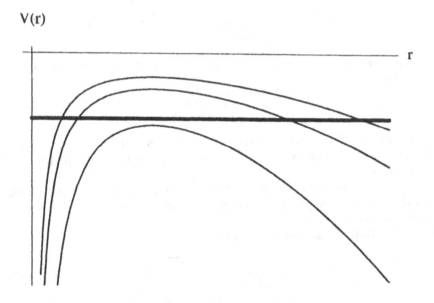

Figure 15.7: A schematic representation of the potential energy
$$V(a) \equiv -\frac{2M}{a} - [2\pi\sigma(a)a]^2.$$
This particular potential corresponds to the material on the throat being described by a classical membrane. Classically, motion is confined to the region below the solid bar.

Consider the classical membrane. The equation of state is $\sigma = \vartheta$, so that $\dot{\sigma} \equiv 0$. It is immediately clear from equation (15.66) that a traversable wormhole built using a classical membrane is dynamically unstable. We need merely observe that the potential in (15.66) is unbounded below. Wormholes of this type either collapse to $a = 0$ or blow up to $a = \infty$ depending on the initial conditions. For example, if $a \gg M$ we may write down the approximate solution

$$a(\tau) \approx \frac{1}{2\pi\sigma} \cosh(2\pi\sigma\tau). \tag{15.68}$$

Even the somewhat *outré* condition that $M < 0$ will only help to stabilize the wormhole against collapse; it will do nothing to prevent the system exploding. Recall that, since the surface energy density is already negative, the possibility of a negative total net gravitational mass is no longer excluded. We have already violated the conditions for the applicability of the positive mass theorem just to construct the wormhole. Since the presence of the wormhole has allowed us to excise the otherwise naked singularity at $r = 0$, this geometry does not violate the cosmic censorship hypothesis even for $M < 0$.

More generally, note that for $M > 0$ the potential near $a = 0$ is unbounded below, regardless of the behavior of $\sigma(a)$. Indeed if $a < 2M$, we see that a runaway solution develops with $a \to 0$ as $\tau \to \infty$. Physically this is a reflection of the fact that if the throat falls within its own Schwarzschild radius, then the wormhole is doomed. Thus if $M > 0$ the best we can hope for is that the wormhole be *metastable* against collapse to $a = 0$.

Even if $M < 0$, one still must require $2\pi \cdot a^{3/2} \cdot \sigma(a) \to k_0 < \sqrt{2|M|}$ as $a \to 0$ in order for the surface density term to not overwhelm the mass term.

On the other hand, if we consider the behavior as $a \to \infty$, we see (regardless of the sign of M) that the wormhole is stable against explosion if and only if $\lim_{a \to \infty}\{2\pi \cdot a \cdot \sigma(a)\} < 1$. If this condition is violated, the wormhole is at best *metastable*. This condition on $\sigma(a)$ will be shown to imply a constraint on the equation of state of the domain wall in the region $\sigma \approx 0$.

15.2.6 Equation of state

To constrain the equation of state, we use the fact that stability against explosion requires $\lim_{a \to \infty}\{2\pi \cdot a \cdot \sigma(a)\} < 1$. Since this implies that $\sigma(a) \to 0$ at spatial infinity, it becomes interesting to expand the equation of state in a Taylor series around $\sigma = 0$:

$$\vartheta(\sigma) = \vartheta_0 + k^2\sigma + O(\sigma^2). \tag{15.69}$$

From this assumed equation of state, and the conservation of stress-energy, one may estimate $\sigma(a)$. Specifically, ignoring $O(\sigma^2)$ terms

$$\vartheta_0 + k^2\sigma = \vartheta = \sigma + \frac{1}{2}a\frac{d\sigma}{da}. \tag{15.70}$$

This differential equation is easily solved

$$\sigma(a) = \frac{\vartheta_0}{1 - k^2} + \sigma_0 \cdot (a/a_0)^{2(k^2-1)}. \tag{15.71}$$

By looking at the $a \to \infty$ behavior we see that $\vartheta_0 = 0$. The pair of Einstein equations now simplify to the single equation

$$\dot{a}^2 - \frac{2M}{a} - [2\pi\sigma_0 a_0]^2 \left(\frac{a_0}{a}\right)^{2-4k^2} = -1; \tag{15.72}$$

So stability requires $k^2 \le \frac{1}{2}$. Thus

$$\vartheta(\sigma) = k^2\sigma + O(\sigma^2); \qquad k^2 \in (-\infty, 1/2]. \tag{15.73}$$

In particular, the case $k = 0$ (corresponding to negative energy dust, $\vartheta = 0$), is stable against explosion.

As a moves in from infinity, $\sigma(a)$ grows and the $O(\sigma^2)$ terms eventually become significant. We cannot really trust the simple linear equation of state for small a (high σ). Nevertheless, if we ignore these warning signs, and also take $M < 0$, then stability against collapse requires $k^2 \in [1/4, \infty)$. Thus the overlap region, in this case the region $k^2 \in [1/4, 1/2]$, gives an equation of state describing a wormhole system that is stable against both collapse and explosion.

Note that the stability analysis has only been carried out for radial perturbations—stability against nonspherical perturbations is completely uncharted territory.

Similar constructions can be carried out in various modified theories of gravity [176, 137].

Exercise: (Long but straightforward) Consider an arbitrary static wormhole of the type discussed above. Perform a linearized stability analysis by Taylor series expanding around the static values of σ_0 and ϑ_0. The result will be a cross between the type of analysis above, and the Brady–Louko–Poisson linearization stability analysis for a thin shell surrounding a black hole [26].

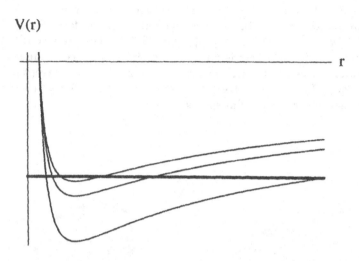

Figure 15.8: A schematic representation of the potential energy
$$V(a) \equiv -\frac{2M}{a} - [2\pi\sigma(a)a]^2.$$
This particular potential corresponds to the material on the throat being described by one of the stability-inducing equations of state discussed in the text. Classically, motion is confined to the region below the solid bar.

15.3 Dynamic loop-based wormholes

15.3.1 Minkowski cut and paste

With a little more work, it is possible to use the thin-shell formalism to construct and investigate "loop-based" wormholes. These are in some sense variants of the polyhedral wormholes previously discussed. Take two copies of flat Minkowski space. For the time being let everything be time independent (static). Because we are dealing with Minkowski space the total overall mass is identically zero.

For convenience, temporarily restrict attention to some fixed time t. Take two copies of flat three-space, each of which contains one of a pair of geometrically identical loops L_1, L_2. Each of these loops can be taken to be the boundary of one of a pair of geometrically identical two-dimensional surfaces S_1, S_2. Each one of these surfaces has two sides, $S_1{}^+$, $S_1{}^-$, and $S_2{}^+$, $S_2{}^-$, respectively.

We can even let the loops L_1, L_2 be knotted. For any arbitrary knot L it is still possible to find a complicated surface S such that the boundary of the surface S is the knot L. Such surfaces are called Seifert surfaces [160, 161].

Now cut the two flat three spaces open along the surfaces S_1, S_2. Identify S_1^+ with S_2^-, and identify S_1^- with S_2^+. These are smooth identifications, and in fact there is no discontinuity in the second fundamental form. Consequently the Riemann tensor is everywhere zero except possibly at the loops L_1, L_2 themselves. This is now our model for a "loop-based" wormhole connecting two flat spaces.

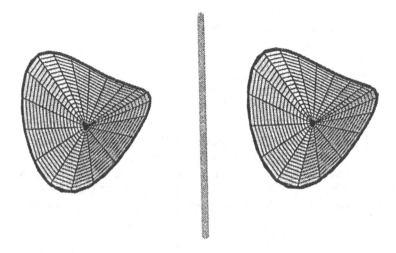

Figure 15.9: Loop-based wormhole: An arbitrary loop may always be spanned by some Seifert surface. Slice the spacetime open along the Seifert surface. Then identify opposite faces of the Seifert surface from different universes.

By power counting (dimensional analysis) we know that the Einstein tensor has dimensions $1/(\text{length})^2$. By construction we know that the Einstein tensor is concentrated on the loop. Thus the Einstein tensor must be proportional to a two-dimensional delta function. Let the position of the loop be given parametrically by $\vec{y}[\xi]$. Then the delta function that has support only on the loop is

$$\delta^2(\vec{x}; L) \equiv \int_L d\xi \, \delta^3(\vec{x} - \vec{y}[\xi]). \qquad (15.74)$$

Here ξ is a single coordinate residing on the loop.

Because this delta function has dimensions $1/(\text{length})^2$, we know that the three-dimensional Einstein tensor is of the form

$$G_{ij}(\vec{x}) = \delta^2(\vec{x}; L) \, f_{ij}. \tag{15.75}$$

Here f_{ij} is a dimensionless tensor that can be constructed only out of the spatial metric and geometrical information about where the loop is located. Now take a particular ansatz for the tensor f_{ij}. Let

$$G_{ij}(\vec{x}) = \delta^2(\vec{x}; L) \, [ah_{ij} + bg_{ij}]. \tag{15.76}$$

(This is a sufficiently general ansatz for the sort of loop-based wormhole of interest in this chapter.) Here h_{ij} is the induced metric on the loop, while g_{ij} is the metric of space. The dimensionless constants a and b are as yet unspecified. For a one-dimensional loop the induced metric is given very simply in terms of the unit tangent vector t_i by $h_{ij} = t_i t_j$.

If we now let the loops L_1, L_2 move around as a function of time we just repeat the whole analysis for each time-slice. By doing this we can promote the entire discussion to $(3+1)$-dimensions. The loop now sweeps out a $(1+1)$-dimensional world-sheet in $(3+1)$-dimensional spacetime. (This implies that, instead of having to do an *ab initio* analysis, one can simply carry over much of the technical discussion developed to handle superstring theories. See, for instance, [123, 127, 238].) Let T denote the time axis. Let the world-sheet be denoted by $\Sigma = T \times L$. The $(3+1)$-dimensional Einstein tensor is

$$G_{\mu\nu}(x) = \delta^2(x; \Sigma) \, [ah_{\mu\nu} + bg_{\mu\nu}], \tag{15.77}$$

subject to the modification

$$\delta^2(x; \Sigma) \equiv \int_{\Sigma} d^2\xi \, \delta^4(x - X[\xi]). \tag{15.78}$$

All this has just been done with dimensional analysis, a simple ansatz, and the observation that away from the loop the space is (by assumption) flat.

The induced metric is simply given by

$$h_{\mu\nu} = -t^0_\mu t^0_\nu + t^1_\mu t^1_\nu. \tag{15.79}$$

Here t^0 is a timelike unit vector tangent to the world-sheet Σ, while t^1 is a orthogonal spacelike unit vector also tangent to the world-sheet [cf. equation (15.20)].

One now has to actually do a minor calculation to show that $b = 0$. [This is just the loop analog of the fact that for a thin shell of matter the normal component of the Einstein tensor vanishes ($G_{nn} = 0$).] Second, one can also show that $a = -2\pi$. The easiest way of deducing these constants

is to take the special case where the loop is flat (confined to a plane) and
to compare with the previous discussion of dihedral wormholes.

The net result is, that for loop-based wormholes constructed using cut
and paste Minkowski spacetimes, the Einstein tensor is

$$G_{\mu\nu}(x) = -2\pi \, \delta^2(x; \Sigma) \, h_{\mu\nu}. \tag{15.80}$$

Note that the logic has been arranged in such a way that I have not yet
made any use of the Einstein field equations—all I've done is to give an
expression for the curvature tensor directly in terms of the location of the
loop without yet specifying what the loop is made of. The only thing we
know for sure about the matter fields is that, by now invoking the Einstein
field equations, the stress-energy tensor must satisfy

$$T_{\mu\nu}(x) = -(1/4G) \, \delta^2(x; \Sigma) \, h_{\mu\nu}. \tag{15.81}$$

This does *not* mean that we know what the matter action is. We can,
however make a few reasonable guesses. Because the stress-energy is concentrated on a loop it's a good bet that the matter Lagrangian should be
concentrated at the same place. This suggests

$$S_m = \int \sqrt{-g} \, \delta^2(x; \Sigma) \, \mathcal{L}(x) d^4 x \tag{15.82}$$

Here $\mathcal{L}(x)$ is some, as yet arbitrary, scalar function that depends on the loop
configuration, and possibly depends on whatever other fields one wishes to
place on the world-sheet swept out by the loop.

For instance: one might place boson or fermion fields on the $(1+1)$-
dimensional world-sheet and couple them to the $(3+1)$-dimensional spacetime metric via a Polyakov type term [216, 217]. On the other hand,
one might add terms depending on the extrinsic curvature of the world-
sheet. Terms depending on extrinsic curvature result in the so-called "stiff
strings" [218].

Warning: A $(1+1)$-dimensional world-sheet propagating in $(3+1)$-
dimensional spacetime has two spacelike normals and hence has two second
fundamental forms (two extrinsic curvatures).

The simplest guess we could make is

$$S_m = \mu \int \sqrt{-g} \, \delta^2(x; \Sigma) d^4 x. \tag{15.83}$$

This is another way of writing the Nambu–Goto action. [Compare with
equation (15.19).] This action defines a Nambu–Goto string with (mass

per unit length) = (tension) = μ. As noted previously, the stress-energy tensor of a Nambu–Goto string is calculated by varying S_m with respect to the metric g. The result is

$$T_{\mu\nu}(x) = \mu\, \delta^2(x; \Sigma)\, h_{\mu\nu}. \tag{15.84}$$

So, a "loop-based" wormhole can be supported, in the sense that the Einstein equations are satisfied, by a length of Nambu–Goto string with

$$\mu = -1/(4G). \tag{15.85}$$

This is a minor variant of the result obtained for polyhedral wormholes. But notice what has not yet been investigated: What are the equations of motion for the matter field? Fortunately, this is easy to answer: Starting from the Nambu–Goto action one can easily derive the usual classical string equations of motion. Note that the string equations of motion are independent of the string tension. The equations of motion do not care whether or not the string is normal or exotic.

What is the net result of all this?

A loop-based wormhole in flat Minkowski space can, but need not, be supported by an exotic piece of Nambu–Goto string. If one chooses the matter content of the wormhole to be described by the Nambu–Goto action the loop will move as though it were a simple string moving in flat Minkowski space; the gravitational interactions can be safely ignored. The exotic nature of the string can be safely ignored; it does not affect the string equations of motion.

Because the string is effectively flapping around in Minkowski space one can construct its "energy" in the usual fashion. This *cannot* be the total energy of the wormhole system since the total energy is *zero*. Nevertheless, one can invoke the usual flat-space Minkowski machinery to define an "energy" which we might profitably think of as the "energy excluding gravitational effects". The length of the string is then of order

$$(\text{length}) \approx E/|\mu| = 4GE. \tag{15.86}$$

This is the distance scale setting the size of the wormhole. Unfortunately, the class of static solutions to the Nambu–Goto string equations is distressingly limited, Nambu–Goto strings always accelerate in the direction of their curvature. For example, simple circular loops of string tend to accelerate inwards and collapse. To get static, stable, loop-based wormholes one needs a different type of string—something more complicated than a Nambu–Goto string. Using an ordinary Polyakov string [216] is no help. Since everything is still classical Nambu–Goto and Polyakov strings are equivalent. Something like a "stiff string" is needed [218].

Nonstatic, oscillating loops are also potentially of interest, though it should be noted that oscillating loops typically have self-intersection events, which in the present context could lead to peculiarities such as topology change.

To summarize:

1. Satisfying the four-dimensional Einstein equations for the gravitational field is easy. We do it by construction.

2. Satisfying the equations of motion for the matter is more problematic. Progress can be made only on a case by case basis once the matter Lagrangian has been specified.

Research Problem 7 *Investigate, in detail, the properties of loop-based wormholes that use stiff strings. Calculate the stress-energy tensor and perform a stability analysis.*

Research Problem 8 *Investigate, in detail, the properties of loop-based wormholes with assorted $(1 + 1)$-dimensional matter fields running around on the world-sheet. Perform a stability analysis.*

Note that while the physical situation here is rather different, one should be able to proceed quickly by borrowing wholesale large chunks of the mathematical machinery developed by the superstring theorists.

15.3.2   Curved-space cut and paste

Once some matter goes through one of these flat-space loop-based wormholes, one of the mouths acquires a positive mass and the other acquires an equal negative mass. Since each of the loops framing the wormhole now lives in a nonflat spacetime the analysis gets much much messier. (For some general comments concerning the difficulties encountered when inserting loop-based sources into general spacetimes, see the opening remarks of Frolov *et al.* [94].)

In principle the analysis follows that of the preceding section; in practice extra attention to details is required. I only know how to do the analysis if the wormhole is symmetric under interchange of asymptotically flat regions.

Take two copies, $\mathcal{M}_1$, $\mathcal{M}_2$, of an arbitrary curved spacetime. For simplicity let both of them be static. (I shall relax this constraint later.)

Let each of these curved spaces contain one of a pair of geometrically identical loops L_1, L_2. Each of these loops can be taken to bound one of a pair of geometrically identical two-surfaces S_1, S_2. We can still let the loops L_1, L_2 be knotted. Now cut the two curved three-spaces open along the surfaces S_1 and S_2. Identify opposite sides of S_1 with opposite sides of S_2. These are smooth identifications, and in fact there is no discontinuity in the second fundamental form. Consequently the Riemann tensor is everywhere smooth except possibly at the loops L_1, L_2 themselves. In particular, the construction guarantees that the Einstein tensor is everywhere smooth, and equal to the Einstein tensor of the original spacetime prior to surgery, except possibly at the loops L_1, L_2 themselves.

Power counting is sufficient to tell us that the Einstein tensor has dimensions $1/(\text{length})^2$. By construction we know that the deviation of the Einstein tensor from that of the presurgical Einstein tensor is concentrated on the loop. Thus the Einstein tensor must again be proportional to a two-dimensional delta function. We can repeat the previous argument to suggest the ansatz

$$G_{ij}(\vec{x}) = G^0_{ij}(\vec{x}) + \delta^2(\vec{x}; L)\,[ah_{ij} + bg_{ij}]. \tag{15.87}$$

Here h_{ij} is the induced metric on the loop, while g_{ij} is the metric of space. Again, a and b are dimensionless constants as yet unspecified. G^0_{ij} is the Einstein tensor of the original manifold $\mathcal{M}$ prior to surgery.

Let the loops L_1, L_2 move around as a function of time. Promote the entire discussion to $(3+1)$-dimensional spacetime. Then

$$G_{\mu\nu}(x) = G^0_{\mu\nu}(x) + \delta^2(x; \Sigma)\,[ah_{\mu\nu} + bg_{\mu\nu}]. \tag{15.88}$$

The dimensionless constants a and b depend only on the local physics. One may thus carry over the previous flat-space results $b = 0$ and $a = -2\pi$. The Einstein tensor is

$$G_{\mu\nu}(x) = G^0_{\mu\nu}(x) - 2\pi\,\delta^2(x; \Sigma)\,h_{\mu\nu}. \tag{15.89}$$

The loop (rather, the two loops) are now embedded in two copies of an arbitrary curved spacetime so this generalization is sufficiently powerful to enable one to handle loop-based wormholes where each mouth is permitted to have nonzero mass. The rest of the flat-space calculation carries over *mutatis mutandis*.

Research Problem 9 *Find a generalization of this construction to the case of unequal masses for the wormhole mouths. (More generally, abandon the requirement that the wormhole connect two identical universes).*

Research Problem 10 *Working with loop-based wormholes in a generic spacetime, make some specific choices for the matter Lagrangian. Analyze Nambu–Goto strings, Polyakov strings, stiff strings, and then add arbitrary $(1+1)$-dimensional matter to the world-sheet. (This should keep you busy for a while.)*

Chapter 16

Topological censorship

The various traversable wormholes considered so far have all been rather special in one way or another—either spherical symmetry or the short-throat (thin-shell) approximation have been invoked to keep the analysis tractable. A general traversable wormhole would be asymmetric, with an arbitrarily long throat, possibly with time-dependent geometry. Analysis of such configurations is rather difficult, and requires the use of so-called global techniques. One major result that can be proved with some degree of confidence is that, loosely

Claim 1 *In any "reasonable" spacetime containing a traversable wormhole the ANEC condition is violated for at least one inextendible null geodesic.*

Note that all the examples considered so far satisfy this claim. Various proofs of this claim, of varying levels of mathematical precision and rigor, have appeared in the literature. The most precise statement equivalent to this result is the topological censorship theorem of Friedman, Schleich, and Witt [92]. Loosely

Claim 2 *If a "reasonable" spacetime satisfies the ANEC on all inextendible null geodesics, then this spacetime does not contain any traversable wormholes.*

I wish to take a few pages to explain this result. I will keep the physical arguments simple, and necessarily imprecise, but will state (without proof) the precise mathematical form of the topological censorship theorem.

16.1 Theorem: Topological censorship

Theorem 13 *In any asymptotically flat, globally hyperbolic spacetime such that every inextendible null geodesic satisfies the ANEC, every causal curve*

195

from past null infinity to future null infinity is deformable to the trivial causal curve.

Technical points: *Asymptotically flat*—this is merely the requirement that at large distances from the highly curved region things settle down to flat Minkowski space. *Globally hyperbolic*—this condition requires the spacetime to be causally well behaved: There is a global Cauchy surface. This implies that the topology of spacetime is of the form $\mathcal{M} \sim \Re \times \Sigma$ and that there can be no closed causal curves. *Trivial causal curve*—a causal curve is trivial if it runs from the infinite past to the infinite future while all the while remaining in the asymptotic region, that is, if it runs from past null infinity to future null infinity by way of spacelike infinity.

The proof of this theorem is highly technical. The import of this theorem is that (assuming global ANEC) it bars any active probing of the topology of space. The theorem does not guarantee that the topology of space Σ is trivial. It does show (assuming global ANEC) that it is impossible to shine a flashlight through (or make a trip through) any nontrivial topology that might be present. Wormholes are permitted. Traversable wormholes are not. A key observation is that the topological censorship theorem can be massaged to give a reasonably general and mathematically precise definition of a traversable wormhole.

Definition 12 *If an asymptotically flat spacetime $\mathcal{M}$ possesses a causal curve γ that stretches from past null infinity to future null infinity, such that γ is not deformable to the trivial causal curve, then $\mathcal{M}$ possesses a traversable wormhole and the curve γ is said to go through the wormhole.*

This definition captures the essence of the previously discussed "traversable in principle" wormholes. No constraints on tidal forces have been formulated. The key point is that a light ray (or traveler) should be able to penetrate the wormhole region and get to the other side. Using this definition, the converse of the topological censorship theorem is

Theorem 14 *Any spacetime containing a traversable wormhole either (1) is not globally hyperbolic, or (2) is such that there exists at least one inextendible null geodesic along which the ANEC is violated.*

Unfortunately, due to technical reasons to do with the way the proof is set up, there is no constructive information on the location of the ANEC violating null geodesic. Presumably at least one of the null geodesics that goes through the wormhole violates the ANEC, but there is as yet no direct general proof of this. Note that it is certainly not true that all null geodesics through the wormhole violate the ANEC. The static polyhedral short-throat wormholes of the preceding chapter provide explicit counterexamples.

16.2   Theorem: ANEC violation

These issues bring home forcefully the importance of any general theorems
that one might be able to prove concerning ANEC violation. There are
certain situations in which we know that the ANEC is guaranteed to hold.
If we could prove that the ANEC was always satisfied in quantum field
theory, then we would have very strong arguments against the existence
of traversable wormholes of the type considered so far. Fortunately it is
known that quantum field theory does sometimes lead to ANEC violation.
We have seen some examples of this already. Some general theorems have
been proved by Klinkhammer [166], by Yurtsever [294], and by Wald and
Yurtsever [277].

Consider the free scalar field with curvature coupling ξ, described by
the Lagrangian

$$\mathcal{L} = \frac{1}{2}(\nabla\phi)^2 + \frac{1}{2}m^2\phi^2 + \xi R\phi^2. \qquad (16.1)$$

Theorem 15 *In* $(3+1)$-*dimensional Minkowski space the vacuum expectation value of the quantum stress-energy tensor for free (massive or massless)
scalar fields satisfies the ANEC along all inextendible null geodesics.*

Theorem 16 *In* $(3+1)$-*dimensional Minkowski space the vacuum expectation value of the quantum stress-energy tensor for free (massive or massless)
scalar fields can violate the ANEC along some inextendible nongeodesic null
curves.*

Theorem 17 *In* $(3+1)$-*dimensional Minkowski space subject to periodic
boundary conditions in the z direction, the vacuum expectation value of the
quantum stress-energy tensor for free (massive or massless) scalar fields
violates the ANEC along all null geodesics (except those perpendicular to
the z axis).*

(This is just the topological Casimir effect considered earlier in this book).
For timelike geodesics the situation depends on the curvature coupling ξ.

Theorem 18 *In* $(3+1)$-*dimensional Minkowski space, if the curvature coupling is in the range* $\xi \in [0, 4]$, *then the vacuum expectation value of the
quantum stress-energy tensor for free (massive or massless) scalar fields
satisfies the ASEC along all inextendible timelike geodesics.*

Theorem 19 *In* $(3+1)$-*dimensional Minkowski space, if the curvature cou-
pling is in the range* $|\xi| \leq 1/4$, *then the vacuum expectation value of the
quantum stress-energy tensor for free (massive or massless) scalar fields
satisfies the AWEC along all inextendible timelike geodesics.*

Stronger results are known in $(1+1)$ dimensions. These results depend critically on the vast simplifications inherent in eliminating two spatial dimensions. For instance [294],

Theorem 20 *In any $(1+1)$-dimensional spacetime that is asymptotically flat and globally conformal to $(1+1)$-dimensional Minkowski space, the expectation value of the quantum stress-energy tensor for a conformally-coupled scalar field in any conformal quantum state satisfies the ANEC along all inextendible null geodesics.*

Another result is [277]

Theorem 21 *In any globally hyperbolic $(1+1)$-dimensional spacetime, the expectation value of the quantum stress-energy tensor for a massless scalar field in any Hadamard quantum state satisfies the ANEC along all inextendible achronal null geodesics.*

Technical points: (1) *Achronal*—a null geodesic is achronal if no two points on the null geodesic can be connected by a timelike curve. A null geodesic is chronal if there exists at least one pair of points on the null geodesic such that this pair of points can be connected by a timelike curve. (2) In any globally hyperbolic $(1+1)$-dimensional spacetime the existence of at least one chronal null geodesic implies that all inextendible null geodesics are chronal and that the spacetime is globally conformal to $\Re \times S^1$. (3) In any globally hyperbolic $(1+1)$-dimensional spacetime: if all inextendible null geodesics are achronal, then the spacetime is globally conformal to $(1+1)$-dimensional Minkowski space (though not necessarily asymptotically flat).

Wald and Yurtsever also present an argument [277, p. 415] that

Theorem 22 *For a massless field the ANEC cannot hold in a general curved $(3+1)$-dimensional spacetime.*

The net result of all of these theorems is to leave the situation rather murky. The theorems we currently have apply only to free fields, and have been derived in the test-field limit—the background spacetime geometry is fixed and the renormalized stress-energy tensor is calculated. No attempt is made to feed this stress-energy tensor back into the Einstein field equations to see what the back-reaction on the geometry might be. No really nice clear-cut criterion is known whereby one could easily decide whether or not quantum field theory permits violations of the ANEC in any particular class of spacetimes. The best result I have been able to obtain to date is

Theorem 23 *In any spacetime with nonzero scale anomaly, the renormalization scale can always be chosen so that the ANEC is violated.*

Since the formulation of this theorem (not to mention the proof), requires technical machinery not currently at hand, I shall defer discussion of this point till later in the book. (See the technical discussion on p. 290ff.) This theorem may be read as saying that "generic" spacetimes violate the ANEC.

Summary: The converse of the topological censorship theorem tells us that any "reasonable" spacetime containing a traversable wormhole and satisfying the Einstein field equations *must* violate the ANEC. The precise conditions under which quantum field theory *permits* this required violation of the ANEC are less than pellucid. This remains an area of active research.

Part IV

Time Travel

Chapter 17

Chronology: Basic notions

From the earliest days of the wormhole renaissance it was realized that the ability to travel to far away places seemed to imply the ability to travel to far away times. The fundamental correspondence between space and time that underlies both the special and general relativistic notions of spacetime indicate that a trip to elsewhere would seem to just as easily imply a trip to "elsewhen". If the whereabouts of the other end of the wormhole are arbitrary, what about its "whenabouts"?

This opens the very disturbing possibility that by recklessly admitting traversable wormholes into the pantheon of physics we have opened Pandora's box by also admitting the curse of time travel. Since notions of causality are fundamental to all physical theories and are basic underpinnings of both logic and most physicists' overall world view, the rather disturbing possibilities inherent in time travel and its associated logical paradoxes deserve to be treated with some care and delicacy.

Aside: The word "paradox" has by now been so abused that I feel it necessary to provide an explicit definition. There are at least two different essentially opposite meanings:

- A logical inconsistency in an apparently plausible argument.

- An apparent inconsistency in a perfectly correct argument.

Many of the more noisy arguments about "paradoxical" aspects of relativity and time travel boil down to the various disputants using these differing definitions without realizing it. The situation is not helped by various sloppy dictionaries that do not include both definitions. In modern usage the phrase "true paradox" and the word "pseudoparadox" are sometimes used to distinguish these two meanings.

Psycho-ceramics warning: To belabor the obvious (at least I hope the reader thinks it's obvious), the "twin paradox" of special relativity is a paradox in the sense of being an "apparent inconsistency" or "pseudoparadox". There are no inconsistencies in special relativity. (Crackpots are politely requested to refrain from reading this paragraph.)

The next order of business is to outline the general strategy for the next few chapters:

- Develop a good working definition of a time machine for use within the context of general relativity. This, thankfully, is a rather easy task.

- Sketch the classical logical paradoxes associated with time travel. These paradoxes fall into two broad classes:

    - Consistency paradoxes.
    - Bootstrap paradoxes.

- Present various elementary examples of universes containing time machines.

- Turning to the issue of traversable wormholes: I shall give a clear exposition of why it is that traversable wormholes, if they exist, *seem* to lead, almost inevitably, to time machine formation.

- Possible responses to the problem of time travel, and various resolutions of the paradoxes, will then be discussed. Responses include

    - The violent and radical rewriting of physics from the ground up.
    - The invoking of "consistency constraints".
    - The denial of the possibility of time travel.
    - The denial of the possibility of traversable wormholes.

All of these options deserve careful scrutiny.

17.1   What is a time machine?

At its most basic, a time machine is any object or system that permits one to travel into the past. Paradoxes arise because once back in the past one should, *a priori*, be able to influence one's own future (which is also one's own past) by either leaving a message or by influencing oneself by some more physical means. To do anything useful with these notions it is necessary to make these ideas mathematically and physically precise. To that end, adopt the following definitions. (See, for instance, [132, pp. 180–199] and [275, pp. 188-200].)

Definition 13 *A causal curve is a curve that is nonspacelike, that is, piecewise either timelike or null (lightlike).*

Physical influences, such as observers and/or the radio messages they send, are constrained to travel along timelike or null curves, respectively. This is nothing more nor less than the assertion that physical influences cannot locally propagate faster than light. Thus if two points are connected by a causal curve it is (in principle) possible to get a signal from one point to the other by a combination of radio and hand-carried messages.

For technical reasons, it is useful to also define

Definition 14 *A chronological curve is a curve that is timelike.*

Chronological curves are appropriate for messengers who are limited to hand-carried messages. Many theorems based on the analysis causal curves have close analogs that use chronological curves. On the other hand, the two notions are logically distinct and one cannot blindly substitute the word chronological for the word causal.

To proceed, one invokes the notion of time-orientability. For convenience, I reiterate the definition

Definition 15 *A Lorentzian spacetime $\mathcal{M}$ is time-orientable if and only if it admits an everywhere nonvanishing continuous timelike vector field.*

This means that an observer placed at any point in spacetime should be able to divide the light cone into a "future light cone" and a "past light cone". All spacetimes discussed in this chapter will be assumed to be time-orientable. There are good physical arguments for this, see the discussion on p. 285. (There are also good physical reasons for expecting spacetime to be space-orientable. We will not need to impose this constraint for the time being.) If spacetime is time-orientable, then, for any point $p \in \mathcal{M}$, one can construct the notion of the causal future and causal past sets.

Definition 16 *If $\mathcal{M}$ is a time-orientable spacetime, then $\forall p \in \mathcal{M}$, the causal future of p, denoted $J^+(p)$, is defined by*

$$J^+(p) \equiv \{q \in \mathcal{M} | \exists \text{ a future-directed causal curve from } p \text{ to } q\}. \quad (17.1)$$

Similarly, the causal past of p is denoted by $J^-(p)$, and is defined in terms of past-directed causal curves.

These causal curves are all assumed to be of nonzero extent, so that under normal circumstances $p \notin J^+(p)$.

Exercise: Define the chronological past, $I^-(p)$, and the chronological future $I^+(p)$, by using chronological curves instead of causal curves.

Exercise: Show that $I^+(p)$ is an open set. Show that $J^+(p)$ need neither be open nor closed. [The causal future $J^+(p)$ is often a closed set, but this is not a universal property.] Show that $\overline{I^+(p)} = \overline{J^+(p)}$, where the bar denotes set closure. Show that boundaries of the chronological and causal futures are equal, $\partial I^+(p) = \partial J^+(p)$. Repeat this exercise, considering the chronological and causal pasts. (After working on this a little, you can find proofs in Hawking and Ellis [132, pp. 182–183].)

The basic definitions of the two fundamental types of time machine are

Definition 17 *If a spacetime $\mathcal{M}$ contains a closed causal curve γ, then $\mathcal{M}$ contains a causality-violating time machine, and the curve γ traverses the time machine.*

Definition 18 *If a spacetime $\mathcal{M}$ contains a closed chronological curve (that is, a closed timelike curve) γ, then $\mathcal{M}$ contains a chronology-violating time machine, and the curve γ traverses the time machine.*

Note that all chronology-violating time machines are causality-violating time machines. It is possible have causality-violating time machines that are not chronology-violating time machines, but as we shall soon see these are rather restricted objects.

Exercise: (Easy) Find an explicit example of a spacetime containing a causality-violating time machine that does not also contain a chronology-violating time machine. [Hints: There are many examples of this behavior. Some particularly simple examples can be constructed by identifying suitable points in an otherwise flat spacetime. A particularly elegant example is everywhere flat (zero Riemann tensor) and singularity free.]

This exercise generalizes to a nice theorem:

Theorem 24 *If a smooth spacetime $\mathcal{M}$ contains a causality-violating time machine, but does not contain a chronology-violating time machine, then the only closed causal curves in the spacetime are closed null geodesics.*

The first step of the proof is easy: a causality-violating time machine that is not chronology violating must by definition contain at least one closed nonspacelike curve, but not contain any closed timelike curves. Next, if any one of these closed nonspacelike curves is not a geodesic, then it can be deformed to produce a closed timelike curve, which is a contradiction. (See Hawking and Ellis [132, Corollary 4.5.1, p. 105; Discussion on p. 183].)

Physically, this means that any causality-violating time machine that is not also chronology violating is of limited interest. While it permits a timelike observer to catch a glimpse of the back of his or her head, it does not give one opportunity to do anything useful with the information.

Exercise: Show that these closed null geodesics are geodesically complete into both the past and future. Hint: see Hawking and Ellis [132, Discussion on p. 190; Corollary 6.4.4, p. 191]. You will need the contrapositive of that result.

On the other hand, one can also show

Theorem 25 *If a smooth spacetime $\mathcal{M}$ contains a causality-violating time machine that is not a chronology-violating time machine, then there exist infinitesimal perturbations of the metric that result in a new spacetime that contains a chronology-violating time machine.*

A spacetime containing a causality-violating time machine is certainly not stably causal. Now adopt the C^0 open topology on the set of all Lorentzian metrics defined on the manifold $\mathcal{M}$. (See Hawking and Ellis [132, p. 198].) Then the lack of stable causality implies that every open set which includes the original metric g must also contain some other metrics that possess closed timelike curves.

These observations suggest that the generic situation of interest is to consider chronology-violating time machines. Unfortunately, to keep things (reasonably) mathematically precise, both notions will have to be carried along.

Definition 19 *The causality-violating region of a spacetime $\mathcal{M}$ is the set of all points x which are connected to themselves by some closed causal curve.*

By definition, each causality-violating region contains a time machine. In causally well-behaved spacetimes the causality-violating region is empty. Note that for points p in the causality-violating region, the intersection between the causal past $J^-(p)$ and the causal future $J^+(p)$ is nonempty. This observation permits one to adopt more refined definitions:

Definition 20 *The causality-violating region associated with the point p is defined as*

$$J^0(p) \equiv J^+(p) \cap J^-(p). \tag{17.2}$$

Definition 21 *The causality-violating region of the spacetime $\mathcal{M}$ is*

$$J^0(\mathcal{M}) \equiv \bigcup_{p \in \mathcal{M}} J^0(p). \tag{17.3}$$

Definition 22 *A spacetime satisfies the causality condition (and is called a causal spacetime), if and only if the causality-violating region is empty.*

The normal assumption in general relativity is that all spacetimes are causal, see for instance [132, p. 189]. We shall for the time being relax this assumption.

Exercise: Use the notion of chronological curve to define chronology-violating regions (also known as dischronal regions) in direct analogy to the above.

Theorem 26 *The causality-violating region is the disjoint union of sets of the form $J^0(q) \equiv J^+(p) \cap J^-(p)$. That is, there exists a set $Q \subset M$ such that*

$$J^0(\mathcal{M}) = \bigcup_{q \in Q} J^0(q) \quad \text{and} \quad q \neq q' \Rightarrow J^0(q) \cap J^0(q') = \emptyset. \quad (17.4)$$

Warning: the set Q is in no sense unique!
(Hawking and Ellis [132, proposition 6.4.3, p. 190].)

Exercise: (Easy) Suppose the causality-violating set consists of a single isolated closed null geodesic. Find all possible candidates for the set Q.

Theorem 27 *The chronology-violating region is the disjoint union of sets of the form $I^0(q) \equiv I^+(p) \cap I^-(p)$. That is, there exists a set $Q \subset M$ such that*

$$I^0(\mathcal{M}) = \bigcup_{q \in Q} I^0(q) \quad \text{and} \quad q \neq q' \Rightarrow I^0(q) \cap I^0(q') = \emptyset. \quad (17.5)$$

Warning: the set Q is in no sense unique!
(Hawking and Ellis [132, proposition 6.4.1, p. 189].)

Exercise 1: (Easy) Find a spacetime such that the chronology-violating set is empty, but the causality-violating set covers the entire spacetime.

Exercise 2: Show that the set closure of the causality-violating region $\overline{J^0(\mathcal{M})}$ differs from the closure of the chronology-violating region $\overline{I^0(\mathcal{M})}$ by at most a disjoint union of geodesically complete closed null geodesics

$$\overline{J^0(\mathcal{M})} = \overline{I^0(\mathcal{M})} \cup \{\gamma\}. \quad (17.6)$$

All the machinery to do this is now in place.

With this technical machinery out of the way, we are finally in a position to define suitable notions of time travel horizons:

Definition 23 *The future causality horizon of a spacetime $\mathcal{M}$ is the boundary of the causal future of the causality-violating region:*

$$H^+(J) \equiv \partial[J^+(J^0(\mathcal{M}))] \qquad (17.7)$$

Definition 24 *The future chronology horizon of a spacetime $\mathcal{M}$ is the boundary of the chronological future of the chronology-violating region:*

$$H^+(I) \equiv \partial[I^+(I^0(\mathcal{M}))] \qquad (17.8)$$

Spacetimes containing time machines will typically be causally well behaved up to a certain point—the causality horizon. Past the causality horizon all manner of evils lurk. Note that the boundary of the causality-violating region and the boundary of the chronology-violating region do not play the central role one might expect—while the boundary of the causality-violating region delimits the "active core" of the time machine, the diseases associated with time travel propagate to the entire future of this region and destroy predictability to the entire future of the causality-violating region.

Theorem 28 *The future causality horizon $H^+(J)$ and future chronology horizon $H^+(I)$ are generated by null geodesic segments. If any of these null geodesic segments possess future end-points, these future end-points will lie within $H^+(J)$ or $H^+(I)$, respectively. If any of these null geodesic segments possess past end-points, these past end-points will lie within $J^0(\mathcal{M})$ or $I^0(\mathcal{M})$, respectively.*

The fact that the horizons are generated by null geodesic segments follows from the fact that they are the boundaries of future sets. The rest of the theorem follows from adaptation of the discussion on pp. 187–188 of Hawking and Ellis [132].

Warning: (1) A causality horizon is not, in general, also an event horizon. (2) A causality horizon is a special case of a Cauchy horizon. All causality horizons are Cauchy horizons; most Cauchy horizons are not causality horizons. (3) The chronology horizon need not be equal to the causality horizon and may not accurately reflect all the predictability problems in the spacetime. (4) In most model time machines presently considered in the literature, the causality and chronology horizons coincide. (5) In many (but not all) of the model time machines considered in the literature the chronology-violating region is coincident with its future, so one need not distinguish the chronology horizon from the boundary of the dischronal region.

Exercise 1: (Easy) Construct a spacetime that has empty chronology horizon but nonempty causality horizon.

Exercise 2: (Easy) Find the chronology-violating region for Politzer's model time machine [214]. Find the boundary of the chronology-violating region. Find the chronology horizon. (They are not the same.) Find a single point q such that $I^0(\mathcal{M}) = I^+(q) \cap I^-(q)$. Show that this point q is not unique, and find all such points. Repeat for the causality-violating region and causal horizon.

Exercise 3: (Easy) Show that if the set closure of the causality-violating region coincides with the closure of the chronology-violating region, then the causality horizon coincides with the chronology horizon.

Observation: In certain situations, where mathematical precision is not paramount, the distinction between chronological curves and causal curves is often ignored.

Several additional concepts in common use in the literature also deserve mention.

Definition 25 *Polarized hypersurface: A point x lies on a polarized hypersurface if and only if there exists a self-intersecting null geodesic that connects the point to itself.*

Note that I adopt the phrase "self-intersecting null geodesic", as distinguished from "closed null geodesic"; this emphasizes the fact that the tangent vector to the null geodesic is permitted to be discontinuous at the point x itself, though it should be continuous elsewhere.

Exercise: Find, schematically, all of the polarized hypersurfaces for the Morris–Thorne–Yurtsever time machine [191]. (After working on this for a while, see [164].) Now find the polarized hypersurfaces for Politzer's model time machine [214]. Discuss.

Lemma 1 *The polarized hypersurfaces all lie within the causality-violating region.*

If the spacetime contains only a single traversable wormhole, then self-intersecting curves (be they causal or noncausal) can be characterized by a single winding number N—this is just the total number of times the curve traverses the wormhole. The fundamental group (first homotopy group) is $\pi_1(\mathcal{M}) = Z$.

Definition 26 *N'th polarized hypersurface: In any spacetime containing a single traversable wormhole a point x is said to lie on the N'th polarized hypersurface if and only if (1) there exists a self-intersecting null geodesic that connects the point to itself, and (2) this curve traverses the wormhole N times.*

Claim 3 *Under certain technical assumptions the chronology horizon is the $N \to \infty$ limit of the N'th polarized hypersurfaces.*

Exercise: (Tricky) Find the precise technical requirements to turn this claim into a theorem. Hint: See [164, p. 3932]. Now apply that argument to Politzer's model time machine [214], and check to see what happens.

If the spacetime contains more than one traversable wormhole then the fundamental group $\pi_1(\mathcal{M})$ is more complicated. In fact, for n traversable wormholes

$$\pi_1(\mathcal{M}) = Z \oplus Z \oplus Z \cdots Z. \qquad (17.9)$$

This denotes the n-fold free sum of the integers. In this case one associates a polarized hypersurface with each element of the fundamental group.

Definition 27 *g-polarized hypersurface: In any spacetime containing multiple traversable wormholes a point x is said to lie on a g-polarized hypersurface if and only if (1) there exists a self-intersecting null geodesic that connects the point to itself, and (2) this curve is homotopic to $g \in \pi_1(\mathcal{M})$.*

The technical analysis of chronology horizons is considerably simplified if they are "compactly generated" [129, 130].

Definition 28 *A chronology horizon is said to be past compactly generated if and only if all inextendible null curves on the chronology horizon tend, in the limit as one goes to negative infinite affine parameter, to a denumerable number of smooth closed null geodesics. These smooth closed null geodesics are called the fountains of the spacetime.*

Many (but not all) of the chronology horizons discussed in this monograph are compactly generated.

Technical point: Since the manifolds I consider are taken to be time-orientable it makes sense to distinguish past fountains (as considered above) from future fountains (sinks?). Future fountains are of interest if one wishes to look at the destruction, rather than creation, of a time machine.

Exercise: Suppose that the only closed null geodesics in the spacetime are past fountains for the chronology horizon. Show that in this case the chronology horizon coincides with the causality horizon. Show that in this case the chronology horizon coincides with the boundary of the chronology-violating region. Show that in this case the chronology horizon is the $N \to \infty$ limit of the N'th polarized hypersurfaces.

Lemma 2 *In any spacetime containing a single traversable wormhole every fountain lies on each one of the N'th polarized hypersurfaces.*

Lemma 3 *In any spacetime containing multiple traversable wormholes, every fountain will lie on at least one of the g-polarized hypersurfaces.*

Lemma 4 *In any spacetime containing multiple traversable wormholes, if a fountain lies on the g-polarized hypersurface, then $\forall j$, the fountain also lies on the g^j-polarized hypersurface.*

17.2   Time travel paradoxes

There are two basic classes of time travel "paradox" that require some analysis. They are the consistency paradoxes and the bootstrap paradoxes. For the time being I will just present a brief discussion of what these paradoxes are. Discussion of some of the possible resolutions of these paradoxes will be postponed until we have seen a few examples of time machines and analyzed the possibility of time machine formation from traversable wormholes.

17.2.1   Consistency paradoxes

Consistency paradoxes arise because time travel to the past, coupled with our notion of free will (autonomy), seems to open the possibility of changing history. A popular but trite place to start the discussion is the so-called "grandfather paradox". Suppose one travels, say, 75 years into the past to encounter either of one's grandfathers before one's own parents were born. Suppose further that one somehow prevents one's grandfather from interacting with one's grandmother. (If you wish to be melodramatic, homicidal techniques can be employed.) What then happens to you, the time traveler? You are now a man or woman without a past since at least one of your parents was never born. How does the universe react to such insult?

Of course, the paradox has nothing to do with grandfathers *per se*. The "grandfather paradox" is merely one example of a general class of consistency paradoxes. A more blunt and immediate version of the paradox is this: Simply take a time machine back to, say, five minutes ago. Then use your second amendment rights to permanently discourage your younger self from any future experimental research into time travel. Now ask: Who killed you? Your future self? But you are now dead, so there is no future self able to come back and kill you. Therefore you cannot be dead. Therefore, in five minutes' time you *can* hop into the time machine and come back to kill your past self. Therefore, you are in fact dead. With a little thought it is possible to come up with all manner of variations on this theme. Consistency paradoxes arise whenever there is a possibility of changing one's own history.

There is certainly an issue here that needs to be addressed. The consistency paradoxes seem to be "true paradoxes" in that they *seem* to indicate an actual logical inconsistency in the notion of time travel. These consistency paradoxes can be resolved, however, by suitably enlarging the universe of discourse. In fact there are *many* ways of dealing with these consistency paradoxes. I shall discuss the various resolutions in due course.

Exercise: Read a few science fiction books.

17.2.2 Bootstrap paradoxes

A second class of logical paradoxes associated with time travel are the bootstrap paradoxes related to information (or objects, or even people?) being created from nothing.

Suppose I travel back in time a year or so and give my younger self a bound and printed copy of this monograph. (Better yet, I give my younger self copies of the final version of the LaTeX files.) My younger self could save himself a lot of work by simply submitting the final version of the manuscript to AIP Press without the bother of actually having to write anything. Information seems to have been created from nothing. (I trust that most readers will agree that there is some information content in this monograph.) Who wrote this book? Variations on the bootstrap theme have been extensively discussed in the science fiction literature and also in mainstream cinema. While there are no logical inconsistencies associated with straightforward bootstrap paradoxes (bootstrap paradoxes are "pseudoparadoxes") the purported effects are certainly weird.

What is more disturbing is that perturbing a bootstrap paradox can give rise to a consistency paradox. Suppose that my future self sends me a copy of the final version of this monograph. This is a bootstrap paradox. Now perturb the bootstrap paradox: It would seem that I (my present self) could make a determined decision to never get near a time machine, and so never send myself the monograph that I have already received from myself. One has now generated a consistency paradox.

Exercise: Read a few more science fiction books.

Exercise: Watch "Star Trek IV—The Voyage Home". Ask yourself: Who discovered "transparent aluminum"?

17.3   Simple time machines

Before turning attention to the temporal aspects of traversable wormholes it is useful to get a feel for some of the simple classical Lorentzian geometries that *seem* to lead to time travel. General relativity is in fact infested with peculiar geometries that *seem* to produce time machines.

17.3.1   The van Stockum time machine

To the best of my knowledge, the earliest explicit example of a spacetime containing closed timelike curves is the van Stockum spacetime [258]. The van Stockum spacetime is a stationary cylindrically symmetric solution to the Einstein field equations that describes an infinitely long rigidly rotating cylinder of dust surrounded by vacuum. The dust is supported against collapse by rotation. The causal properties of the van Stockum spacetime have been explicated by Tipler [251].

By using the assumed symmetries, the metric can be cast into the form [258, 22, 149, 170]

$$ds^2 = -F(r)dt^2 + 2M(r)d\varphi dt + L(r)d\varphi^2 + H(r)[dz^2 + dr^2]. \qquad (17.10)$$

The range of the coordinates (t, φ, r, z) is

$$t \in (-\infty, +\infty); \quad \varphi \in [0, 2\pi]; \quad r \in (0, +\infty); \quad z \in (-\infty, +\infty). \qquad (17.11)$$

In matrix form, the metric can be written as

$$g_{\mu\nu} = \begin{bmatrix} -F & M & 0 & 0 \\ M & L & 0 & 0 \\ 0 & 0 & H & 0 \\ 0 & 0 & 0 & H \end{bmatrix}. \qquad (17.12)$$

The determinant of this matrix is $g = -(FL + M^2)H^2$. Therefore the metric has Lorentzian signature provided $FL + M^2 > 0$.

Consider an azimuthal curve of fixed (r, z, t). That is, consider an integral curve of the coordinate φ. By construction, this is a closed curve of invariant length

$$s_\gamma^2 \equiv L(r)[2\pi]^2. \qquad (17.13)$$

If $L(r)$ ever becomes negative, then this azimuthal curve is a closed timelike curve. If $L(r)$ is ever zero, then this azimuthal curve is a closed null curve.

Alternatively, consider a null azimuthal curve. By this I mean a null curve in the (φ, t) plane that remains at fixed r, z. This curve need not, and in general will not be a geodesic. Neither will it be, in general, a closed curve. The null condition $ds^2 = 0$ implies

$$0 = -F + 2M\dot{\varphi} + L\dot{\varphi}^2. \qquad (17.14)$$

(Here $\dot{\varphi} \equiv d\varphi/dt$.) Solving the quadratic,

$$\dot{\varphi} = \frac{-M \pm \sqrt{M^2 + FL}}{L}. \qquad (17.15)$$

In view of the Lorentzian signature constraint, the roots will always be real. If $L(r) < 0$ then the light cones are tipped over sufficiently far to permit a trip to the past. By going once around the azimuthal direction the total backward time-jump for a null curve is

$$\Delta T = \frac{2\pi|L|}{M + \sqrt{M^2 - F|L|}}. \qquad (17.16)$$

Loosely speaking, the tipping over (that is, tilting over) of light cones is a generic feature of spacetimes containing closed causal curves. Remember that the definition of a causal curve is one that is always locally traveling at or below the speed of light. So if all the light cones are pointing in roughly the same direction, closed causal curves cannot arise. It is only when the light cones are sufficiently disorganized, so that at some locations they point back into what would naively be thought of as the past, that closed timelike curves arise.

If $L(r) < 0$ for even a single value of r, then there is a closed causal curve passing through every point of the spacetime. To see this, start at an arbitrary point x. Follow an ordinary null curve to some value of r such that $L(r) < 0$. Now take the null curve that wraps around the azimuth a total of N times. The total backward time-jump is then $N\Delta T$, which can be made as large as desired. Finally, follow an ordinary null curve back to the starting point x.

That is, if $L(r) < 0$ for even a single value of r, the chronology-violating region covers the entire spacetime.

The argument presented so far has relied solely on symmetry considerations without yet invoking the Einstein equations. Now let the surface of the dust cylinder be located at $r = R$, and let the cylinder rotate with angular velocity Ω. Then the interior solution ($r < R$) is relatively easy to write down [258, 22]. The metric is

$$ds^2 = -dt^2 + 2\Omega r^2 d\varphi dt + r^2(1 - \Omega^2 r^2)d\varphi^2 + e^{-\Omega^2 r^2}(dz^2 + dr^2). \quad (17.17)$$

Applying the Einstein field equations, the density and four-velocity of the dust are given by

$$8\pi G\rho = 4\Omega^2 e^{+\Omega^2 r^2}; \qquad V^\mu = (1, 0, 0, 0). \qquad (17.18)$$

Note that this coordinate system corotates with the dust. From the form of the interior solution it is already clear that closed timelike curves arise for $\Omega R > 1$. Because the source is simply positive density dust, it is clear that all the energy conditions are satisfied.

Exercise: Calculate the vorticity of the dust.

Considerably more can be said by also considering the exterior solutions to the Einstein equations ($r > R$). Exterior solutions to the vacuum Einstein equations exist for all values of the parameter ΩR and can be smoothly matched to the interior solution given above. Unfortunately, the exterior solutions are rather messy. If $\Omega R > 1/2$, then it can be shown that the function $L(r)$ is negative for some values of r. The first zero of $L(r)$ occurs at

$$r_0 = R \, \exp\left\{ \frac{\pi - 3\tan^{-1}\sqrt{4R^2\Omega^2 - 1}}{\sqrt{4R^2\Omega^2 - 1}} \right\}. \tag{17.19}$$

Exercise: (Tedious) Prove this assertion. Use the explicit form of the exterior metrics given in [22].

Note that $r_0 \to \infty$ as $\Omega R \to 1/2$, while $r_0 \to R$ as $\Omega R \to 1$. That is, azimuthal curves are first timelike for $\Omega R = (1/2)^+$, and the innermost timelike azimuthal curves move in toward the dust cylinder as ΩR rises to unity. At $\Omega R = 1$ the interior and exterior solutions agree that the innermost null azimuthal curve lies at $r = R$, the surface of the dust cylinder.

In summary, the van Stockum spacetime contains closed causal curves provided $\Omega R > 1/2$. The chronology-violating region covers the whole spacetime.

The typical reaction to this observation is to dismiss the van Stockum solution as "unphysical", at least for $\Omega R > 1/2$. At least three excuses can be made for this reaction:

1. The van Stockum solution applies only to an infinitely long cylinder. Cylinders of finite length are believed not to induce closed timelike curves.

2. The van Stockum solution may be viewed as a particular instance of the general principle "garbage in—garbage out". If physically unreasonable assumptions are built into the definition of the model, then one should not be surprised by the fact that physically unreasonable consequences can then be deduced from the model. Indeed the van Stockum solutions are not asymptotically flat, and so cannot be physically realized in our universe.

3. Furthermore, the van Stockum solutions are solutions to the field equations only in the sense that we have not asked what the appropriate initial data are. Solving a set of differential equations makes no sense unless one specifies the relevant initial data, or more generally, relevant boundary conditions. If the boundary conditions are themselves diseased, the problem can be ill-posed and the resulting "solution", if any, is itself problematic.

17.3.2 The Gödel universe

In its original formulation, the Gödel universe [117] is interpreted as a universe that contains both (rotating) dust and a nonzero cosmological constant. The total stress-energy tensor may be written as

$$T^{\mu\nu}_{\text{total}} = \rho V^\mu V^\nu - \frac{\Lambda}{8\pi G}\, g^{\mu\nu}. \tag{17.20}$$

By redefining

$$T^{\mu\nu}_{\text{total}} = (\bar\rho + \bar p)V^\mu V^\nu + \bar p g^{\mu\nu}, \tag{17.21}$$

with

$$\bar p = -\frac{\Lambda}{8\pi G}; \qquad \bar\rho = \rho + \frac{\Lambda}{8\pi G}, \tag{17.22}$$

this stress-energy tensor can be reinterpreted as that of a (rotating) perfect fluid of density $\bar\rho$ and pressure $\bar p$ in a universe with zero cosmological constant. The metric of the Gödel solution is rather messy,

$$ds^2 = -dt^2 - 2e^{\sqrt{2}\Omega y}dt dx - \frac{1}{2}e^{2\sqrt{2}\Omega y}dx^2 + dy^2 + dz^2. \tag{17.23}$$

Here the parameter Ω is related to the stress-energy by

$$4\pi G\rho = \Omega^2 = -\Lambda \qquad \text{or equivalently} \qquad \bar\rho = \bar p = \frac{\Omega^2}{8\pi G}. \tag{17.24}$$

This implies that the null, weak, strong, and dominant energy conditions are all satisfied. The dominant energy condition, though satisfied, is on the verge of being violated. The four-velocity and vorticity of the fluid are simply

$$V^\mu = (1,0,0,0); \qquad \omega^\mu = (0,0,0,\Omega). \tag{17.25}$$

Consider the alternative set of coordinates $(\bar t, r, \varphi, z)$ defined by

$$\Omega x \exp(\sqrt{2}\Omega y) = \sin\varphi \sinh(\sqrt{2}\Omega r),$$
$$\exp(\sqrt{2}\Omega y) = \cosh(\sqrt{2}\Omega r) + \cos\varphi \sinh(\sqrt{2}\Omega r),$$
$$\tan\{[\varphi + \Omega(t - \bar t)]/2\} = \exp(-\sqrt{2}\Omega r)\tan\{\varphi/2\}. \tag{17.26}$$

Using these coordinates the metric may be written as

$$
\begin{aligned}
ds^2 &= -d\bar{t}^2 + 4\Omega^{-1}\sinh^2(\sqrt{2}\Omega r)d\varphi d\bar{t} \\
&\quad + 2\Omega^{-2}\sinh^2(\sqrt{2}\Omega r)\left[1 - \sinh^2(\sqrt{2}\Omega r)\right]d\varphi^2 \\
&\quad + dr^2 + dz^2.
\end{aligned} \tag{17.27}
$$

Exercise: (Tedious) Prove this. These coordinates are a minor variant of those discussed by Gödel [117, p. 449]. See also Hawking and Ellis [132, p. 168].

The metric is now in the same format as was used for discussing the van Stockum solutions. (The rotational symmetry of the Gödel universe is now manifest.) Azimuthal curves [curves of fixed $(\bar{t}, r, z)$] will be closed timelike curves provided

$$
\sinh(\sqrt{2}\Omega r) > 1; \quad \text{that is, provided} \quad \Omega r > \frac{\ln(1 + \sqrt{2})}{\sqrt{2}}. \tag{17.28}
$$

These are just the most obvious closed timelike curves. There are many others. By the preceding analysis, the chronology-violating region covers the entire spacetime. In fact, despite appearances, the Gödel universe is homogeneous (all points are equivalent under translational symmetry).

Exercise: (Tedious) Prove this. This is not supposed to be obvious.

The typical reaction to time travel in the Gödel universe is, again, to dismiss this solution as unphysical. The reasons given for this reaction are more limited in scope than in the case of the van Stockum spacetime. Since Gödel's spacetime is homogenous it is no longer possible to blame the unphysical behavior on the assumed presence of "unphysical" infinitely long rotating dust cylinders. While observationally the Gödel universe is not acceptable as a model for our own universe, condemning it on those grounds is somewhat of a cheap shot in that it leaves important issues of principle unanswered.

More honestly one falls back on the "garbage in—garbage out" principle. In deriving the Gödel solution one has solved the Einstein field equations. So what? Diseased boundary conditions lead to diseased physics. For instance: The Gödel solution is certainly not foliated by spacelike hypersurfaces. What, pray tell, are the "initial data" that evolve (via the Einstein differential equations) into the Gödel universe? In fact, there are no such initial data. The Gödel universe cannot be analyzed by the initial data formalism. (See, for example, [186, 275] for discussions of the initial data formalism.)

To put the issue more properly in perspective: Consider a piece of Minkowski spacetime that has been compactified in the time direction by applying periodic boundary conditions (period T). The metric is

$$ds^2 = -dt^2 + dx^2 + dy^2 + dz^2. \qquad (17.29)$$

And the coordinate t is limited to the range $t \in [0, T]$, with $t = 0$ being identified with $t = T$. This spacetime certainly contains closed timelike curves. It is also a solution of the Einstein field equations. (After all, locally it's just flat Minkowski space.) But one would not (or at least, should not) claim that the existence of this solution to the Einstein field equations proves the reality of time travel.

Most physicists therefore view the Gödel universe as a curiosity of dubious relevance to the real world. It is sometimes claimed that the discovery of the Gödel solution proves the existence of time travel as a real physical phenomenon. I view such claims as gross over enthusiasm.

Exercise 1: (Trivial) Consider the compactified Minkowski space discussed above. Find the chronology-violating region and causality-violating region. Find the fountains, if any.

Exercise 2: (Easy) Find a solution to the vacuum Einstein equations which is everywhere flat (zero Riemann tensor) but is nevertheless not time-orientable. Does this solution contain any closed causal curves? Find the chronology-violating region and causality-violating region. Find the fountains if any. What lesson should you deduce from the existence of this solution?

Exercise 3: After reviewing the definitions of space-orientability and spacetime-orientability, find a solution to the vacuum Einstein equations which is everywhere flat (zero Riemann tensor) but nevertheless is not space-orientable. Repeat this exercise, now finding (several) solutions that are not spacetime-orientable. What lesson should you deduce from the existence of these solutions? (Hint: think of a variation on the theme of the Möbius strip.)

17.3.3 The Kerr time machine

The Kerr black hole (a rotating vacuum black hole) is another example of a spacetime possessing closed timelike curves. The presence of closed timelike curves is intimately related to the ringlike structure of the singularity. (See p. 75.) The details of the discussion are rather different depending on whether or not this singularity is naked.

Case $a < M$: The ring singularity is surrounded by an event horizon (see p. 75). Consider azimuthal curves of constant (r, θ, t), that is, integral curves of φ. From the explicit form of the metric [equation (7.5)] it is easy to see that these are closed timelike curves for $\theta = \pi/2$ and $r \in (-\sqrt[3]{2Ma^2}, 0)$. To establish that the region $r < 0$ is actually physically sensible and part of the maximally extended Kerr solution requires a little work.

With a little more work it can be seen that every point inside the inner horizon $(r_- = M - \sqrt{M^2 - a^2})$ lies on some closed timelike curve [132, p. 164]. The chronology-violating region is thus contiguous with the interior of the inner horizon and the chronology horizon coincides with the inner horizon. (The inner horizon is also both a Cauchy horizon and an apparent horizon.) The chronology horizon is not compactly generated. More details concerning the causal structure of Kerr spacetime are given in the series of papers by Clarke, de Felice, and co-workers [31, 54, 53, 40, 41].

The chronology violations are decently hidden behind an event horizon and are therefore deemed to be not disturbing. If an outside observer can't see it, don't worry about it. There are also other reasons for not being too worried by chronology violation in the Kerr black hole: We have no clear idea of what the interior metric is for a rotating star, let alone what a rotating star collapses to after black hole formation. We have no particular reason to believe that the part of the Kerr solution inside the event horizon has anything to do with astrophysical black holes formed from stellar collapse. Additionally, the inner horizon (which is a Cauchy horizon) is known to be generically unstable [212, 153, 180].

Exercise: (Straightforward) Work through the details of the maximally extended Kerr spacetime. Use any of the standard textbooks.

Case $a > M$: The ring singularity is now naked (not surrounded by an event horizon, see p. 75). As previously, azimuthal curves of constant (r, θ, t) are closed timelike curves for $\theta = \pi/2$ and $r \in (-\sqrt[3]{2Ma^2}, 0)$. In this case these closed timelike curves can be deformed to pass through every point in the (maximally extended) spacetime [132, p. 162]. The chronology-violating region thus covers the entire spacetime.

The fact that the singularity is naked is already a problem in its own right. Normally one appeals to the cosmic censorship conjecture to dismiss this spacetime as unphysical.

17.3.4 Spinning cosmic strings

Consider an infinitely long straight piece of "string" that lies on and spins about the z axis. The symmetries of the system are identical to those for the van Stockum spacetimes but the asymptotic behavior is different.

The source is a delta-function source confined to the z axis, characterized by a mass per unit length μ, a tension τ, and an angular momentum per unit length J. For so-called cosmic strings the mass per unit length is equal to the tension, $\mu = \tau$. In this case the Einstein equations can be easily solved [55, 56, 111, 120, 136, 159, 177]. In cylindrical coordinates the metric is

$$ds^2 = -[d(t + 4GJ\varphi)]^2 + dr^2 + (1 - 4G\mu)^2 r^2 d\varphi^2 + dz^2. \tag{17.30}$$

The coordinates cover the ranges

$$t \in (-\infty, +\infty); \quad \varphi \in [0, 2\pi]; \quad r \in (0, +\infty); \quad z \in (-\infty, +\infty). \tag{17.31}$$

Adopt a new set of coordinates

$$\bar{t} = t + 4GJ\varphi; \qquad \bar{\varphi} = (1 - 4G\mu)\varphi. \tag{17.32}$$

The metric may be rewritten as

$$ds^2 = -d\bar{t}^2 + dr^2 + r^2 d\bar{\varphi}^2 + dz^2. \tag{17.33}$$

The new coordinates cover the ranges

$$\bar{t} \in (-\infty, +\infty); \quad \bar{\varphi} \in [0, (1 - 4G\mu)2\pi]; \quad r \in (0, +\infty); \quad z \in (-\infty, +\infty). \tag{17.34}$$

They are subject to the identification

$$(\bar{t}, r, \bar{\varphi}, z) \equiv (\bar{t} + 8\pi GJ, r, \bar{\varphi} + 2\pi[1 - 4G\mu], z). \tag{17.35}$$

It is now obvious that, outside the core at $r = 0$, the metric is locally flat (the Riemann tensor is zero). The geometry is that of flat Minkowski space subject to a somewhat peculiar set of identifications. On traveling once around the string, one sees that the spatial slices are "missing a wedge of angle $8\pi G\mu$". This defines the so-called deficit angle (cf. p. 168)

$$\Delta\theta \equiv 8\pi G\mu. \tag{17.36}$$

Furthermore, on traveling once around the string, one undergoes a backward time-jump of size

$$\Delta\bar{t} = 8\pi GJ. \tag{17.37}$$

To investigate the issue of closed timelike curves, consider azimuthal curves. In (t, r, φ, z) coordinates these are just the integral curves of φ. Repeating the by now familiar analysis, azimuthal curves are closed timelike curves whenever

$$r < \frac{4GJ}{1 - 4G\mu}. \tag{17.38}$$

If one prefers the $(\bar{t}, r, \bar{\varphi}, z)$ coordinates, the azimuthal curves are given by

$$X(\bar{\varphi})\Big|_{(t,r,z)} = \left(t + \frac{4GJ\bar{\varphi}}{1 - 4G\mu}, r, \bar{\varphi}, z\right), \qquad (17.39)$$

subject to the identifications

$$(t, r, 0, z) \equiv (t + 8\pi GJ, r, 2\pi[1 - 4G\mu], z). \qquad (17.40)$$

These azimuthal curves are timelike for $r < 4GJ/(1 - 4G\mu)$. (The two different coordinate systems agree on this point, as indeed they must.)

As was the case in the previously considered examples, these closed time-like curves can be deformed to cover the entire spacetime. Consequently the chronology-violating region covers the whole manifold.

While perfectly valid as solutions to the Einstein equations, these geometries are typically deemed to be unphysical for the sorts of reasons discussed in earlier examples.

More realistic models for spinning cosmic strings replace the delta-function source with some complicated spinning core of finite thickness. The van Stockum interior solution is not appropriate. For a cosmic string we need unequal principal pressures $\rho = p_z \neq 0$, but $p_r = p_\varphi = 0$. For the van Stockum solution $\rho \neq 0$, but $p_r = p_\varphi = p_r = 0$. An appropriate class of interior solutions has been written down by Jensen and Soleng [159]. If the core of the spinning string happens to be of radius $R > J$ then the closed timelike curves have been abolished. Unfortunately there is no general requirement to this effect.

17.3.5   The Gott time machine

The Gott time machine is a variation on the theme of the spinning cosmic string. Two cosmic strings are employed, each of which is non-spinning [120] (zero intrinsic angular momentum). The idea is to let the strings be precisely parallel to each other, and to move with respect to each other. The orbital angular momentum of the pair is now used to induce closed timelike curves [121]. (More complicated, non-parallel geometries are also possible.)

Each of the two cosmic strings is parameterized by a two-dimensional velocity $(\vec{\beta}_1, \vec{\beta}_2)$, a mass per unit length (μ_1, μ_2), and a two-dimensional location $(\vec{x}_1, \vec{x}_2)$. The relative motion of the pair is parameterized by the impact parameter a. The strings are taken to lie parallel to the z axis, and explicit mention of this third dimension is often suppressed.

Outside the region containing the two cosmic strings the geometry is a locally flat vacuum solution to the Einstein equations—and so must be characterized by the total mass per unit length μ_{tot}, the total angular momentum per unit length J_{tot}, and the location of the center of the compound system. The geometry exterior to the region containing the two cosmic strings is described by the spinning cosmic string previously discussed. If both strings simultaneously approach within a distance $4GJ_{\text{tot}}/(1-4G\mu_{\text{tot}})$ of the center of the compound system, then some closed timelike curves are formed [273]. Unfortunately $J_{\text{tot}}(\mu_1, \beta_1, \mu_2, \beta_2, a)$ is an extremely messy function of its parameters.

If we temporarily confine attention to very light strings $(4G\mu \ll 1)$, then the global geometry is very close to that of Minkowski space and we expect to be able to use ordinary relativistic mechanics. For rapidly moving objects it is convenient to define the so-called rapidity parameters by

$$\beta_i = \tanh\zeta_i; \qquad \gamma_i = \cosh\zeta_i; \qquad \beta_i\gamma_i = \sinh\zeta_i. \qquad (17.41)$$

For instance, for sufficiently light strings one expects the special relativistic result:

$$\mu_{\text{tot}}^2 \approx \mu_1^2 + \mu_2^2 + 2\mu_1\mu_2\cosh(\zeta_1 - \zeta_2). \qquad (17.42)$$

Exercise: Prove this.

Now go to the center of momentum frame,

$$\mu_1\beta_1\gamma_1 = -\mu_2\beta_2\gamma_2, \qquad (17.43)$$

that is,

$$\mu_1\sinh\zeta_1 = -\mu_2\sinh\zeta_2. \qquad (17.44)$$

The orbital angular momentum is then approximately

$$J \approx (\mu_1\beta_1\gamma_1)a = (\mu_1\sinh\zeta_1)a. \qquad (17.45)$$

For sufficiently light cosmic strings, the center of the compound system is in the obvious place, and at the point of closest approach both strings are a distance $a/2$ from the center. Thus closed timelike curves form only if

$$\frac{a}{2} \lesssim \frac{4GJ}{1-4G\mu} \approx (4G\mu_1\sinh\zeta_1)a, \qquad (17.46)$$

that is, provided

$$(8G\mu_1)\sinh\zeta_1 \gtrsim 1. \qquad (17.47)$$

This implies that closed timelike curves *cannot* form if the masses and velocities are both small. The only hope for obtaining closed timelike curves is with heavy rapidly moving cosmic strings. An algebraically messy general analysis is called for.

If the two cosmic strings have equal mass per unit length, and equal but opposite velocities, then Gott has shown [121] that closed timelike curves form if and only if

$$\tan(4\pi G\mu_1)\sinh\zeta_1 > 1. \qquad (17.48)$$

This exact result generalizes the low mass approximation given in equation (17.47) above. The exact result is equivalent to

$$\sin(4\pi G\mu_1)\cosh\zeta_1 > 1. \tag{17.49}$$

Exercise: Check the equivalence of these two inequalities.

The closed timelike curves can be deformed to penetrate infinitely far into the past and future [197, 50]. The chronology-violating region does not cover the whole spacetime and the location of the chronology horizon is a very messy function of the position, mass per unit length, and velocities of the cosmic strings [197, 50]. The chronology horizon is not compactly generated.

The Gott time machine is now believed be unphysical for a number of reasons [56, 35, 36].

1. The use of infinitely long strings is as always disturbing.

2. The fact that closed timelike curves penetrate infinitely far into the past indicates that this solution of the field equations is yet another example of the "garbage in—garbage out" principle. If closed timelike curves are (effectively) built into the initial data at $t = -\infty$, the fact that the rest of the spacetime is unpleasantly behaved should not be a surprise.

3. Furthermore, the existence of closed timelike curves requires an "unphysical" value for the total mass of the system. A brief digression is in order.

An excruciating calculation yields the total mass of a pair of moving cosmic strings

$$\begin{aligned}\cos(4\pi G\mu_{\text{tot}}) \;=\; & \cos(4\pi G\mu_1)\cos(4\pi G\mu_2) \\ & - \sin(4\pi G\mu_1)\sin(4\pi G\mu_2)\cosh(\zeta_1 - \zeta_2). \end{aligned} \tag{17.50}$$

(As long as this total mass is small, the total mass can be defined in terms of the total angular deficit of the spacetime. More generally the total mass is defined in terms of the identification procedure used to construct the Gott spacetime.)

Exercise: (Hard) Prove this. That is: Prove equation (5.13) of reference [55].

If the two cosmic strings have equal mass per unit length, and equal but opposite velocities (rapidities) the total mass per unit length simplifies to

$$\sin(2\pi G\mu_{tot}) = \sin(4\pi G\mu_1)\cosh(\zeta_1). \qquad (17.51)$$

But the inequality that must be satisfied to generate closed timelike curves then implies

$$\sin(2\pi G\mu_{tot}) \geq 1. \qquad (17.52)$$

This can only be satisfied for a complex value of the total mass! (This is an indication of the fact that the overall four-momentum of a Gott time machine is in fact spacelike, rather than timelike. See references [56, 35, 36, 273]. For other features of the Gott spacetime see [122, 197].)

17.3.6 Summary

All of these classical geometries that *seem* to lead to closed timelike curves are arguably unphysical in one form or another. A rule of thumb seems to be that closed timelike curves are associated with sufficiently rapid rotation (where "sufficiently rapid" seems to imply "unphysical"). In the next chapter we shall see examples of putative time machines based on traversable wormholes. This new class of time machines does not rely on the rotation-induced tipping of the light cones. Discussion of the physical acceptability or unacceptability of these spacetimes must be based on other issues.

Chapter 18

From wormhole to time machine

18.1 Isolated wormholes

It is one of the more entertaining features of wormhole physics that, given a traversable wormhole, it *appears* to be very easy to build a time machine [190, 191]. Indeed so easy is the construction that it *seems* that the creation of a time machine might be the generic fate of a traversable wormhole [95, 268]. To see the underlying physics behind this claim it is useful to perform a *gedanken experiment* that clearly and cleanly separates the various steps involved in constructing a time machine:

1. Acquire a traversable wormhole.

2. Induce a "time-shift" between the two mouths of the wormhole.

3. Bring the wormhole mouths close together.

It is only the induction of the "time-shift" in step 2 that requires intrinsically relativistic effects. The *apparent* creation of the time machine in step 3 can take place arbitrarily slowly in a nonrelativistic manner [268].

18.1.1 Step 1: Acquire a wormhole

To get the discussion started, assume that an arbitrarily advanced civilization has by whatever means necessary acquired and continues to maintain a traversable wormhole. This stage of the process is rather reminiscent of the recipe for dragon stew: First, find a dragon ...

It is sufficient, for the purposes of this chapter, to take an extremely simple idealized model for a traversable wormhole. Start by considering a short-throat wormhole embedded in flat Minkowski space—one of the "cut and paste" wormholes described by the thin-shell formalism. (This approximation throws away all of the complications having to do with the thickness of the wormhole throat.) Now make a further approximation, and take the radius of the wormhole throat to be zero. (This throws away all of the complications having to do with the finite transverse extent of the wormhole mouth.) This approximation is useful for deducing gross features of time machine formation—but we will certainly go beyond this approximation in subsequent chapters.

The wormhole is now mathematically modeled by Minkowski space with two timelike worldlines identified. For simplicity one may further assume that initially the two mouths of the wormhole are at rest with respect to each other, and also that the wormhole mouths connect "equal times" as viewed from the mouth's rest frame. Mathematically this means that one is considering $(3 + 1)$-dimensional Minkowski space with the following two lines identified:

$$\ell_1^\mu(\tau) = \bar{\ell}_0^\mu + \frac{1}{2}S^\mu + V^\mu\tau, \qquad \ell_2^\mu(\tau) = \bar{\ell}_0^\mu - \frac{1}{2}S^\mu + V^\mu\tau, \qquad (18.1)$$

here V^μ is an arbitrary timelike vector, S^μ is perpendicular to V^μ and so is spacelike, and $\bar{\ell}_0^\mu$ is completely arbitrary. The center of mass of the pair of wormhole mouths follows the line

$$\bar{\ell}^\mu(\tau) = \bar{\ell}_0^\mu + V^\mu\tau, \qquad (18.2)$$

and the vector S^μ describes the separation of the wormhole mouths. Notation: $s = \sqrt{S^\mu S_\mu}$. Thus s is the invariant interval that gives the distance from a point on the wormhole throat to itself, as measured by going the long way around in normal space.

The present construction of course does not describe a time machine, but one shall soon see that very simple manipulations *appear* to be able to take this "safe" wormhole and turn it into a time machine.

18.1.2   Step 2: Induce a time-shift

As the penultimate step in one's construction of a time machine it is necessary to induce a (possibly small) "time-shift" between the two mouths of the wormhole. For clarity, work in the rest frame of the wormhole mouths. After a suitable translation and Lorentz transformation the simple model wormhole of step 1 can, without loss of generality, be described by the identification

$$(t, 0, 0, 0) \equiv (t, 0, 0, s). \qquad (18.3)$$

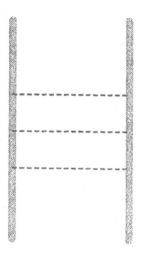

Figure 18.1: Idealized wormhole: safe variety. The gray lines indicate the worldlines of the wormhole mouths, while the dashed lines indicate which points on the worldline are to be identified. This wormhole connects "equal times".

Exercise: (Easy) Prove this. Hints: V^μ is a timelike unit vector, so without loss of generality one can take $V^\mu = (1,0,0,0)$. But S^μ is a spacelike vector perpendicular to V^μ, so without loss of generality one can take $S^\mu = s(0,0,0,1)$. The line ℓ_1 is now $\ell_1(\tau)^\mu = (\tau,0,0,-s/2)$, while the line ℓ_2 is now $\ell_2(\tau)^\mu = (\tau,0,0,+s/2)$. Change notation $\tau \to t$, and shift everything a distance $s/2$ along the z axis.

One wishes to change this state of affairs to obtain a wormhole of type

$$(t,0,0,0) \equiv (t+T,0,0,\ell), \tag{18.4}$$

where T is the time-shift and ℓ is the distance between the wormhole mouths. Mathematically one wishes to force the vectors V^μ and S^μ to no longer be perpendicular to each other so that it is possible to define the time-shift to be $T = V^\mu S_\mu$, while the distance between the mouths is

$$\ell = \|(\delta^\mu{}_\nu + V^\mu V_\nu)S^\nu\|. \tag{18.5}$$

Physically this may be accomplished in a number of different ways:

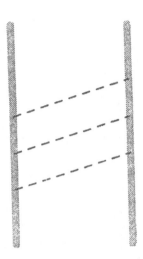

Figure 18.2: Idealized wormhole: with time-shift. The wormhole has now been modified so that traveling through the wormhole results in a time-shift. This is not a time machine unless the time-shift is larger than the (long way around) distance from one mouth to the other.

The relativity of simultaneity

Apply identical forces to the two wormhole mouths so that they suffer identical four-accelerations (as seen by the initial rest frame). It is clear that if the mouths initially connect equal times (as seen by the initial rest frame), and if their four-accelerations are equal (as seen by the initial rest frame), then the paths followed by the wormhole mouths will be identical up to a fixed translation by the vector S^μ. Consequently the wormhole mouths will always connect equal times—as seen by the initial rest frame. Mathematically this describes a situation where S^μ is a constant of the motion while V^μ is changing. Of course, after the applied external forces are switched off the two mouths of the wormhole have identical four-velocities V_f^μ not equal to their initial four-velocities V_i^μ. And by the relativity of simultaneity equal times as viewed by the initial rest frame are not the same as equal times as viewed by the final rest frame. If one takes the center of motion of the wormhole mouths to have been accelerated from four-velocity $(1, 0, 0, 0)$ to four-velocity $(\gamma, 0, 0, \gamma\beta)$, then the relativity of simultaneity induces a time-shift

$$T = \gamma \beta s. \tag{18.6}$$

As seen in the final rest frame the distance between the wormhole mouths is

$$l = \gamma s. \tag{18.7}$$

Exercise: Check this.

Special relativistic time dilation

Another way of inducing a time-shift in the simple model wormhole of step 1 is to simply move one of the wormhole mouths around and to rely on special relativistic time dilation. This can either be done by using rectilinear motion [191] or by moving one of the mouths around in a large circle [195]. Assuming for simplicity that the two mouths are finally returned to their initial four-velocity, the induced time-shift is simply given by

$$T = \int_i^f (\gamma_1 - \gamma_2) dt. \tag{18.8}$$

Mathematically, V^μ is unaltered, while T and hence S^μ are changed by this mechanism.

Psycho-ceramics warning: The presentation given here has the advantage of avoiding the phrase "twin paradox". Of course, the discussion could easily be rephrased in terms of the language of the special relativistic twin paradox but that would only serve to put the lunatic fringe into an excited state. (Again, crackpots are politely requested to refrain from reading this paragraph.)

General relativistic time dilation

As an alternative to relying on special relativistic time dilation effects, one may instead rely on the general relativistic time dilation engendered by the gravitational redshift [95]. One merely places the wormhole mouths in different positions in a gravitational potential for a suitable amount of time to induce a time-shift:

$$T = \int_i^f \left[\sqrt{g_{00}(x_1)} - \sqrt{g_{00}(x_2)} \right] dt. \tag{18.9}$$

This particular process has a number of additional interesting features and will be discussed in more detail in a subsequent chapter. (See pp. 239ff.)

Comment

Of course, this procedure has not yet built a time machine. The discussion so far has merely established that given a traversable wormhole it is trivial to arrange creation of a traversable wormhole that exhibits a time-shift upon passing through the throat—this time-shift can be created by a number of different mechanisms so that the ability to produce such a time-shift is a very robust result.

18.1.3   Step 3: Bring the mouths together

Having constructed a traversable wormhole with time-shift, the final stage of time machine construction is deceptively simple: Merely push the two wormhole mouths towards one another (this may be done as slowly as is desired). A time machine forms once the physical distance between the wormhole mouths l is less than the time-shift T. Once this occurs, it is clear that closed timelike geodesics have formed—merely consider the closed geodesic connecting the two wormhole mouths and threading the wormhole throat. As an alternative to moving the mouths of the wormhole closer together one could of course arrange for the time-shift to continue growing while keeping the distance between the mouths fixed—such an approach conflates steps 2 and 3.

The advantage of clearly separating steps 2 and 3 in one's mind is that it is now clear that the wormhole mouths may be brought together arbitrarily slowly—in fact adiabatically—and still *appear* to form a time machine. If one makes the approach adiabatic (so that one can safely take the time-shift to be a constant of the motion), then the wormhole may be mathematically modeled by identifying the lines

$$(t, 0, 0, z(t)) \equiv (t + T, 0, 0, l_0 - z(t)). \qquad (18.10)$$

The physical distance between the wormhole mouths is

$$l(t) = l_0 - 2z(t). \qquad (18.11)$$

A family of closed "almost" geodesics is defined by the set of straight lines, labeled by t, that connect $l_1{}^\mu(t)$ with $l_2{}^\mu(t)$. Specifically, with $\sigma \in [0, 1]$:

$$\begin{aligned}
X^\mu(t, \sigma) &= \sigma l_1{}^\mu(t) + \{1 - \sigma\} l_2{}^\mu(t) \\
&= \left([t + \{1 - \sigma\}T], \, 0, \, 0, \, [\{1 - \sigma\}l_0 - \{1 - 2\sigma\}z(t)] \right).
\end{aligned}$$
$$(18.12)$$

Here σ is a parameter along the closed geodesics, while t labels the particular closed geodesic under consideration. These curves are true geodesics everywhere except at the point $(\sigma = 0) \equiv (\sigma = 1)$, the location of the wormhole mouths, where there is a "kink" in the tangent vector. This "kink" is induced by the relative motion of the wormhole mouths. The invariant length of the members of this family of closed geodesics is

$$s(t) = \|\ell_1{}^\mu(t) - \ell_2{}^\mu(t)\| = \|(T, 0, 0, \ell(t))\| = \sqrt{\ell(t)^2 - T^2}. \qquad (18.13)$$

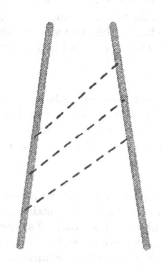

Figure 18.3: Idealized wormhole: time machine formation. If one moves the wormhole mouths closer together, keeping the time-shift fixed, a time machine forms once the (long way around) distance from one mouth to the other is less than the time-shift. The first closed null geodesic that forms is a fountain for the chronology horizon.

Once $\ell(t) < T$ a time machine has formed. Prior to time machine formation the vector $(T, 0, 0, \ell(t))$ is, by hypothesis, spacelike. Therefore it is possible to find a Lorentz transformation to bring it into the form $(0, 0, 0, s)$ with $s = \sqrt{\ell^2 - T^2}$. A brief calculation shows that this will be accomplished by a Lorentz transformation of velocity

$$\beta = T/\ell(t) = T/\sqrt{T^2 + s^2(t)}, \qquad (18.14)$$

so that $\gamma = \ell/s$ and $\gamma\beta = T/s$.

Exercise: Check this.

In this new Lorentz frame the wormhole connects "equal times", and this frame can be referred to as the frame of simultaneity (synchronous frame). Note in particular, that as one gets close to building a time machine $s(t) \to 0$, so that $\beta(t) \to 1$. The velocity of the frame of simultaneity approaches the speed of light.

The construction process described above clearly separates the different effects at work in the putative construction of time machines. Given a traversable wormhole arbitrarily small special and/or general relativistic effects can be used to generate a time-shift. Given an arbitrarily small time-shift through a traversable wormhole, arbitrarily slow adiabatic motion of the wormhole mouths towards each other *appears* to be sufficient to construct a time machine. This immediately leads to all of the standard paradoxes associated with time travel and is very disturbing for the state of physics as a whole.

Exercise 1: Show that this wormhole based construction of a time machine leads to a compactly generated chronology horizon. (Hint: the first closed null geodesic is the fountain.) Find the chronology-violating region. Find the chronology horizon. Find the polarized hypersurfaces. (Hint: see [164].)

Exercise 2: Now suppose that the wormhole mouths continue to move past each other, while the time-shift remains fixed. Show that there exists a final closed null geodesic. This implies that the time machine is destroyed after a finite interval. Show that this final closed null curve terminates the chronology-violating region. Find, qualitatively, the chronology horizon and compare it with the boundary of the chronology violating region.

18.2 Multiple wormholes

The analysis of multiple wormhole configurations is a topic of considerable additional complexity. It is relatively easy to think up a putative time machine built out of two or more wormholes, each of which taken in isolation is not itself a time machine. The simplest such configuration was considered by Roman, and uses two wormholes [190, 250, 264, 272, 178]. The phrase "Roman configuration" is used to denote any multi-wormhole collection wherein the individual wormholes in the collection do not of themselves form time machines.

For example, for the time being ignore the internal structure of the two wormholes and simply model them by identifying two world-lines. Let wormhole number 1 be defined by the identification

$$(t, 0, 0, 0) \equiv (t + T_1, 0, 0, \ell). \tag{18.15}$$

As long as the distance between the wormhole mouths ℓ is greater than the time-shift T_1, wormhole number 1 does not, in and of itself, constitute a time machine. Now add a second wormhole. Let wormhole number 2 be offset a distance ζ along the x axis. That is

$$(t, \zeta, 0, \ell) \equiv (t + T_2, \zeta, 0, 0). \qquad (18.16)$$

Again, as long as $\ell > T_2$, wormhole number 2 does not, in and of itself, describe a time machine.

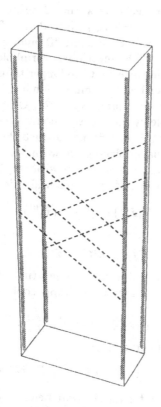

Figure 18.4: Idealized wormholes: Roman configuration. Two wormholes, each of which (considered in isolation) is not a time machine. The dashed lines indicate the result of "jumping" through the wormhole mouths. The compound system may still be a time machine.

Now consider the null trajectory

$$(0, 0, 0, 0) \equiv (T_1, 0, 0, \ell) \rightarrow$$

$$(T_1 + \zeta, \zeta, 0, \ell) \equiv (T_1 + \zeta + T_2, \zeta, 0, 0) \rightarrow$$
$$(T_1 + \zeta + T_2 + \zeta, 0, 0, 0). \tag{18.17}$$

Notation: Here $x \rightarrow y$ means "travel from the point x to the point y along a straight line trajectory" (travel from x to y at constant speed without any detours). This future-pointing null trajectory has returned to its spatial starting point in a total time

$$\Delta T = T_1 + T_2 + 2\zeta. \tag{18.18}$$

This total time-shift can easily be arranged to be negative. (Example: $\ell = 3\zeta$, $T_1 = T_2 = -2\zeta$, $\Delta T = -2\zeta$.) Note that $\Delta T > -2\ell + 2\zeta$, so that a necessary condition for this compound "Roman configuration" to form a time machine is $\ell > \zeta$. (The individual wormholes should permit spatial "jumps" across the universe that are longer than the "offset" distance between wormhole number 1 and wormhole number 2.) The maximum possible size of the backward time-jump is $2\ell - 2\zeta$. One can add some internal structure to the wormhole by making the mouths spherically symmetric of radius R. As long as $R \ll \zeta$ this will not disturb the discussion.

Generalize the analysis by placing the wormhole mouths in general positions. All mouths are taken to be at rest with respect to one another, and wormhole number 1 is described by the identification

$$(t, \vec{x}_1^{in}) \equiv (t + T_1, \vec{x}_1^{out}). \tag{18.19}$$

Define

$$\ell_1 \equiv \|\vec{x}_1^{out} - \vec{x}_1^{in}\|. \tag{18.20}$$

Since one does not wish this wormhole to be a time machine in its own right $|T_1| < \ell_1$. For wormhole number 2 one simply copies all of these definitions, for example,

$$(t, \vec{x}_2^{in}) \equiv (t + T_2, \vec{x}_2^{out}). \tag{18.21}$$

Now consider the closed curve $\gamma(t_0)$

$$(0, \vec{x}_1^{in}) \equiv (T_1, \vec{x}_1^{out}) \rightarrow (t_0, \vec{x}_2^{in}) \equiv (t_0 + T_2, \vec{x}_2^{out}) \rightarrow (0, \vec{x}_1^{in}). \tag{18.22}$$

Here t_0 is a parameter that is for the time being arbitrary. We shall adjust t_0 in such a manner as to make γ a geodesic. The arc length along the curve $\gamma(t_0)$ is

$$s[\gamma] = \sqrt{(\vec{x}_1^{out} - \vec{x}_2^{in})^2 - (T_1 - t_0)^2} + \sqrt{(\vec{x}_2^{out} - \vec{x}_1^{in})^2 - (t_0 + T_2)^2}. \tag{18.23}$$

To simplify life, introduce the notation

$$\begin{aligned} \ell_{1\rightarrow2} &\equiv \|\vec{x}_1^{out} - \vec{x}_2^{in}\|, \\ \ell_{2\rightarrow1} &\equiv \|\vec{x}_2^{out} - \vec{x}_1^{in}\|. \end{aligned} \tag{18.24}$$

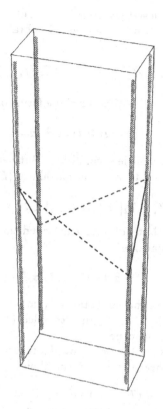

Figure 18.5: Idealized wormholes: Roman configuration. Example of a closed timelike curve through a Roman configuration. The solid lines represent ordinary trips in ordinary flat space.

Then

$$s[\gamma] = \sqrt{(\ell_{1\to 2})^2 - (T_1 - t_0)^2} + \sqrt{(\ell_{2\to 1})^2 - (t_0 + T_2)^2}. \qquad (18.25)$$

The arc length of the curve γ is extremized when

$$\frac{\partial s}{\partial t_0} \equiv \frac{T_1 - t_0}{\sqrt{(\ell_{1\to 2})^2 - (T_1 - t_0)^2}} - \frac{T_2 + t_0}{\sqrt{(\ell_{2\to 1})^2 - (T_2 + t_0)^2}} = 0. \qquad (18.26)$$

This equation has the parametric solution

$$\begin{aligned}
T_1 - t_0 &= k\ell_{1\to 2}, \\
T_2 + t_0 &= k\ell_{2\to 1}, \\
T_1 + T_2 &= k(\ell_{1\to 2} + \ell_{2\to 1}).
\end{aligned} \qquad (18.27)$$

The arc length of the closed geodesic of current interest, one that wraps once through the two wormhole system, is obtained by setting t_0 to its critical value. Thus

$$s[\gamma] = \sqrt{(\ell_{1\to 2} + \ell_{2\to 1})^2 - (T_1 + T_2)^2}. \tag{18.28}$$

And one sees that a putative time machine forms once

$$|T_1 + T_2| > \ell_{1\to 2} + \ell_{2\to 1}. \tag{18.29}$$

This is a necessary and sufficient condition. But because one does not wish the individual wormholes to be time machines, $|T_1| < \ell_1$, and $|T_2| < \ell_2$. Consequently

$$|T_1 + T_2| < |T_1| + |T_2| < \ell_1 + \ell_2. \tag{18.30}$$

So another necessary (but not sufficient) condition on the formation of a "two-wormhole time machine" is

$$\ell_{1\to 2} + \ell_{2\to 1} < \ell_1 + \ell_2. \tag{18.31}$$

This means that the net spatial distance "jumped" through the two wormholes has to exceed the total distance traveled (in normal space) in going from one wormhole to the other.

One may also consider the behavior of a future-pointing null curve (not a closed curve) that threads the two wormholes

$$(0, \vec{x}_1^{in}) \equiv (T_1, \vec{x}_1^{out}) \to$$
$$(T_1 + \ell_{1\to 2}, \vec{x}_2^{in}) \equiv (T_1 + \ell_{1\to 2} + T_2, \vec{x}_2^{out}) \to$$
$$(T_1 + \ell_{1\to 2} + T_2 + \ell_{2\to 1}, \vec{x}_1^{in}). \tag{18.32}$$

This curve returns to its spatial starting point in total time,

$$\Delta T \equiv T_1 + T_2 + \ell_{1\to 2} + \ell_{2\to 1}. \tag{18.33}$$

As per the preceding discussion this can easily be negative. The virtue of this type of analysis is that it is now clear that the maximum possible size of the (single trip) backward time-jump is

$$\Delta T_{\max} \equiv \ell_1 + \ell_2 - \ell_{1\to 2} - \ell_{2\to 1}. \tag{18.34}$$

Exercise: Check this.

Additional complications can be added by:

1. Extending the analysis to include the effect of closed curves in the "crossed" configuration. ($\vec{x}_1^{in} \equiv \vec{x}_1^{out} \to \vec{x}_2^{out} \equiv \vec{x}_2^{in} \to \vec{x}_1^{in}$.)

2. Extending the analysis to include multiple trips through the wormholes.

3. Permitting a nonzero relative velocity for the two wormholes and their four mouths. One such configuration has been discussed by Lyutikov [178]. Since this adds time dependence to the analysis the complications are enormous.

4. Adding yet more wormholes: With three, four, or many wormholes there are clearly additional levels of complexity.

The lesson to be learned is this: Multiple wormholes can interact in complicated and unexpected ways. To check for the nonexistence of closed timelike curves one has to check each topologically inequivalent class of curves separately. Technically, for each element $\{\gamma\}$ of the fundamental group of the manifold $\pi_1(\mathcal{M})$, one has to check that none of the curves γ in the equivalence class $\{\gamma\}$ are closed timelike curves.

Exercise: Check that for n wormholes there are (up to topological deformation) $2^{(n-1)} \times (n-1)!$ equivalence classes of single-trip self-intersecting curves that pass through the entire complex. (There are $n!$ ways of ordering the list of n wormholes, but the starting point in the list is irrelevant. For each ordering there are 2^n choices of which wormhole mouth to enter, but one overall orientation is irrelevant.) Similarly, there are a total of $\sum_{j=1}^n \binom{n}{j} 2^{(j-1)} \times (j-1)!$ equivalence classes of self-intersecting curves that do not visit any particular wormhole twice. Each of these equivalence classes should be checked independently for the onset of time travel.

18.3 Locally-static spacetimes

In this section I will explore in more detail the suggestion that general relativistic time dilation can be used to turn a generic traversable wormhole into a time machine. The analysis follows that of Frolov and Novikov [95, 93], and seems to indicate that time machines will almost always form. In an earlier chapter we looked at the general metric for a spherically symmetric static wormhole (see p. 101):

$$ds^2 = -e^{2\phi(l)}dt^2 + dl^2 + r^2(l) \left[d\theta^2 + \sin^2\theta \, d\varphi^2\right]. \tag{18.35}$$

It was pointed out at that time that there is no particular reason to demand that time run at the same rate on either side of the wormhole. More precisely, it is perfectly acceptable to have $\phi(l = +\infty) \neq \phi(l = -\infty)$. This is fine for inter-universe wormholes but leads to interesting results if we try to build an intra-universe wormhole along these lines.

Because both sides of the wormhole reside in an asymptotically flat universe we know that with negligible local geometric distortion we can embed the two mouths in a single asymptotically flat universe. Though the local effects are minor, the influence on the global geometry is profound. We shall see that closed timelike curves develop at sufficiently late and sufficiently early times [95].

Roughly speaking, because time is running at different rates for the two wormhole mouths they will be getting further and further out of step with each other. To build a time machine is then simply a matter of waiting until they are sufficiently far out of step with each other.

Consider a $(3 + 1)$-dimensional intra-universe wormhole. Concentrate attention on the $(1 + 1)$-dimensional surface swept out by the central geodesic connecting the two mouths. The spacetime metric on this $(1 + 1)$-dimensional surface is easily obtained by throwing away the angular variables of the general traversable wormhole to define a simple Lorentzian metric in the (t, l) plane:

$$ds^2 = -e^{2\phi(l)}dt^2 + dl^2. \tag{18.36}$$

The range of the l coordinate is $(-L/2, +L/2)$ and the coordinate values $l = -L/2$ and $l = +L/2$ are to be identified. The total "long way around" distance from one wormhole mouth to the other is L. For convenience define

$$\phi_\pm \equiv \phi(l = \pm L/2); \qquad \Delta\phi \equiv \phi_+ - \phi_-. \tag{18.37}$$

At the junction $l = \pm L/2$, the metric has to be smooth.

Warning: It is the *metric* that has to be smooth. This does not mean that the *components* of the metric have to be smooth, only that any discontinuity in the components be matched with a discontinuity in the coordinates so that the combination $ds = \sqrt{g_{\mu\nu}dx^\mu dx^\nu}$ is smooth.

This implies

$$d\tau = e^{\phi_-}\, dt_- = e^{\phi_+}\, dt_+. \tag{18.38}$$

If we define the origin of the time coordinate by identifying the points

$$(0, -L/2) \equiv (0, +L/2), \tag{18.39}$$

then the discontinuity of the time coordinate is

$$t_+ = t_-\, \exp(\phi_- - \phi_+) = t_-\, \exp(-\Delta\phi) \tag{18.40}$$

This means that in general we should identify the points

$$(t_-, -L/2) \equiv (t_-\, \exp[-\Delta\phi], +L/2). \tag{18.41}$$

(So time t_+ on one side of the wormhole mouth is to be identified with time t_- on the other side.) With this identification, the metric is smooth across the seam at $l = \pm L/2$.

Now ask what happens to a null geodesic (light ray) that starts out from $l = -L/2$ at some initial time $[t_i]_-$. The condition for a null curve is that $ds = 0$, or equivalently that

$$\frac{dl}{dt} = \pm e^{+\phi(l)}. \tag{18.42}$$

Here the $+$ sign corresponds to a right-moving null curve, the $-$ sign to a left-moving null curve.

Integrating, a right-moving null curve arrives at $l = +L/2$ at coordinate time

$$[t_f]_+ = [t_i]_- + \int_{-L/2}^{+L/2} e^{-\phi(l)} dl. \tag{18.43}$$

Using the matching condition across the coordinate discontinuity at $l = \pm L/2$, this means that the light ray returns to its starting point at coordinate time

$$[t_f]_- = [t_f]_+ \exp(\Delta\phi) = \left[[t_i]_- + \oint e^{-\phi(l)} dl \right] \exp(\Delta\phi). \tag{18.44}$$

So the null curve can easily return to its starting point before it left. A right-moving closed null curve exists if $[t_f]_- = [t_i]_-$. This is equivalent to

$$[t_i]_-^R = \frac{\oint e^{-\phi(l)} dl}{\exp(-\Delta\phi) - 1}. \tag{18.45}$$

On the other hand, a left-moving null curve returns to its starting point at time

$$[t_f]_- = \exp(-\Delta\phi)[t_i]_- + \oint e^{-\phi(l)} dl. \tag{18.46}$$

Exercise: Prove this.

If $[t_f]_- = [t_i]_-$, one has obtained a left-moving closed null curve. This is equivalent to

$$[t_i]_-^L = \frac{\oint e^{-\phi(l)} dl}{1 - \exp(-\Delta\phi)} = -[t_i]_-^R. \tag{18.47}$$

To simplify life, define

$$T_- \equiv \frac{\oint e^{-\phi(l)} dl}{|1 - \exp(-\Delta\phi)|}; \qquad T_+ \equiv \frac{\oint e^{-\phi(l)} dl}{|\exp(\Delta\phi) - 1|}. \tag{18.48}$$

The two closed null curves pass through the points

$$(\pm T_-, -L/2) \equiv (\pm T_+, +L/2). \qquad (18.49)$$

Closed timelike curves form at sufficiently late and sufficiently early times. There is only a limited temporal region into which closed timelike curves do *not* penetrate. In the $(1+1)$-dimensional subspace just analyzed, the past and future chronology horizons are precisely the closed null curves just discussed. In the full $(3+1)$-dimensional spacetime, which I shall now analyze, these two closed null curves will be seen to be the fountains of the past and future chronology horizons.

There is a minor puzzle here that may cause some confusion. The metric we started from was completely time independent. So how did time dependence sneak into the analysis? Via the boundary conditions! The identification [equation (18.41)] that we were forced to make to keep the metric smooth has explicitly broken the time translation invariance of the geometry.

The analysis can easily be extended off the $(1+1)$-dimensional subspace spanned by the central geodesic. (In fact, we can even avoid the simplifying assumption of spherical symmetry that was previously made.) Consider a time-independent but otherwise arbitrary metric of the form

$$ds^2 = e^{2\phi(x)}dt^2 + g_{ij}(x)dx^i dx^j. \qquad (18.50)$$

This would under normal circumstances be called a static spacetime, but we are about to mutilate it in an interesting manner.

Consider the spacelike two-surfaces defined by looking at the equipotentials of $\phi(x)$. Suppose that somewhere in the spacetime, some two of these equipotential surfaces (say $\partial\Omega_+$, $\partial\Omega_-$) have the same induced two-metric. Then we can cut out the interiors of these two surfaces and paste the geometry together by identifying the two surfaces $\partial\Omega_+ \equiv \partial\Omega_-$. This is enough to keep the geometry on the spatial slices continuous, so that we can think about applying the extended thin-shell formalism defined in previous chapters. Unfortunately, the job is not yet done, we also need to make the spacetime geometry continuous. Note that we have not needed to demand $\phi|_{\partial\Omega_+} = \phi|_{\partial\Omega_-}$ to keep the spatial geometry continuous across the identification.

If we permit $\phi|_{\partial\Omega_+} \neq \phi|_{\partial\Omega_-}$ then the spacetime geometry is continuous if and only if we identify the time coordinates according to the prescription

$$t_+ \exp(+\phi|_{\partial\Omega_+}) = t_- \exp(+\phi|_{\partial\Omega_+}). \qquad (18.51)$$

This identification process has explicitly broken the time translation symmetry. The resulting geometry is an example of what Frolov and Novikov call a "locally-static" spacetime [95]. With this identification process now complete, we have a continuous spacetime metric and can apply the extended thin-shell formalism in the usual manner. By smoothing out the geometry at the throat $\partial\Omega_+ \equiv \partial\Omega_-$ we can make the delta-function contribution to the stress-energy vanish and so construct a completely smooth "locally-static" spacetime.

Locally-static spacetimes can also be characterized in a more abstract manner:

Definition 29 *A spacetime $\mathcal{M}$ is locally static if and only if there exist two globally defined vector fields u and ω such that*

1. $u_\mu u^\mu = -1$. *(In particular, the vector field u is everywhere timelike.)*

2. $\nabla_\mu u_\nu + \nabla_\nu u_\mu = -\omega_\mu u_\nu - \omega_\nu u_\mu$.

3. $\omega_{[\mu;\nu]} = 0$.

In terms of Lie derivatives and exterior derivatives the last two conditions may be rephrased as

$$\mathcal{L}_u g = -\omega \otimes u - u \otimes \omega; \qquad d\omega = 0. \tag{18.52}$$

Now because the one-form ω is closed, it can locally be written as a gradient. That is, $\omega = d\alpha$, at least on simply-connected regions. This permits one to rewrite the first condition as

$$\mathcal{L}_{(e^\alpha u)} g = 0. \tag{18.53}$$

In more prosaic coordinate-based notation

$$\nabla_\mu(e^\alpha u_\nu) + \nabla_\nu(e^\alpha u_\mu) = 0. \tag{18.54}$$

This implies that, at least locally, the vector field $\xi \equiv e^\alpha u$ is a timelike Killing vector.

Exercise: Read a little about Killing vectors in any of the standard textbooks. See, for example, [186, 275].

Unfortunately, there is in general no guarantee that this vector field ξ exists globally over the whole spacetime. We deduce

Lemma 5 *A locally-static spacetime is static if and only if the one-form ω is exact.*

An alternative definition is

Definition 30 *A spacetime $\mathcal{M}$ is locally static if and only if (1) on each simply connected region there exists a timelike Killing vector field ξ, and (2) whenever two simply connected regions overlap the relevant Killing vector fields are parallel.*

Using this alternative definition one can, in any simply connected region, uniquely define fields α and u by $e^{2\alpha} = -g(\xi, \xi)$ and $\xi = e^{\alpha} u$, so that $g(u, u) = -1$. Because the Killing vector fields are parallel in overlap regions it follows that the vector field u is globally defined. From the Killing equation applied to the two Killing vector fields in any overlap region it follows that α changes by at most a constant in going from one simply connected region to another. This permits one to uniquely define a one-form by $\omega = d\alpha$.

This abstract formalism can be connected back to the previous version by taking the timelike Killing almost-vector $\xi \equiv e^{-\alpha} u$, which is defined only on simply connected regions, and extending it to cover almost the whole manifold. One can certainly extend it to cover everything except the throat of the wormhole, where some matching conditions will be needed. By doing this the metric can be put in the (locally static) form

$$ds^2 = -e^{2\phi(x)} dt^2 + g_{ij}(x)\, dx^i\, dx^j. \tag{18.55}$$

Thus the discontinuity in ϕ is related to the discontinuity in α by

$$\Delta\phi \equiv \phi_+ - \phi_- = \alpha_+ - \alpha_- = \oint \omega. \tag{18.56}$$

Exercise: Check this. Calculate $g(\xi, \xi)$ two different ways.
First $g(\xi, \xi) = e^{2\alpha} g(u, u) = -e^{2\alpha}$. Second $g(\xi, \xi) \equiv g_{\mu\nu}\, \xi^\mu \xi^\nu = -e^{2\phi}$.

Now pick any time slice of fixed t. Within such a time slice consider some spacelike geodesic γ connecting the two mouths. Then the change in coordinate time required for a light signal to get from one mouth to the other along this spatial curve is

$$\Delta t = \oint_\gamma e^{-\phi} dl. \tag{18.57}$$

One can define a minimum time curve γ_0 by minimizing Δt. Closed timelike curves form once the time coordinates at the mouths satisfy

$$|t_-| > T_- \equiv \frac{\oint_{\gamma_0} e^{-\phi(l)} dl}{|1 - \exp(-\Delta\phi)|},$$

$$|t_+| > T_+ \equiv \frac{\oint_{\gamma_0} e^{-\phi(l)} dl}{|\exp(\Delta\phi) - 1|}. \tag{18.58}$$

Exercise: Check this.

This is the natural generalization of the $(1+1)$-dimensional analysis to arbitrary locally-static spacetimes.

If the "potential" $\phi(x)$ and its discontinuity are small, this may be approximated as indicating the onset of chronology violation once

$$|t| \gtrsim \frac{L}{|\Delta\phi|}. \tag{18.59}$$

The key point is that it is *apparently* very easy to build a time machine in this manner.

- Mathematically: Start with an inter-universe wormhole that has time running at different rates in the two universes (this is still a globally static spacetime). Now put both mouths into the same universe. This cannot be done in a globally static manner—one is forced to construct a locally-static spacetime. Sit back and wait, a time machine forms automatically.

- Physically: Start with an intra-universe traversable wormhole that is globally static, and place a large mass next to one mouth but not the other. Doing this will change the gravitational potential at one mouth and convert a globally static intra-universe traversable wormhole to a locally-static intra-universe traversable wormhole. Since clocks at the two mouths now run at different rates it is simply a matter of waiting long enough for the chronology horizon to form. (This is simply a more detailed exposition of the idea sketched earlier, see p. 231.)

Another interesting feature of locally-static traversable wormholes is that they appear, at first glance, to be perpetual motion machines (of the first kind). From the normalization condition $u^\mu u_\mu = -1$ one deduces $u^\mu \nabla_\nu u_\mu = 0$. Twice contracting the equation $\nabla_\mu u_\nu + \nabla_\nu u_\mu = -\omega_\mu u_\nu - \omega_\nu u_\mu$ with the vector u shows that $u^\mu \omega_\mu = 0$. Now, consider the four-acceleration of an observer who moves with four-velocity u. One has

$$a^\mu \equiv u^\nu \nabla_\nu u^\mu = \omega^\mu. \tag{18.60}$$

Thus in a locally-static spacetime the gravitational force is nonconservative. For a particle of total mass-energy m the "work done" by the gravitational field is simply

$$\frac{dm}{dt} = -ma\frac{dx}{dt} = -m\omega\frac{dx}{dt} = -m\frac{d\alpha}{dt}. \tag{18.61}$$

Exercise: Perform all the ritual incantations needed to make this argument rigorous.

Now, integrating the work done along some path,

$$m_f = m_i \, \exp(-[\alpha_f - \alpha_i]) = m_i \, \exp\left(-\int_i^f \omega\right). \qquad (18.62)$$

The same result follows from a more careful treatment using the Killing vector directly. Let V be the four-velocity of the particle that traverses the wormhole. On simply connected patches (assuming geodesic motion) we know that the inner product $g(\xi, V) = e^\alpha g(u, V) = k$ is conserved. But the energy of the particle is then

$$m \equiv m_0 \, \gamma = -m_0 \, g(u, V) = -e^{-\alpha} \, m_0 \, k. \qquad (18.63)$$

This reproduces the previous result.

Exercise: Check that this also works, *mutatis mutandis*, for massless particles (null geodesics).

Dump a certain amount of mass energy m_i into the "+" mouth of a locally-static traversable wormhole. The nonconservative nature of the gravitational field then implies that the mass-energy coming out the "−" mouth is

$$m_f = m_i \, \exp(\Delta\phi) = m_i \, \exp\left(\oint \omega\right). \qquad (18.64)$$

By repeating this process many times (N trips) it seems, at first glance, that we can recover a total energy

$$m_f = m_i \, \exp(N\Delta\phi), \qquad (18.65)$$

and thereby extract arbitrary amounts of energy from the wormhole system. This *seems* to be a perpetual motion machine of the first kind. This analysis is too naive and misses two crucial issues:

Back-reaction: Certainly the object gains (or loses) energy upon traversing the wormhole. It does so at the cost of altering the mass of the wormhole mouths themselves. For a single trip

$$M_f^+ = M_i^+ + m_i; \qquad M_f^- = M_i^- - m_i \, \exp(\Delta\phi). \qquad (18.66)$$

While for repeated trips

$$
\begin{aligned}
M_f^+ &= M_i^+ + m_i \, \frac{1 - \exp(N\Delta\phi)}{1 - \exp(\Delta\phi)}; \\
M_f^- &= M_i^- - m_i \, \exp(\Delta\phi) \left[\frac{1 - \exp(N\Delta\phi)}{1 - \exp(\Delta\phi)}\right].
\end{aligned}
\qquad (18.67)
$$

The total energy gained by the object traversing the wormhole is matched by a corresponding mass loss for the wormhole mouths

$$\Delta m \equiv m_f - m_i = - \left[(M_f^+ + M_f^-) - (M_i^+ + M_i^-) \right]. \qquad (18.68)$$

Chronology violation: In a sense the possibility of perpetual motion is the least of one's worries. The formation of closed timelike curves with the concomitant possibility of time travel is much more disturbing.

Chapter 19

Response to the paradoxes

By now the reader has seen many examples of the relative ease with which it *seems* to be possible to build a time machine. We have also very briefly sketched the reasons why time travel is considered problematic, if not downright repugnant, from a physics point of view. How should we respond to such concerns? One can either learn to live with it or do something about it—as a matter of logical necessity precisely *one* of the following alternatives must hold:

1. The radical rewrite conjecture.

2. The Novikov consistency conjecture.

3. The Hawking chronology protection conjecture.

4. The boring physics conjecture.

19.1 The radical rewrite conjecture

"All hell breaks loose".

19.1.1 General considerations

The most profoundly radical response to issues of time travel is simply to abandon the causal underpinnings of known physics, unreservedly admit time travel into the pantheon of physical effects and attempt to rewrite all of physics from the ground up. Attempts at doing this have led to the complex of ideas I refer to as the radical rewrite conjecture.

Warning and sanity check: In case anyone has by now lost track of
empirical reality, it cannot be strongly enough emphasised that there is not
the slightest shred of experimental evidence to support the notion that time
travel ever occurs anywhere. The radical rewrite proposals are very highly
speculative "what if" proposals.

To fully accommodate the full panoply of time travel effects (consistency
paradoxes, bootstrap paradoxes) one somehow has to give the universe the
possibility of having more than one *actual* past history. It seems that somehow the chronology-violating region behind the chronology horizon should
contain multiple actual histories (multiple "timelines") for the universe—
the different histories being connected to each other by time travel effects.
The ordinary notion of spacetime as a (Hausdorff) differentiable manifold
is clearly inadequate to the task.

19.1.2   Non-Hausdorff manifolds

A mathematically well defined model that appears to be sufficiently general
to accommodate time travel (at least at the classical level) is that of a non-
Hausdorff differentiable manifold with a Lorentzian signature metric [206,
264, 268]. The property of being Hausdorff is a technical restriction on the
acceptable topology of spacetime.

Definition 31 *A topology is Hausdorff if and only if for any two points*
$x_1 \neq x_2$ *there exist open sets* $\mathcal{O}_1$ *and* $\mathcal{O}_2$ *such that* $x_1 \in \mathcal{O}_1$; $x_2 \in \mathcal{O}_2$; *and*
$\mathcal{O}_1 \cap \mathcal{O}_2 = \emptyset$.

Normally manifolds are assumed Hausdorff by definition. Relaxing this
condition permits the existence of "branched" manifolds. Because the ideas
are subtle, a few explicit examples will be given.

Example 1:   Define a set $\mathcal{E}_1$ by considering the ordinary real line $\Re$. Remove the point 0, and replace it by *two* points 0_1 and 0_2.
   Define a basis for the topology on $\mathcal{E}_1$:

1. Any open set on $\Re$ that does not contain 0 is also an open set in $\mathcal{E}_1$.

2. $\forall \epsilon > 0$, $(-\epsilon, 0) \cup \{0_1\} \cup (0, \epsilon)$ is an open set in $\mathcal{E}_1$ that contains 0_1.

3. $\forall \epsilon > 0$, $(-\epsilon, 0) \cup \{0_2\} \cup (0, \epsilon)$ is an open set in $\mathcal{E}_1$ that contains 0_2.

In this topology, every open set containing 0_1 has a non-empty intersection with every open set containing 0_2. This is an elementary example
of "point doubling".

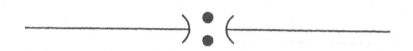

Figure 19.1: A non-Hausdorff topology: point doubling.

Example 2: Define a set $\mathcal{E}_2$ by considering the ordinary real line $\Re$. Remove the half-closed interval $I = [0, +\infty)$, and replace it by *two* copies $I_1 = [0_1, +\infty_1)$ and $I_2 = [0_2, +\infty_2)$.

Define a basis for the topology on $\mathcal{E}_2$:

1. Any open interval on $(-\infty, 0)$ is also an open set in $\mathcal{E}_2$.

2. Any open interval in I_1 is also an open set in $\mathcal{E}_2$.

3. Any open interval in I_2 is also an open set in $\mathcal{E}_2$.

4. $\forall \epsilon > 0$, $(-\epsilon, 0) \cup [0_1, 0_1 + \epsilon)$ is an open set in $\mathcal{E}_2$ that contains 0_1.

5. $\forall \epsilon > 0$, $(-\epsilon, 0) \cup [0_2, 0_2 + \epsilon)$ is an open set in $\mathcal{E}_2$ that contains 0_2.

This topology fails to be Hausdorff for the pair of points $\{0_1, 0_2\}$. Because of its explicit construction in terms of the real line, this set can be given a manifold structure in the obvious fashion. This is an elementary example of a one-dimensional manifold that "branches" into two.

Example 3: Define a set $\mathcal{E}_3$ by considering ordinary $(3 + 1)$-dimensional Minkowski space $M(3, 1)$. Remove the set F containing the point 0 and its entire future. (That is, everything inside and on the future light-cone with vertex at 0.) Replace F by *two* copies F_1 and F_2.

The topology is constructed in the obvious manner. Define a basis for the topology on $\mathcal{E}_3$:

1. Any open set in $[M(3, 1) - F] \cup F_1$ is an open set in $\mathcal{E}_3$.

2. Any open set in $[M(3, 1) - F] \cup F_2$ is an open set in $\mathcal{E}_3$.

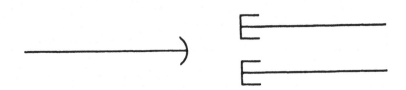

Figure 19.2: A non-Hausdorff topology: line splitting.

This topology fails to be Hausdorff for any pair of doubled points on the light cone. If $x_1 \in L_1 \equiv \partial F_1$ and $x_2 \in L_2 \equiv \partial F_2$ are a pair of double points, then the topology fails to be Hausdorff for the pair $\{x_1, x_2\}$. This example is our prototype for a universe that splits into two. It is now clear how to extend this construction to arbitrary spacetimes.

Example 4: Define a branched spacetime $\mathcal{M}(x)$ by considering an ordinary (3+1)-dimensional Hausdorff spacetime $\mathcal{M}_0$. Pick some point $x \in \mathcal{M}_0$. Remove the set $F(x)$ containing the point x and its entire future. [Technical point: $F(x)$ is the set closure of the causal future, $J^+(x)$, of the point x.]

Replace $F(x)$ by *two* copies $F_1(x)$ and $F_2(x)$. Thus

$$\mathcal{M}(x) \equiv [\mathcal{M}_0 - F(x)] \cup F_1(x) \cup F_2(x). \qquad (19.1)$$

The topology is constructed in the obvious manner. Define a basis for the topology on $\mathcal{M}(x)$:

1. Any open set in $[\mathcal{M}_0 - F(x)] \cup F_1(x)$ is an open set in $\mathcal{M}(x)$.

2. Any open set in $[\mathcal{M}_0 - F(x)] \cup F_2(x)$ is an open set in $\mathcal{M}(x)$.

If $x_1 \in \partial F_1$ and $x_2 \in \partial F_2$ are a pair of double points, then the topology fails to be Hausdorff there. What does it mean physically for spacetime to be non-Hausdorff? While coordinate patches remain four-dimensional in such a spacetime, the manifold itself can be arbitrarily complicated. Local physics remains tied to nicely behaved four-dimensional coordinate patches. Thus one can, for instance, impose the Einstein field equations in the usual manner. Every now and then, however, a "temporal anomaly"

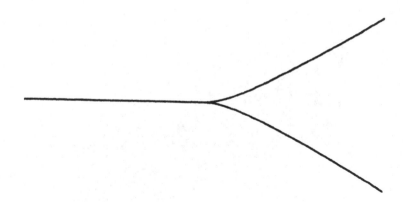

Figure 19.3: Alternative representation of the line splitting topology.

induces a "non-Hausdorff wavefront" which duplicates the whole universe. (There is nothing sacred about splitting the universe into *two* branches. Any arbitrary natural number suffices.)

This structure is now sufficiently rich to be able to offer a "multiple history" resolution of the consistency paradoxes. This is *one* possible resolution of the consistency paradoxes; there are others. Suppose a time traveler goes back into the past. He/she/it leaves behind a perfectly acceptable universe from which he/she/it has now departed. The moment x that he/she/it steps out of the time machine into the past, one might suppose that a "temporal anomaly" occurs. A new history (a new timeline) is initiated. This new history can diverge from the old history only in the causal future of x. This is exactly the sort of situation that the mathematical formalism of non-Hausdorff manifolds is capable of describing. In this formalism, the traveler can change history all he/she/it likes—the alterations are merely alterations to a new history, the old history proceeding completely unaffected.

Note: Both the old and new history share a common asymptotic region. In particular both histories share the same spacelike infinity. Thus both histories have the same ADM mass.

Warning: While we *can* use non-Hausdorff spacetimes to analyze time travel there is nothing to tell us that we *must* do it this way. In particular one can have non-Hausdorff manifolds without having time travel (one might wish to use the notion of universes splitting for other nefarious purposes), and one can have time travel without invoking non-Hausdorff

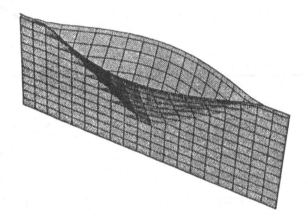

Figure 19.4: A non-Hausdorff topology: universe splitting.

manifolds (e.g., via the Novikov consistency conjecture).

More generally, let $\mathcal{M}_0$ be any Lorentzian spacetime with non-empty causality-violating region. Let x be any point of the causality-violating region. Any attempt to change history at the point x converts $\mathcal{M}_0$ into a non-Hausdorff manifold $\mathcal{M}(x)$.

Warning: Because $\mathcal{M}_0$ contains a time machine, the future of x might not be where you naively expect it to be.

One might now wish to iterate this construction. At each point in the chronology-violating region one could in principle initiate a new timeline— a new $\mathcal{M}(x)$—the resulting conglomeration consists of an infinite collection of timelines. The only relatively pleasant feature of the whole business is that the multiplicity of different timelines is restricted to lie in the future of the causality violating region (that is, inside the causality horizon).

If one wishes an even more bizarre model of reality, one could question the naive notion that the "present" has a unique fixed "past history". After all, merely by adding a time-reversed "branching event" to our non-Hausdorff spacetime one obtains a "merging event" where two universes merge into one. Not only is predictability more than somewhat dubious in such a universe, but one appears to have lost retrodictability as well. Even worse than time travel, such a cognitive framework would render the universe unsafe for historians, as it would undermine the very notion of the existence of a unique "history" for the historians to describe!

These notions are still firmly classical; even though there are many histories for the universe (many "timelines") the whole conglomeration is assembled into what at this stage is still a classical though non-Hausdorff spacetime. The various "temporal anomalies" ("branching events") that switch the universe from one time track to another can be treated at the classical level without directly invoking quantum effects. (One does indirectly invoke semiclassical quantum effects. The best way of building a time machine seems to be via a traversable wormhole. To hold the traversable wormhole open one has to violate the ANEC. The most reasonable way of violating the ANEC seems to be via semiclassical quantum effects.)

19.1.3 Exercise: Non-Hausdorff topologies

The examples of non-Hausdorff manifolds discussed in the preceding section are really minor variations on a particular general theme for constructing non-Hausdorff topologies.

Start with an arbitrary Hausdorff topological space X. Let $\Omega \subset X$ be an open set with closure $\bar{\Omega}$. Let the boundary of Ω be nonempty; $\partial \Omega \equiv \bar{\Omega} - \Omega \neq \emptyset$. Then consider the set $\tilde{X}$ obtained from X by excising $\bar{\Omega}$ and replacing it by N copies

$$\tilde{X}_N \equiv [X - \bar{\Omega}] \bigcup \left[\bigcup_{i=1}^{N} \bar{\Omega}_i \right]. \tag{19.2}$$

Define a basis for the topology on $\tilde{X}$: $\forall i \in [1, N]$, any open set in $[X - \bar{\Omega}] \cup \bar{\Omega}_i$ is an open set in $\tilde{X}$.

Check that this does in fact define a basis for a topology. Check that this topology is non-Hausdorff.

19.2 The Novikov consistency conjecture

"Suffer not an inconsistency to exist".

19.2.1 General considerations

The consistency conjecture championed by Novikov (and many others) is an alternative way of making one's peace with the notion of time travel. A primitive version of the Novikov consistency conjecture starts with the assumption that there is only *one* history for the universe, and that this one history *must* be consistent no matter what. This assumed consistency then enforces unexpected constraints on possible events in the vicinity of closed timelike curves. These consistency constraints are deemed by many to be disturbing in that they seem to affect the idea of free will or autonomy. (See [132, p. 189]. But see Deutsch [57] for a dissenting viewpoint.) More sophisticated versions of Novikov's consistency conjecture (also known as the principle of self-consistency) attempt to *derive* self-consistency, typically by invoking some microphysical assumptions regarding what it means to quantize gravity.

Warning: Consistent histories in the sense of the consistency constraints have nothing to do with the consistent histories sometimes used in studying the metaphysical interpretations of quantum mechanics.

Detailed analyses of the consistency constraints in certain model space-times (time machines plus classical billiard balls) have been carried out by various consortia of physicists [88, 75, 196]. This type of analysis has been extended to considerations of the behavior of the wave equation as one crosses the chronology horizon into the chronology-violating region [87]. Note: these particular analyses apply only to the "test particle" or "test field" limit. No attempt is made to self consistently analyze the gravitational back-reaction.

At a less technical level, the consistency constraints can be summarized by the phrase: "You can't change recorded history" (this theme has been extensively discussed in science fiction). Loosely speaking, any attempt to change past history will fail, because it must fail, because history already tells one (at least in broad outline) what did happen. It is easy to think up failure modes that are so contrived that the consistency condition takes on the air of a consistency conspiracy.

For example, consider the previously discussed situation in which a time traveler attempts to jump back into the past to kill his/her former self. The consistency conjecture demands that any such attempt *must* fail. Either the gun will misfire, or the time machine will be maladjusted and deposit the future self on the wrong side of town, or (any of an endless list of contorted and convoluted excuses for why the future self does not succeed in killing his/her former self) ... The implied conspiratorial nature of reality is, to say the least, unsettling. In the presence of closed timelike curves the consistency conjecture forces certain low probability events to become virtual certainties.

The sometimes expressed viewpoint that the Novikov consistency conjecture is the unique answer to the causality paradoxes is manifestly untenable. The Novikov consistency is *one* possible resolution of the consistency paradoxes. Indeed the consistency conjecture is still firmly wedded to the notion of spacetime as a four-dimensional Hausdorff differentiable manifold.

One may make life even more entertaining (and complicated) by introducing quantum mechanics into this volatile mix.

19.2.2 Quantum timelines?

As long as one is dealing with strictly classical physics the Novikov consistency conjecture and the multiple timeline version of the radical rewrite conjecture can be clearly separated as logically different resolutions of the consistency paradoxes. Once quantum effects are included the situation becomes more blurred. Indeed, there are certain formal similarities between a classical universe consisting of multiple interrelated timelines and Feynman's quantum mechanical "sum over histories".

Note that there are *at least* three levels at which quantum physics can be introduced:

1. Keep the background spacetime manifold fixed and Hausdorff (single timeline), and study the quantum mechanics of a particle that is permitted to traverse the chronology-violating region an arbitrary number of times. This is the test particle limit. See [214, 118, 119].

2. Keep the background spacetime manifold fixed and Hausdorff, and study the quantum field theory of a single-valued quantum field in the presence of a chronology-violating region. This is the test field limit. See [25, 91, 90, 215].

3. Permit spacetime itself to participate in the quantum fluctuations. The only way I know of formalizing this is via non-Hausdorff manifolds.

For instance, Deutsch [57, 58] has explored the suggestion that a branching spacetime of the multiple timeline type might be interpretable in terms of the Everett–DeWitt "many-worlds" interpretation of quantum mechanics [80, 69].

Warning: (1) Because Deutsch wishes his version of multiple timelines to do two jobs—to fix the time travel paradoxes *and* to "explain" quantum mechanics—the structure of his multiple timelines will be somewhat more complicated than those considered in the previous discussion. (2) The complications arise due to EPR correlations—since Einstein–Podolsky–Rosen correlations are nonlocal the picture of universes splitting at a point, followed by the split propagating at lightspeed, has too much a flavor of "local realism" to be compatible with quantum mechanics. Presumably (in the absence of time travel) one would wish to modify the non-Hausdorff construction by picking an entire spacelike hypersurface and splitting off an entire new universe at once. In the presence of closed timelike curves it is far from clear what happens to EPR correlations. (3) Deutsch does not actually use the word "non-Hausdorff" but formalizing his notion of multiple histories will need something along these lines.

The more radical proposal that the universe might not even have a unique past might further be bolstered by the observation that if one takes Feynman's "sum over paths" notion of quantum mechanics seriously, then all possible past histories of the universe should contribute to the present "state" of the universe. Some preliminaries are in order.

19.2.3   Interpretations of quantum mechanics

While essentially all physicists agree on what the rules of quantum mechanics are, there is very little agreement on what it all means. Interpretational issues are still murky (despite many loud, forceful, and mutually incompatible claims to the contrary). Major interpretational schemes are

- Copenhagen interpretation: Bohr's original interpretation—this is the official party line [157].

- Everett's relative state interpretation: This has metamorphosed over the years into the Everett–DeWitt "many-worlds" interpretation. Popular among the high priests of quantum cosmology [80, 69].

- Bohm's hidden variable interpretation: This one is a wild card. Works well for non-relativistic systems but does not seem to have a clean relativistic extension. Highly non-local. Not really popular with anybody [20, 21, 4].

- Cramer's transactional interpretation [48].

- Decoherence [103, 295, 204].

- The "shut-up-and-calculate" non-interpretation: Extremely popular when teaching.

- Various variations on these themes. [Consistent histories (in the sense of quantum mechanical consistent histories), pilot waves, etc.]

The single most important thing to know about these various interpretations of quantum mechanics is that they are *all* compatible with experiment. Applied to ordinary quantum mechanics these interpretations are in fact experimentally indistinguishable.

Many physicists will be utterly mortified by the realization that there is a consistent hidden variable interpretation of ordinary (non-relativistic) quantum mechanics that is experimentally indistinguishable from the more usual interpretations. Read Bohm [20, 21] and the commentary by Bell [14]. Note that Bell's theorem deals with *local* hidden variable theories. (Warning: Do not confuse Bohm's 1952 papers [20, 21] with his later work on "implicate order". These are very different theories.)

Since the various interpretations of quantum mechanics are experimentally indistinguishable they are by definition metaphysical constructs of no physical relevance. So why bring these issues up in a physics book? The various metaphysical interpretations of ordinary quantum mechanics become physically interesting if and only if one tries to alter or extend quantum mechanics in some fashion. Different metaphysical interpretations lead to different ideas of what a "natural" extension of quantum mechanics might be.

In summary: An interpretation of quantum mechanics has no physical consequences and is merely an additional layer of metaphysics added to the theory. If one ever finds a situation where different "interpretations" of quantum mechanics lead to different physical results then one has *not* "interpreted" quantum mechanics—rather one has *modified* quantum mechanics to produce a different physical theory.

Warning: the relevant scientific literature is vast, inconclusive, and internally contradictory.

19.2.4 Parameterized post-classical formalism?

When discussing experimental tests of general relativity it is extremely useful to invoke the parameterized post-Newtonian (PPN) formalism. The PPN formalism is a general phenomenological framework for describing weak-field gravity that contains a number of adjustable parameters. Depending on the values of these parameters, the PPN formalism describes the weak-field limits of Newtonian gravity, Einstein gravity, and many other *a priori* reasonable alternatives to ordinary general relativity. It is then a matter of experiment to measure the various parameters in the PPN formalism to see whether or not Einstein gravity is the theory that describes physical gravity. It is. (See Will [289, 290].)

So far, no similar unified "parameterized post-classical" (PPC) formalism exists for studying quantum mechanics. What would be desired is a phenomenological model with many adjustable parameters. One choice of parameters should correspond to classical Newtonian mechanics. Another choice of parameters should correspond to ordinary quantum mechanics. Present tests of quantum mechanics are pretty much constructed on an *ad hoc* basis—a quantum mechanics calculation is performed and the results checked experimentally. The closest we currently have to a PPC formalism is the general complex of ideas associated with the Bell inequalities. Studying the metaphysical interpretations of ordinary quantum mechanics *might* give us a hint on how to flesh out a PPC formalism.

Research Problem 11 *Develop a flexible, coherent, and unified parameterized post-classical (PPC) formalism.*

19.2.5   Modifying quantum mechanics

Adding time travel to physics is certainly a significant modification that might be expected to affect quantum mechanics in interesting ways. Several authors have looked at such issues. For example, it has been argued that, for interacting fields, there will be a breakdown of quantum unitary evolution in the vicinity of closed timelike curves. Within the context of the Novikov consistency conjecture this loss of unitarity has been discussed by Politzer [214, 215],[1] Friedman, Papastamatiou, and Simon [91, 90], and Goldwirth, Perry, Piran, and Thorne [119, 118].

For readers familiar with certain technical aspects of quantum field theory, there is a quick way to see that time travel implies unitarity violation:

- Perturbative unitarity of a quantum field theory is equivalent to the existence of Cutkosky rules for the Feynman diagrams.

- The existence of Cutkosky rules is equivalent to the imposition of Feynman's "$i\epsilon$" prescription for the propagators.

- The "$i\epsilon$" prescription is equivalent to the interpretation of propagators in terms expectation values of time-ordered products of quantum fields.

- But, it is impossible to define time-ordering in the presence of a time machine.

- More precisely, it is simply not possible to define time-ordering in the causality-violating region.

With certain additional assumptions (basically adopting the radical rewrite conjecture) Deutsch [57, 58] has suggested a possible experimental test of the Everett–DeWitt "many worlds" interpretation. (The only reason that there is any hope of seeing an effect, even *in principle*, is that after these modifications one is very definitely *not* dealing with ordinary quantum mechanics.)

The Deutsch test of the many-worlds interpretation[2] requires one to

1. Acquire a time machine.

---

[1] Politzer's model time machines have interesting curvature singularities [38].

[2] More precisely: The test of that particular modification of ordinary quantum mechanics that Deutsch felt was most natural based on the Everett–DeWitt "many world" metaphysics ...

2. Adopt a particular version of the radical rewrite conjecture (many different possible timelines for the universe).

3. Adopt the assumption that the switching of timelines can be identified with the splitting of universes in the "many-worlds" interpretation. This is not obvious: Just because timelines are branching in a non-Hausdorff spacetime, and universes are splitting in "many world" metaphysics does not mean that the two phenomena necessarily have anything to do with one another.

4. Accept the loss of quantum unitarity in the presence of closed timelike curves. As pointed out above, this loss of unitarity can be rigorously proved, in the test field limit, if one has a time machine in a universe with a single timeline subject to consistency constraints [91, 90]. The situation when multiple timelines are present is less clear. (It is not particularly clear what unitarity means in a situation with multiple timelines. Should probabilities add to unity in each timeline separately, or should one perform a sum over all timelines weighted by the probability of existence of the timeline?)

5. Despite the fact that one has assumed that the universe suffers from a multiplicity of timelines, the quantum physics of the entire class of possible timelines is assumed to be described by a single quantum density matrix that is subject to Novikov style consistency constraints.

With all these assumptions in place, Deutsch showed that one particular set of reasonable self-consistent resolutions of the consistency and bootstrap paradoxes can be constructed.

Warning: When Deutsch uses the phrase "unmodified quantum theory" he means quantum mechanics with the Everett–DeWitt "many-worlds" metaphysics [57]. I think it fair to say that when most physicists use this phrase they mean either "shut-up-and-calculate" or possibly ordinary quantum mechanics with the Copenhagen metaphysics.

On the other hand, Politzer [215] has investigated a somewhat different generalization of quantum mechanics to causality-violating spacetimes. This particular approach is based on a functional integral over single-valued histories—that is, the consistency conditions are applied to the individual histories before functional integration. In the absence of chronology-violating regions this model and the Deutsch density matrix model reduce to the same limit—ordinary quantum mechanics. In the presence of chronology violating regions the two models describe different physics.

Yet another generalization of quantum mechanics capable of dealing with causality-violating spacetimes is based on the decoherence approach to quantum physics [103].

Finally, I wish to point out that Copenhagen metaphysics suggests yet more possibilities. Based on the Copenhagen metaphysics, Penrose has suggested modifying quantum mechanics by raising the "collapse of the wave function" to the status of a physical phenomenon, as opposed to a metaphysical convenience [208, 209]. Penrose has furthermore argued that the physical collapse of the wave function is intrinsically a quantum gravitational effect. If this suggestion is correct, life gets very entertaining. In the absence of a chronology-violating region one could assert that the every-now-and-then physical collapse of the wave function leads to the development of some single unique spacetime manifold encoding the history of the universe. In the presence of a chronology-violating region one can either

1. assert that the existence of the chronology-violating region forces all the collapse events in the chronology-violating region to be consistent with each other, or

2. assert that collapse events within the chronology-violating region lead to new timelines.

Trying to prove either of these possibilities from first principles will be extremely difficult—not least because there is no agreement as to what the relevant first principles might be. One issue in particular that will have to be addressed is whether or not this collapse process is local or nonlocal, and what exactly this implies for the EPR correlations.

Comment: (Disclaimer) I do not raise all of these issues because I particularly believe that this is the way the universe works. I do so because in order to evaluate the various different ideas advocated by various authors it is necessary to have a reasonably clear idea of what is being claimed. Once one has opened Pandora's box by permitting time travel, I see no particular reason to believe that the damage done to our notions of reality would be at all mild. Something rather radical is probably required to shore things up.

Research Problem 12 *Once you have satisfactorily developed a parameterized post-classical (PPC) formalism, use the PPC formalism to analyze different ways of extending quantum mechanics to deal with issues of time travel.*

19.3 The chronology protection conjecture

"Suffer not a closed causal curve to exist".

19.3.1 General considerations

Hawking's chronology protection conjecture is a considerably more conservative method for dealing with temporal paradoxes. This conjecture permits the existence of traversable wormholes but forbids the existence of time travel [129, 130]. Given the current experimental situation (absolutely *zero* experimental evidence for time travel) it is entirely acceptable to adopt the nonexistence of time travel as a working hypothesis, or better yet, as an axiom. (Some things one should not try to prove mathematically, instead one assumes them as axiomatic and then checks by experiment whether or not the assumption corresponds to physical reality. However, if the experimental situation changes one should not hesitate to reject such an axiom.)

To paraphrase Hawking: the observed fact that we are not hip-deep in tourists from the future can be interpreted as experimental evidence in support of some form of the notion of chronology protection.

If the chronology protection conjecture is to hold, then there must be something wrong with the *apparently* simple manipulations described in previous chapters that *appeared* to lead to the formation of time machines. Investigating the physics of the putative formation of time machines is certainly interesting for the light it may throw on the physical mechanisms that one assumes enforce the chronology protection conjecture. There is a whole raft of strange and peculiar effects that come into play near the chronology horizon. Aside: The relative importance of various effects depends critically on the dimensionality of spacetime. For instance, certain classical effects that are important in $(1 + 1)$ dimensions are in general negligible in $(3 + 1)$ dimensions.

For the purposes of this chapter let me focus on one particular issue—the effect of the spacetime geometry on the renormalized vacuum expectation value of the quantum stress-energy tensor associated with various quantized fields. These calculations are most commonly restricted to the case of conformally-coupled scalar fields, though sometimes spinor fields are also discussed. (Most technical details will be deferred until subsequent portions of the book—see p. 333ff.—the present discussion is an overview.) It appears that any attempt to transform a single isolated wormhole into a time machine results in large vacuum polarization effects. These vacuum polarization effects seem sufficient to disrupt the internal structure of the wormhole long before the onset of Planck scale physics, and before the onset of time travel.

Many authors have calculated the vacuum expectation value of the quantum stress-energy tensor in model spacetimes containing traversable wormholes [93, 116, 169, 163, 164, 167, 268, 270, 114, 178]. This may equally well be viewed as calculating the vacuum polarization induced by the presence of the wormhole. Generically one encounters expressions of the form

$$\langle 0|T_{\mu\nu}(x)|0\rangle \approx \hbar \sum_{\gamma}{}' \frac{\Delta_\gamma(x,x)^{1/2}}{\pi^2 s_\gamma(x,x)^4}\, t_{\mu\nu}(x;\gamma). \qquad (19.3)$$

The prime indicates that the sum runs over all nontrivial geodesics connecting the point x to itself. The symbol $s_\gamma(x,x)$ denotes the length of this self-intersecting geodesic. The tensor $t_{\mu\nu}(x;\gamma)$ is a dimensionless object that is built up out of the spacetime metric and the tangent vectors to γ [270]. Furthermore, $\Delta_\gamma(x,x)$ denotes the object known as the van Vleck determinant [259, 189, 202, 68, 59, 64, 66]. The overall size of this prefactor is a key ingredient in governing whether or not semiclassical quantum effects are sufficient to enforce Hawking's chronology protection conjecture [129, 130, 164, 268, 270, 114, 178]. To get a handle on what is going on, note that a good parameterization of the magnitude of the quantum back-reaction is obtained by considering the scalar invariant

$$\mathcal{T}_0 \equiv \sqrt{\langle 0|T_{\mu\nu}(x)|0\rangle\, \langle 0|T^{\mu\nu}(x)|0\rangle}. \qquad (19.4)$$

Then, if one of the closed self-linking geodesics (say γ') dominates over the others, one may approximate

$$\mathcal{T}_0 \approx \hbar\, \frac{\Delta_{\gamma'}(x,x)^{1/2}}{\pi^2 s_{\gamma'}(x,x)^4}. \qquad (19.5)$$

(This requires $t^{\mu\nu}t_{\mu\nu} \approx 1$, as will be justified on p. 336.)

The essential points are: (1) the semiclassical quantum back-reaction is directly proportional to the square root of the van Vleck determinant, and (2) the semiclassical quantum back-reaction diverges as s^{-4}. While the above formula is valid at any point x in the spacetime, it is most useful to consider a point x on the throat of the wormhole. The size of the back-reaction should be compared to $\mathcal{T}$ the quantity of stress-energy required to hold the wormhole throat open in the first place. At the throat of a wormhole whose mouth is of radius R the Einstein equations imply

$$\mathcal{T} \equiv \sqrt{T_{\mu\nu}\, T^{\mu\nu}} \approx \frac{\hbar}{\ell_P^2 R^2}. \qquad (19.6)$$

Once $\mathcal{T}_0 \gg \mathcal{T}$ one has what I feel is clear and convincing evidence that the semiclassical vacuum polarization disrupts the internal structure of the wormhole [270]. (Others may choose to disagree with me on this issue.) This disruption occurs at

$$s|_{\text{disrupt}} \approx \Delta_{\gamma'}^{1/8} \sqrt{\pi \ell_P R}. \qquad (19.7)$$

Since $s_\gamma(x,x) \to 0$ is the signal for the onset of time travel, it will *always* be the case that this occurs before the time machine has a chance to form. The real issue is whether the overwhelming of the wormhole's internal structure occurs before or after one reaches the Planck regime. One of the issues adding to the complications is a certain lack of agreement as to where exactly the onset of the Planck scale physics is.

If $T_0 \gg T$ while $s_\gamma(x,x) \gg \ell_P$, then semiclassical vacuum polarization effects overwhelm the wormhole's internal structure in a parameter regime where one still expects the semiclassical approximation to hold. One may thus safely assert that at least in this parameter regime, semiclassical vacuum polarization effects enforce Hawking's chronology protection conjecture.

If $T_0 \gg T$ does not occur until $s_\gamma(x,x) < \ell_P$, then one cannot conclude that time machine formation succeeds. All that one can safely conclude is that the full and hideous glory of full-fledged quantum gravity must be invoked.

A central issue in investigating these issues is thus the evaluation of $\Delta_\gamma(x,x)$ for closed self-intersecting geodesics γ with base point x located on the throat of the wormhole. For the case of a putative time machine built out of a single wormhole it can be shown, in the short-throat limit, that this van Vleck determinant will always equal unity. For putative time machines built out of multiple wormholes it appears possible to make this factor small [178, 270]. If the wormhole sizes and separations are human scale, then it can be shown that wormhole disruption occurs long before one enters the Planck regime [270]. For suitably obtuse (not traversable by humans) choices of wormhole size and location one can push the disruption scale down below the Planck slop. Even begging the question of whether or not one can build or acquire traversable wormholes in the first place, this does *not* mean that multiple wormhole configurations can be used to build a time machine. All it means is that one's limited ability to calculate in the semiclassical regime has been completely obviated by entry into the Planck slop.

Summary: The stress-energy tensor is driven to infinity as one gets close to building a time machine. Modulo arguments concerning the onset of Planck scale physics, this strongly suggests that wormholes will be gravitationally disrupted by back-reaction effects before the onset of chronology violation.

No fully convincing proof of the chronology protection conjecture is currently extant. Part of the problem is that it is far from clear what the technical assumptions and precise statement of any chronology protection theorem should be. In this regard it is useful to consider the historic situation with regard to the cosmic censorship conjecture. While it is relatively easy to give a crude physicist's description of cosmic censorship it took a good five years of work to formulate careful mathematically precise statements of the purported content of the cosmic censorship hypothesis (of course, we have still not managed to *prove* the cosmic censorship hypothesis). In a similar manner, it is to be expected that some time will pass before there is any clear agreement in the community as to what exactly one has to do in order to "prove" (or disprove!) chronology protection. In particular, since current arguments make use of semiclassical quantum effects it is already clear that the classical physics of the Einstein equations and classical energy conditions cannot be used as input hypotheses to any putative chronology protection theorem. It is possible that the chronology protection conjecture could be settled entirely within the context of semiclassical quantum gravity. It is equally possible that resolution of this issue will require use of some as yet undeveloped theory of full-fledged quantum gravity.

An alternative point of view is that notions of causality are so fundamental to our picture of the physical universe that one should adopt the chronology protection as an axiom. If one chooses this option the task is now to build only those theories of gravity that are compatible with chronology protection. This limits the class of acceptable theories of gravity. In particular, one now has to address questions such as: Is classical Einstein gravity without energy conditions compatible with the chronology protection axiom? (Answer: "No".) Is classical Einstein gravity with energy conditions compatible with the chronology protection axiom? [Answer: "Not necessarily" (it depends on your precise definition of classical general relativity).] Is semiclassical Einstein gravity compatible with the chronology protection conjecture? [Answer: "Maybe".] Is quantum gravity compatible with the chronology protection conjecture? [Answer: "First you have to decide on your theory of quantum gravity".] Adopting this point of view implies that the status of chronology protection has been elevated to an empirical experimental question. The chronology protection axiom can be overturned only by the empirical construction of a time machine—a task that we are unlikely to accomplish in the immediate future.

19.3.2   A toy model

To get a feel for some of the physical processes that might enforce Hawking's chronology protection conjecture it is useful to consider a highly idealized toy model. In the context of this toy model a "proof" of the chronology protection conjecture is relatively easy—unfortunately this "proof" does not carry over to more realistic models. Nevertheless the toy model is useful in that it gives analytical insight into the basic physics.

Periodic boundary conditions with time-shift

The toy model I have in mind requires certain brutal approximations to the traversable wormhole spacetime. Take an asymptotically flat spacetime containing a single traversable wormhole. Assume most of the curvature is concentrated near the two mouths of the wormhole. Assume that the throat of the wormhole is short. Now let the radius of the wormhole mouths R go to infinity, while keeping the distance between the wormhole mouths ℓ and the time-shift T fixed. As $R \to \infty$ the stress-energy localized at the throat tends to zero.

The toy model, obtained in the limit $R \to \infty$, is simply Minkowski space with two hyperplanes identified:

$$(t, x, y, 0) \equiv (t + T, x, y, \ell). \tag{19.8}$$

The toy model is described by two parameters ℓ and T, both of which may be assumed to be varied adiabatically. A somewhat more complicated model—Misner space—is discussed by Hawking [129, 130]. That model avoids the adiabatic approximation but, on the other hand, forces ℓ to be related to T in a very specific way. Another variation on this theme is the Grant spacetime [122], further discussed by Laurence [175].

Inspecting the toy model, one is seen to be working with a minor generalization of the ordinary Casimir effect geometry—a Casimir effect geometry with a time-shift.

Note that the "once-through-the-wormhole" interval is

$$s = \| (T, 0, 0, \ell) \| = \sqrt{\ell^2 - T^2}. \tag{19.9}$$

Now perform a Lorentz transformation of velocity $\beta = T/\ell$. In this new Lorentz frame the wormhole connects "equal times". In this frame of simultaneity (synchronous frame) the wormhole is described by identifying the two hyperplanes

$$(\gamma t, x, y, \gamma \beta t) \equiv (\gamma t, x, y, \gamma \beta t + s). \tag{19.10}$$

Classical considerations

For the sake of argument (this is *not* a realistic possibility) assume that the toy model wormhole contains a single classical photon of energy E_0 that travels directly from $z = 0$ to $z = \ell$, loops back to $z = 0$, and repeats this trip indefinitely.

Suppose that the photon starts from $z = 0$ at time $t = 0$. Then the photon will arrive at $z = \ell$ at time $t = \ell$ (since it travels at speed $c \equiv 1$). But time $t = \ell$ at $z = \ell$ is the same as time $t = \ell - T$ at $z = 0$. So the photon passes through the plane $z = 0$ with periodicity $\ell - T$, requiring only a time $\ell - T$ to complete each orbit of the wormhole.

If one takes a time slice at any constant time, the photon will be at many places at the same time. For instance at time $t = 0$ the photon is at $z = 0$, but also at $z = \ell - T$, $z = 2(\ell - T)$, etc. (However, this replication effect cuts off at $z = \ell$.) This implies, in particular, that the total energy is enhanced by a replication factor

$$E_{\text{tot}} = E_0 \, \frac{\ell}{\ell - T}. \tag{19.11}$$

Thus there is a classical energy barrier acting to keep the wormhole mouths apart, and this classical energy barrier becomes infinitely large as ℓ is adiabatically reduced to T. (Essentially the multiple copies of the photon pile up on top of one another as one approaches time machine formation.) This prevents the formation of a time machine.

Unfortunately this simple example, relying on only a single photon, is unphysical in (3+1) dimensions. A real wormhole throat of finite radius always has a defocusing effect on incident particles [190]. Thus in real life, after a finite number of traversals of the wormhole system, the photon will be scattered out of the path joining the wormhole mouths and the associated energy barrier will vanish. [However: The analogous argument in (1 + 1) dimensions is watertight. This is simply because there are no transverse directions for the photon to get lost in.]

Casimir effect with time-shift

A particularly nice feature of this toy model is that it is possible to analytically calculate the renormalized value of the stress-energy tensor. This is done via a minor variation of the calculation used for the ordinary Casimir effect. (For a nice survey article on the Casimir effect see [134]. See also the textbooks [17, 99].)

Working in the frame of simultaneity (synchronous frame), the manifest invariance of the boundary conditions under rotation and/or reflection in the x-y plane restricts the stress-energy tensor to be of the form

$$\langle 0|T_{\mu\nu}|0\rangle = \begin{bmatrix} T_{tt} & 0 & 0 & T_{tz} \\ 0 & T_{xx} & 0 & 0 \\ 0 & 0 & T_{xx} & 0 \\ T_{tz} & 0 & 0 & T_{zz} \end{bmatrix}. \tag{19.12}$$

Consider a scalar quantum field . By considering the effect of the boundary conditions one may write eigenmodes of the d'Alambertian in the separated form

$$\phi(t, x, y, z) = e^{-i\omega t} \, e^{ik_x x} \, e^{ik_y y} \, e^{ik_z z}, \tag{19.13}$$

subject to the constraint

$$\phi(t, x, y, z) = \phi(t, x, y, z + s). \tag{19.14}$$

Therefore the boundary condition on k_z implies the classical quantization

$$k_z = \pm \frac{2\pi n}{s}, \tag{19.15}$$

while k_x and k_y are unquantized (i.e., arbitrary and continuous). The equation of motion constrains ω to be

$$\omega = \sqrt{\left(\frac{2\pi n}{s}\right)^2 + k_x^2 + k_y^2}. \tag{19.16}$$

Now, recall that

$$\langle 0|T_{tz}|0\rangle \propto \langle 0|\partial_t \phi \, \partial_z \phi |0\rangle. \tag{19.17}$$

(See, for instance, p. 120, or Birrell and Davies [17].)

Performing a mode sum over n, it is clear that the positive values of k_z exactly cancel the negative values so that

$$\langle 0|T_{tz}|0\rangle = 0. \tag{19.18}$$

Furthermore, one may note that the quantization conditions on ω and k_z depend only on s, not on β or γ. Thus, without loss of generality one may immediately apply the result for $\beta = 0$ to the case $\beta \neq 0$ and obtain

$$\langle 0|T_{\mu\nu}|0\rangle = -\frac{\hbar k}{s^4}\left(\eta_{\mu\nu} - 4 n_\mu n_\nu\right). \tag{19.19}$$

Here n_μ is the tangent to the closed self-intersecting spacelike geodesic threading the wormhole throat. In the frame of simultaneity (synchronous frame)

$$n_\mu = (0,0,0,1), \tag{19.20}$$

while in the rest frame

$$n_\mu = (T/s, 0, 0, \ell/s). \tag{19.21}$$

The constant k is a dimensionless numerical factor. For a scalar field [99]

$$k = -\frac{\pi^2}{1440}. \tag{19.22}$$

In a more general context the argument given above applies not only to scalar fields but also to fields of arbitrary spin that are (approximately) conformally-coupled. In general k will depend on the nature of the applied boundary conditions (twisted or untwisted), the spin of the quantum field theory under consideration, etc. Furthermore, as s decreases, massive particles may be considered as being approximately massless once $\hbar c/s \gg mc^2$. (That is, once the length of the self-intersecting closed geodesic is less than the appropriate Compton wavelength.) Thus k should be thought of as changing in a roughly stepwise fashion as $s \to 0^+$. (This behavior is entirely analogous to the behavior of the R parameter in electron–positron annihilation. See, for instance, [154, pp. 668–671].) A complete specification of k would thus require complete knowledge of the asymptotic behavior of the spectrum of elementary particles, and it is for this reason that one declines further precision in the specification of k.

The form of the stress-energy tensor can also be determined by noting that the identifications (19.10) and (19.8) describing the wormhole can be easily solved by considering an otherwise free field in Minkowski space subject to the constraint

$$\phi(x^\mu) \equiv \phi(x^\mu \pm jsn^\mu); \qquad j = 0, 1, 2, \ldots \qquad (19.23)$$

Since these constraints do not depend on V, the four-velocity of the wormhole mouth, one may directly apply the results obtained for the ordinary Casimir effect [134, 17, 99].

The fact that the renormalized stress-energy goes to infinity as one gets close to forming a time machine ($s \to 0^+$) is our signal for chronology protection. It is important to realize that the sign of k is largely immaterial. If k is positive then one observes a large positive Casimir energy which tends to repel the wormhole mouths, and so tends to prevent the formation of a time machine. On the other hand if k is negative one observes a large positive stress threading the wormhole throat. If this stress is allowed to back-react on the geometry via the Einstein field equations it tends to act to collapse the wormhole, and so also tends to prevent the formation of a time machine.

The limitations inherent in the type of toy model currently under consideration should also be made clear: the model problem is optimally designed to make it easy to thwart the formation of a time machine (i.e., closed time-like curves). It is optimal for thwarting in two senses: (1) Idealizing the wormhole mouths by a pair of flat planes prevents defocusing effects. (2) The adiabatic approximation requires the relative motion of the wormhole mouths to be arbitrarily slow. This is an idealized limit; and again, it is the case that this idealization makes it easier for nature to thwart the time machine formation, because it gives the growing vacuum polarization an arbitrarily long time to act back on the spacetime and distort it.

k positive—repulsive Casimir energy

For the toy model one may calculate the total four-momentum associated with the stress-energy tensor. Compactify the toy model in the x-y directions by adding periodic boundary conditions that fix the area A in the x-y

directions to be macroscopically large. Then

$$P^\mu = \oint \langle 0|T^{\mu\nu}|0\rangle d\Sigma_\nu = \langle 0|T^{\mu\nu}|0\rangle \, A \, s \, n_\perp^\mu. \qquad (19.24)$$

In the frame of simultaneity (synchronous frame)

$$n_\perp^\mu = (1,0,0,0). \qquad (19.25)$$

That is, $n_\perp$ is perpendicular to n. Thus

$$P^\mu = \frac{\hbar k}{s^4} \, A \, s \, n_\perp^\mu. \qquad (19.26)$$

By going back to the rest frame one obtains

$$P^\mu = \frac{\hbar k}{s^4} A s \left(\frac{\ell}{s}, 0, 0, \frac{T}{s}\right), \qquad (19.27)$$

so that one may identify the Casimir energy as

$$E_{\text{Casimir}} = -P^\mu V_\mu = \frac{\hbar k A \ell}{s^4}. \qquad (19.28)$$

Take $k > 0$ for the sake of discussion; then this Casimir energy, considered in isolation, would by itself give an infinitely repulsive hard core to the interaction between the wormhole mouths. However, one should exercise some care. The calculation is certainly expected to break down for $s < \ell_P$ (once one enters the Planck slop) so that one may safely conclude only that there is a finite but large potential barrier to entering the full quantum gravity regime. Fortunately this barrier is in fact very high. For $T \gg T_P$ one can estimate

$$E_{\text{barrier}} \approx E_P \frac{A}{\ell_P^2} \left(\frac{T}{T_P}\right), \qquad (19.29)$$

while for $T \ll T_P$ (i.e., $T \approx 0$) one may estimate

$$E_{\text{barrier}} \approx E_P \frac{A}{\ell_P^2}. \qquad (19.30)$$

For macroscopic wormholes these barriers are truly enormous. For microscopic wormholes one really doesn't care what happens. First because it is not too clear whether truly microscopic wormholes ($R \ll \ell_P$) can even exist. Second, because even if such wormholes do exist, they would in no sense be traversable [264], and so would be irrelevant to the construction of usable time machines.

k negative—wormhole disruption

At first glance, should the constant k happen to be negative, one would appear to have a disaster on one's hands. In this case the Casimir force seems to act to help rather than hinder time machine formation. Fortunately yet another physical effect comes into play. The infinitely large stress-energy tensor associated with time machine formation should, via back-reaction effects, distort the geometry of the original toy model.

Consider linearized fluctuations around the locally flat background metric describing the toy model. Using the notation developed on p. 12ff:

$$g_{\mu\nu} = \eta_{\mu\nu} + h_{\mu\nu} = \eta_{\mu\nu} + \bar{h}_{\mu\nu} - \frac{1}{2}\bar{h}\,\eta_{\mu\nu}. \tag{19.31}$$

By going to the Hilbert gauge (Lorentz gauge), $\partial_\nu \bar{h}^{\mu\nu} = 0$, one may write the linearized Einstein field equations as

$$\Box \bar{h}_{\mu\nu} = \frac{-16\pi \ell_P^2}{\hbar} \langle 0|T_{\mu\nu}|0\rangle. \tag{19.32}$$

Now, working in the rest frame, the time translation invariance of the geometry implies

$$\partial_t \bar{h}_{\mu\nu} = 0. \tag{19.33}$$

Similarly, the translation symmetry in the x and y directions implies

$$\partial_x \bar{h}_{\mu\nu} = 0 = \partial_y \bar{h}_{\mu\nu}. \tag{19.34}$$

Thus the linearized Einstein equations reduce to

$$\partial_z^2 \bar{h}_{\mu\nu} = \frac{16\pi k \ell_P^2}{s^4} \left(\eta_{\mu\nu} - 4n_\mu n_\nu\right). \tag{19.35}$$

Note that the tracelessness of $T_{\mu\nu}$ implies the tracelessness of $\bar{h}_{\mu\nu}$ so that one automatically has $\bar{h}_{\mu\nu} = h_{\mu\nu}$. Using the boundary condition that

$$h_{\mu\nu}|_{(z=0, t=0)} = h_{\mu\nu}|_{(z=\ell, t=T)} \tag{19.36}$$

the linearized Einstein field equations integrate to

$$h_{\mu\nu}(z) = \frac{16\pi}{2} (z - \ell/2)^2 \frac{k\ell_P^2}{s^4} \left(\eta_{\mu\nu} - 4n_\mu n_\nu\right). \tag{19.37}$$

To estimate the maximum perturbation of the metric calculate

$$\delta g_{\mu\nu} = h_{\mu\nu}|_{(z=0)} = h_{\mu\nu}|_{(z=\ell)} = \frac{16\pi k \ell_P^2 \ell^2}{8s^4} \left(\eta_{\mu\nu} - 4n_\mu n_\nu\right). \tag{19.38}$$

Define

$$\|\delta g\| \equiv \sqrt{(\delta g_{\alpha\beta})\, \eta^{\beta\gamma}\, (\delta g_{\gamma\delta})\, \eta^{\delta\alpha}}. \tag{19.39}$$

Then

$$\|\delta g\| = \sqrt{12}\, \frac{16\pi k \ell_P^2 \ell^2}{8 s^4}. \tag{19.40}$$

This metric perturbation certainly becomes large well before one enters the Planck regime, thus indicating that the gravitational back-reaction becomes large and important long before one needs to consider full quantum gravity effects.

Because of the delicacy of these particular issues, it may be worthwhile to belabor the point. Suppose one redoes the entire computation in the frame of simultaneity (synchronous frame). One still has

$$\Box \bar{h}_{\mu\nu} = \frac{-16\pi \ell_P^2}{\hbar}\, \langle 0|T_{\mu\nu}|0\rangle, \tag{19.41}$$

where now the boundary conditions are

$$h_{\mu\nu}|_{(z=\beta t, t)} = h_{\mu\nu}|_{(z=\beta t + s, t)}. \tag{19.42}$$

This immediately implies that $h_{\mu\nu}$ is a function of $(z - \beta t)$. Indeed the linearized Einstein equations integrate to

$$h_{\mu\nu}(z, t) = \frac{16\pi}{2}\, \frac{(z - \beta t - s/2)^2}{(1 - \beta^2)}\, \frac{k\ell_P^2}{s^4}\, (\eta_{\mu\nu} - 4 n_\mu n_\nu). \tag{19.43}$$

So that at the mouths of the wormhole, $(z = \beta t) \equiv (z = \beta t + s)$, factors of s and $\gamma = 1/\sqrt{1 - \beta^2}$ combine to yield

$$\delta g_{\mu\nu} = h_{\mu\nu}|_{(z=\beta t, t)} = h_{\mu\nu}|_{(z=\beta t + \ell, t)} = \frac{16\pi k \ell_P^2 \ell^2}{8 s^4}\, (\eta_{\mu\nu} - 4 n_\mu n_\nu), \tag{19.44}$$

which is the same tensor as the result calculated in the rest frame (as of course it must be). While the four-velocity of the wormhole throat does not enter into the computation of the stress-energy tensor it is important to realize that the four-velocity of the throat does, via the naturally imposed boundary conditions, have an important influence on the gravitational back-reaction.

In any event, the gravitational back-reaction will radically alter the spacetime geometry long before a time machine has the chance to form. Note that for k negative, so that the Casimir energy is negative and attractive, the *sign* of δg_{tt} obtained from this linearized analysis hints at the formation of an event horizon should s become sufficiently small. Unfortunately a full nonlinear analysis would be necessary to establish this with any certainty.

Well before s shrinks down to the Planck length, the vacuum polarization will have severely disrupted the internal structure of the wormhole—presumably leading to wormhole collapse. It should be pointed out that there is no particular need for the collapse to proceed all the way down to zero radius and subsequent topology change. It is quite sufficient for the present discussion if the collapse were to halt at a radius of order the Planck length ℓ_P. In fact, there is evidence, based on minisuperspace calculations (see pp. 347–366 herein, and also [262, 263, 265, 266]), that this is indeed what happens. If indeed collapse is halted at a radius of order the Planck length by quantum gravity effects, then the universe is still safe for historians since there is no reasonable way to get a physical probe through a Planck scale wormhole [264].

Summary

Vacuum polarization effects become large as $s \to 0^+$. That is, as the closed self-intersecting spacelike geodesics which must be present in wormhole spacetimes of nontrivial topology shrink to invariant length zero, the stress-energy diverges. Depending on an overall undetermined sign either (1) there is an arbitrarily large force pushing the wormhole mouths apart, or (2) there are wormhole disruption effects at play which presumably collapse the wormhole throat down to the size of a Planck length. Either way, usable time machines are avoided.

Unfortunately, because the toy model currently under consideration is optimally designed to make it easy to thwart the formation of a time machine, it is unclear to what extent one may draw generic conclusions from these arguments. More general arguments will be given in subsequent chapters. These more general arguments will involve rather technical calculations of the singularities exhibited by the stress-energy tensor at the onset of time travel, and will focus in particular on the quantity known as the van Vleck determinant.

19.4   The boring physics conjecture

"Suffer not these obnoxious difficulties to pervert physics"
Another way of dealing with the problems associated with time travel is simply to abolish from consideration all of the physical mechanisms that might lead to closed timelike curves. I call this the boring physics conjecture. A first draft of this conjecture might roughly be formulated as: "suffer not traversable wormholes to exist".

Merely by asserting the nonexistence of traversable wormholes all causal problems of the particular type associated with intra-universe traversable wormholes go quietly away. It had been hoped that this abolition could be achieved through strict enforcement of the averaged null energy condition (though it was never clear how to reconcile such strict enforcement with the relatively benign counterexample on p. 125). It is now realized that the ANEC will fail in generic spacetimes, so that this avenue does not lead to anti-wormhole legislation. (See [277], the general discussion on p. 196, and the technical discussion on p. 290.) Worse, there are many forms of chronology violation that are not associated with traversable wormholes, and some circumstances where chronology violation is not associated with violations of the energy conditions [198, 199]. Furthermore, there are some circumstances (inter-universe wormholes) in which traversable wormholes do not lead to chronology violation.

To completely forbid time travel requires additional assumptions—such as the nonexistence of cosmic strings and limitations on the tipping over of light cones. For instance, requiring the triviality of the fundamental group $\pi_1(\mathcal{M})$ (the first homotopy group) implies that the manifold possesses no closed noncontractible loops. Such a restriction would preclude both (1) traversable wormholes (more precisely, intra-universe traversable wormholes and multiple connections between universes via multiple inter-universe wormholes), and (2) cosmic strings. (Aside: This depends on whether one assumes the core of the string to be a true singularity, and so a homotopy obstruction. If one smears the string core out one can argue that the fundamental group in universes containing physical cosmic stings is in fact trivial.) This takes care of most instances of chronology violation.

But even this is not enough—time travel can seemingly occur in universes of trivial topology [129, 130, 117], so even stronger constraints should be imposed. In this regard Penrose's version of the strong cosmic censorship conjecture [206] implies the causality protection conjecture via the equivalence

$$(\text{strong cosmic censorship}) \Leftrightarrow (\text{global hyperbolicity})$$
$$\Rightarrow (\text{strong causality}) \Leftrightarrow (\exists \text{ global time function}). \quad (19.45)$$

The "boring physics conjecture" might thus best be formulated as a corollary to the strong cosmic censorship conjecture. (But note that the theoretical evidence in favor of strong cosmic censorship is weak. Classically there is a [non-generic] counterexample [180]; quantum effects are believed to destabilize this counterexample.) There is certainly no experimental evidence against the boring physics conjecture, but in many ways this is a relatively uninteresting possibility. In particular, considering the relatively benign conditions on the stress-energy tensor required to support a traversable wormhole it seems to be overkill to dispose of the possibility of wormholes merely to avoid problems with time travel.

Part V

Quantum Effects

Chapter 20

Semiclassical quantum gravity

This part of the book collects an assortment of discussions and comments on several rather technical issues. Beginning graduate students and readers unfamiliar with the technical intricacies of quantum field theory on curved spacetimes may wish to avoid this part of the monograph on first reading.

I shall start with a chapter on basic techniques and issues of general importance. Subsequent chapters shall investigate the van Vleck determinant, and apply this object to discussing issues surrounding the chronology protection conjecture. Finally a chapter on minisuperspace wormholes exhibits ways in which it is possible to go beyond semiclassical quantum gravity.

20.1   Background

Pick some fixed classical spacetime manifold with metric $g_{\mu\nu}$. Suppose that spacetime has residing on it some collection of quantum fields $\Phi(x)$ (collectively called the matter fields). The matter fields are assumed to be in some quantum state $|\psi\rangle$. There are at least two obvious and interesting things to calculate:

- The Einstein tensor, $G_{\mu\nu}(g_{\alpha\beta})$.

- The expectation value of the quantum stress-energy tensor,

$$\langle\psi|T^{\mu\nu}(\Phi, g_{\alpha\beta})|\psi\rangle.$$

If the Einstein tensor is related to the expectation value of the stress-energy tensor by

$$G_{\mu\nu}(g_{\alpha\beta}) = 8\pi G\langle\psi|T^{\mu\nu}(\Phi, g_{\alpha\beta})|\psi\rangle, \tag{20.1}$$

then the spacetime is said to satisfy the semiclassical Einstein equations of semiclassical quantum gravity. Finding such solutions is extremely difficult. The only exact solutions I know of are trivial:

- Minkowski space with all the quantum fields in the vacuum state (and with the cosmological constant renormalized to zero by fiat).

- de Sitter space with all the quantum fields in the vacuum state (and with the renormalized cosmological constant fine tuned by fiat).

- Starobinsky inflation—this is a special case of de Sitter space with all the quantum fields in the vacuum state. If one restricts attention to massless conformally-coupled quantum fields. One can then *calculate* the renormalized cosmological constant in a self-consistent manner via the trace (conformal) anomaly [245, 241].

- Classical solutions obtained by taking the limit $\hbar \to 0$.

I (and the rest of the community) would dearly love to see a self-consistent wormhole solution of the semiclassical field equations. The strategy would go something like this: We know that spherically symmetric traversable wormholes are uniquely described by two functions of the radial parameter—the shape function $b(r)$ and the redshift function $\phi(r)$. The Einstein tensor, as a function of shape and redshift functions, has already been calculated. More complicated, but in principle doable, is the evaluation of

$$\langle 0|T^{\mu\nu}(\Phi, g[\phi(r), b(r)])|0\rangle. \tag{20.2}$$

(Doing this will probably involve some numerical techniques.) Now adjust the shape and redshift functions until a solution of the semiclassical equations is found. Note that numerical techniques require $\phi(r)$ and $b(r)$ to be fully specified—at best one could hope to adopt an iterative approach.

Research Problem 13 *Use this strategy to find a self-consistent wormhole solution of the semiclassical field equations (or show that no such solution exists).*

To continue the discussion, it is useful to define

$$T^{\mu\nu}_{\text{classical}} \equiv \lim_{\hbar \to 0} \langle \psi|T^{\mu\nu}(\Phi, g_{\alpha\beta})|\psi\rangle, \tag{20.3}$$

$$T^{\mu\nu}_{\text{quantum}} \equiv \langle \psi|T^{\mu\nu}(\Phi, g_{\alpha\beta})|\psi\rangle - T^{\mu\nu}_{\text{classical}}. \tag{20.4}$$

Note that this automatically implies that $T^{\mu\nu}_{\text{quantum}} = O(\hbar)$. A good working definition of a quantum vacuum state $|0\rangle$ is any state for which $T^{\mu\nu}_{\text{classical}} = 0$. Since exact solutions to semiclassical quantum gravity are so difficult to come by, the next best thing is to adopt some approximation techniques.

- Linearize around flat Minkowski space: $g_{\mu\nu} = \eta_{\mu\nu} + h_{\mu\nu}$. Put the quantum field Φ in some vacuum state $|0\rangle$, so that $T^{\mu\nu}_{\text{classical}} = 0$. Now solve the linearized Einstein equation

$$\Box\left(h_{\mu\nu} - \frac{1}{2}h\eta_{\mu\nu}\right) = -\frac{16\pi G}{c^2}\, T^{\text{quantum}}_{\mu\nu} + O(h^2). \qquad (20.5)$$

This technique has been explored, for instance, by Horowitz [142] and Horowitz and Wald [145]. Note that because $T^{\text{quantum}}_{\mu\nu}$ is of order $\hbar$, the metric perturbation $h_{\mu\nu}$ will also be of order $\hbar$, and is therefore presumably small.

- Take the test field limit: Let the background metric be any solution of the classical vacuum Einstein equations. Again, put the quantum field Φ in some vacuum state $|0\rangle$, so that $T^{\mu\nu}_{\text{classical}} = 0$. Now calculate $T^{\text{quantum}}_{\mu\nu}$. Since this is automatically of order $\hbar$, and so presumably small, ignore the back-reaction this stress-energy induces on the spacetime geometry. This is done for instance in all current calculations of $\langle T^{\mu\nu}\rangle$ on Schwarzschild, Casimir, Grant, and Misner spacetimes, and is also the approximation adopted when trying to prove or disprove the ANEC.

- Back-reaction (a bootstrap approach): Let the background field be any solution of the classical vacuum Einstein equations. Put the quantum field Φ in some vacuum state $|0\rangle$, so that $T^{\mu\nu}_{\text{classical}} = 0$. Now calculate $T^{\text{quantum}}_{\mu\nu}$. Use this as input to the linearized Einstein equations, where one is now linearizing around a curved background metric g_0 by setting $g_1 = g_0 + h_1$. The linearized Einstein equations are qualitatively of the form

$$\Box h_1 \propto T^{\text{quantum}}(g_0). \qquad (20.6)$$

Solve for the shift in the metric. Use the new metric, $g_1 = g_0 + h_1$, to calculate a new stress-energy tensor $T^{\text{quantum}}_{\mu\nu}(g_1)$. Linearize around this new metric $g_2 = g_1 + h_2$. Then solve for

$$\Box h_2 \propto T^{\text{quantum}}(g_1) - T^{\text{quantum}}(g_0). \qquad (20.7)$$

Repeat the previous performance. After n steps

$$\Box h_n \propto T^{\text{quantum}}(g_{n-1}) - T^{\text{quantum}}(g_{n-2}). \qquad (20.8)$$

This is equivalent to recursively defining g_n by solving

$$G(g_n) \propto T^{\text{quantum}}(g_{n-1}). \qquad (20.9)$$

If one is lucky the whole procedure converges ($h_n \to 0$; $g_n \to g_\infty$; $T_n \to T_\infty$). No complete analysis along these lines is currently extant. Note that I have suppressed many technical details to keep the discussion simple. Successfully completing this program is not going to be an easy task. Some initial steps in this direction are presented in references [8, 139, 141, 293].

- Adiabatic expansions: Use adiabatic techniques (for example, the DeWitt–Schwinger expansion) to *approximate* the quantum stress-energy tensor $T_{\mu\nu}^{\text{quantum}}$. Then adopt any one of the previous techniques. Adiabatic approximations are often surprisingly good. For instance, in calculating $\langle T^{\mu\nu} \rangle$ for Schwarzschild spacetime the Page approximation [200] is an adiabatic approximation that retains only part of $\langle T^{\mu\nu} \rangle$ but is nevertheless good to within 1%. Another example is the computation of the divergences in $\langle T^{\mu\nu} \rangle$ as one gets close to building a time machine. The divergences are adequately described by keeping only the dominant term in the adiabatic expansion.

Though these are all approximation techniques, this is the best that can currently be done.

Research Problem 14 *Push all of these techniques as far as possible.*

20.2   Adiabatic techniques

Adiabatic techniques are a standard tool that I shall repeatedly invoke in the next few chapters. I will not attempt to derive these techniques from scratch, but will content myself with a brief exposition of key results.

20.2.1   The Green function: Hadamard Form

It is a standard result of curved-space quantum field theory that Green functions (propagators) can be put into the so-called Hadamard form:

$$
\begin{aligned}
G(x,y) &\equiv \langle 0|\{\Phi(x), \Phi(y)\}|0\rangle \\
&= \hbar \sum_\gamma \frac{\Delta_\gamma(x,y)^{1/2}}{4\pi^2} \left[\frac{1}{\sigma_\gamma(x,y)} \right. \\
&\qquad\qquad \left. + v_\gamma(x,y)\ln|\sigma_\gamma(x,y)| + \varpi_\gamma(x,y) \right]. \quad (20.10)
\end{aligned}
$$

Such Green functions also arise in solving the classical wave equation, cf. [68]. There are many relevant references, including [68, 59, 64, 66, 17, 99].

1. The summation, $\sum_\gamma$, runs over all distinct geodesics connecting the point x to the point y. In general, even in the absence of wormholes, there can be more than one geodesic going from x to y. However, in the absence of wormholes, and for x and y sufficiently close to each other (so as to avoid conjugate points and caustics), there will be only one such geodesic.

2. The symbol $\Delta_\gamma(x,y)$ denotes the van Vleck determinant [259, 189]. The fact that the van Vleck determinant appears in the adiabatic expansion of the Green function is ultimately the reason that the van Vleck determinant is so important for the purposes of this book. The reader will soon encounter two chapters devoted solely to this topic.

3. The symbol $\sigma_\gamma(x,y)$ denotes the geodetic interval. That is, $\sigma_\gamma(x,y) \equiv \pm(1/2)[s_\gamma(x,y)]^2$ is half the signed square of the geodesic distance from x to y. More details will be provided shortly.

4. The functions $v_\gamma(x,y)$ and $\varpi_\gamma(x,y)$ are known to be smooth in the coincidence limit. That is, as $s_\gamma(x,y) \to 0$. See [68, 59, 64, 66, 17, 99].

20.2.2 Point splitting

The basic idea behind point splitting techniques [68, 59, 64, 66, 17, 99] is to define formally infinite objects in terms of a suitable limiting process such as

$$\langle 0|T_{\mu\nu}(x)|0\rangle = \lim_{y \to x} \langle 0|T_{\mu\nu}(x,y;\gamma_0)|0\rangle. \tag{20.11}$$

The point-split stress-energy tensor $T_{\mu\nu}(x,y;\gamma_0)$, which I have not yet carefully defined, is constructed to be a symmetric tensor at the point x and a scalar at the point y. It is not supposed to be obvious that such a point-splitting procedure is internally consistent. The point-split stress-energy tensor will depend on a particular choice of geodesic γ_0 that connects x to y and the limit $y \to x$ is to be taken as y moves along this geodesic.

The contribution to the point-split stress-energy tensor associated with a particular quantum field is generically calculable in terms of covariant derivatives of the renormalized Green function G_R of that quantum field. Schematically

$$\begin{aligned}\langle 0|T_{\mu\nu}(x,y;\gamma_0)|0\rangle &= D_{\mu\nu}(x,y;\gamma_0)\{G_R(x,y;\gamma_0)\}, \\ G_R(x,y;\gamma_0) &\equiv \langle 0|\{\Phi(x),\Phi(y)\}|0\rangle_R.\end{aligned} \tag{20.12}$$

Here $D_{\mu\nu}(x,y;\gamma_0)$ is some second-order differential operator, built out of covariant derivatives at x and y. The covariant derivatives at y must be parallel transported back to the point x so as to ensure that $D_{\mu\nu}(x,y;\gamma_0)$

defines a proper geometrical object. This parallel transport means that the object $D_{\mu\nu}(x, y; \gamma_0)$ depends on the particular choice of geodesic used to parallel transport from x to y.

Renormalization of the Green function consists of removing the short distance singularities. To this end, consider a scalar quantum field $\Phi(x)$. This subtraction depends on the choice of a particular geodesic γ_0 from x to y. A particularly brutal, though simple, subtraction scheme is to define

$$G_R(x, y; \gamma_0) \equiv G(x, y) - \hbar \frac{\Delta_{\gamma_0}(x, y)^{1/2}}{4\pi^2} \left[\frac{1}{\sigma_{\gamma_0}(x, y)} \right.$$
$$\left. + v_{\gamma_0}(x, y) \ln |\sigma_{\gamma_0}(x, y)| \right]. \quad (20.13)$$

These subtractions correspond to a wave-function renormalization and a mass renormalization, respectively. Note that any such renormalization prescription is always ambiguous up to further finite renormalizations. Since, for the purposes of this monograph, the point of the discussion is the infinities that occur at the onset of time machine formation, worrying about possible finite terms is not profitable. For other applications, more careful treatment of the finite renormalizations is essential. See, for instance, [17, 99, 200]. The scheme described above may profitably be viewed as a modified minimal subtraction scheme. Using the Hadamard form of the Green function this implies

$$G_R(x, y; \gamma_0) = \hbar \frac{\Delta_{\gamma_0}(x, y)^{1/2}}{4\pi^2} \varpi_{\gamma_0}(x, y)$$
$$+ \hbar \sum_{\gamma \neq \gamma_0} \frac{\Delta_{\gamma}(x, y)^{1/2}}{4\pi^2} \left[\frac{1}{\sigma_{\gamma}(x, y)} \right.$$
$$\left. + v_{\gamma}(x, y) \ln |\sigma_{\gamma}(x, y)| + \varpi_{\gamma}(x, y) \right]. \quad (20.14)$$

In particular, if one now let $y \to x$ along the geodesic γ_0, then (for a free scalar quantum field in a curved spacetime), this renormalization prescription is sufficient to render $\langle 0|\Phi^2(x)|0\rangle$ finite:

$$\langle 0|\Phi^2(x)|0\rangle_R = G_R(x, x)$$
$$= \hbar \frac{\Delta_{\gamma_0}(x, x)^{1/2}\varpi_{\gamma_0}(x, x)}{4\pi^2}$$
$$+ \hbar \sum_{\gamma \neq \gamma_0} \frac{\Delta_{\gamma}(x, x)^{1/2}}{4\pi^2} \left[\frac{1}{\sigma_{\gamma}(x, x)} \right.$$
$$\left. + v_{\gamma}(x, x) \ln |\sigma_{\gamma}(x, x)| + \varpi_{\gamma}(x, x) \right]. \quad (20.15)$$

In this limit γ_0 devolves to the trivial degenerate curve that always remains at the point x while the summation now runs over nontrivial geodesics that connect the point x to itself.

For the particular case of a conformally-coupled massless scalar field the differential operator in question is [166, 167, 270]

$$
\begin{aligned}
D_{\mu\nu}(x,y;\gamma_0) \equiv\ & \frac{1}{6}\left(\nabla_\mu^z\, g_\nu{}^\alpha(x,y;\gamma_0)\nabla_\alpha^y + g_\mu{}^\alpha(x,y;\gamma_0)\nabla_\alpha^y\nabla_\nu^z\right) \\
& -\frac{1}{12}g_{\mu\nu}(x)\left(g^{\alpha\beta}(x,y;\gamma_0)\nabla_\alpha^z\nabla_\beta^y\right) \\
& -\frac{1}{12}\left(\nabla_\mu^z\nabla_\nu^z + g_\mu{}^\alpha(x,y;\gamma_0)\nabla_\alpha^y\, g_\nu{}^\beta(x,y;\gamma_0)\nabla_\beta^y\right) \\
& +\frac{1}{48}g_{\mu\nu}(x)\left(g^{\alpha\beta}(x)\nabla_\alpha^z\nabla_\beta^z + g^{\alpha\beta}(y)\nabla_\alpha^y\nabla_\beta^y\right) \\
& -R_{\mu\nu}(x) + \frac{1}{4}g_{\mu\nu}(x)R(x).
\end{aligned}
\tag{20.16}
$$

As required, this object is a symmetric tensor at x and is a scalar at y. The bi-vector $g_\mu{}^\nu(x,y;\gamma_0)$ parallel-transports a vector at y to a vector at x, the parallel transport being taken along the geodesic γ_0. The detailed effects of this parallel transport can often be safely ignored, *vide* [167, equation (3)], and [93, equation (2.40)].

Point-splitting is a general purpose technique of wide applicability. It is one of the standard tricks of the trade that every practitioner must at some level be familiar with. After looking at a few other side issues, and constructing some more technical machinery, I shall return to the use of point-splitting techniques to calculate the vacuum polarization near the onset of time travel.

20.3 Non-orientable wormholes?

Insofar as one is solely concerned with the classical physics of point particles moving along timelike paths in Lorentzian spacetime (typically geodesics), there is nothing in the classical physics of point particles to force spacetime to be an orientable manifold. I mentioned earlier (see p. 109 and p. 168) that macroscopic wormholes of the Morris–Thorne or thin-shell variety could with equal ease, be taken to be either orientable or non-orientable. At a deeper level, the possible existence of non-orientable Wheeler wormholes in the spacetime foam has been investigated by Friedman, Papastamatiou, Parker, and Zhang [89]. (For basic background on non-orientable space-times see, for example, [70, 288]) While non-orientable spacetimes are acceptable at the classical level, the situation changes markedly once fermions are considered:

Theorem 29 *Non-orientable spacetimes are incompatible with the standard model of particle physics.*

To understand why this is so, I shall first have to present some definitions (see, for example, [275, p. 60] or [132, pp. 181–182]).

Definition 32 *A spacetime is orientable (that is, "space-time-orientable") if and only if there exists a continuous and everywhere nonzero globally defined four-form* $\epsilon_{[\mu\nu\lambda\rho]}$*.*

Terminology: This nonzero globally defined four-form, if it exists, is called the volume form.

Definition 33 *A spacetime is space-orientable if (1) there exists a continuous and everywhere nonzero globally defined three-form* $e_{[\mu\nu\lambda]}$*, and (2) there exists an everywhere nonzero timelike vector field* V^μ *that is continuous up to possible sign reversals, and (3)* $V^\mu e_{[\mu\nu\lambda]} = 0$*.*

If spacetime is topologically of the form $\mathcal{M} \sim \Re \times \Sigma$, and if the Σ can all be chosen to be spacelike hypersurfaces, then space-orientability of the spacetime $\mathcal{M}$ reduces to (ordinary) orientability of the spacelike hypersurfaces Σ.

For completeness, I also reiterate:

Definition 34 *A spacetime is time-orientable if there exists a continuous and everywhere nonzero globally defined timelike vector field* V^μ*.*

Note that if spacetime is space-orientable, then $V_{[\mu}e_{\nu\lambda\rho]}$ is an everywhere nonzero four-form which is continuous if and only if V is continuous. Thus space-orientability plus time-orientability implies space-time-orientability. Similarly, any two of these conditions imply the third.

Exercise: Prove this.

The key fact to note about the standard model of particle physics is that the weak interactions are chiral. This means that the fermion fields can be taken to be in definite eigenstates of chirality:

$$\gamma_5 \psi = \pm \psi. \tag{20.17}$$

Particle physicists are used to seeing the γ_5 matrix defined as

$$\gamma_5 \equiv i\, \gamma^0 \gamma^1 \gamma^2 \gamma^3 = i\, \frac{\text{sign}[abcd]}{4!}\, \gamma^a \gamma^b \gamma^c \gamma^d. \tag{20.18}$$

When promoting the standard model of particle physics onto a curved spacetime manifold, the definition of γ_5 has to be generalized to

$$\gamma_5 \equiv i\, \frac{\epsilon_{[\mu\nu\lambda\rho]}}{4!}\, e^\mu{}_a e^\nu{}_b e^\lambda{}_c e^\rho{}_d\, \gamma^a \gamma^b \gamma^c \gamma^d. \tag{20.19}$$

Here $e^\mu{}_a$ is the vierbein (tetrad), while $\epsilon_{[\mu\nu\lambda\rho]}$ is a totally antisymmetric four-index tensor, a four-form. (I have not and will not define the vier-beins/tetrads; look up, for instance, Landau and Lifshitz [174, pp. 291–294], Wald [275, pp. 49–53], or Weinberg [279, pp. 365–373].) The crucial point is that chiral fermions exist if and only if globally defined four-forms exist. (Technical point: The second Stiefel–Whitney class also has to vanish, but this is not where the interesting physics is.) Thus (see, for example, [89])

Theorem 30 *Chiral fermions exist if and only if spacetime is orientable (and the second Stiefel–Whitney class vanishes).*

To get a deeper feel for this result, consider what happens if spacetime is *not* orientable. One can always go to the double cover $\mathcal{D}(\mathcal{M})$, which *is* orientable, and define chiral fermions on the double-cover. If one now pulls this chiral fermion field back from the double-cover to the spacetime one obtains *two* fermion fields defined on the spacetime itself. Furthermore, while one cannot (by the assumption of non-orientability) define a global chirality for either one of these fermion fields one can, on any contractible open region of the spacetime, define a "local" chirality operator

$$\text{``}\gamma_5\text{''} \equiv i\sqrt{-g}\,\frac{\text{sign}[\mu\nu\lambda\rho]}{4!}\,e^\mu{}_a\,e^\nu{}_b\,e^\lambda{}_c\,e^\rho{}_d\,\gamma^a\gamma^b\gamma^c\gamma^d. \tag{20.20}$$

In terms of this "local" chirality operator, a globally chiral fermion on the double-cover pulls back, on spacetime, to a *pair* of "local" chiral fermions of *opposite* "local" chirality. This fermion-doubling problem completely destroys the standard model of particle physics, or more precisely, completely destroys the standard model of weak interactions. (Fermion-doubling problems arise with chiral fermions also in the context of lattice models of particle physics. The genesis of the problem is different here but the effects are similar.)

It has been suggested that the assumed occurrence of non-orientable wormholes in the spacetime foam could be used to generate fermion masses. (See, for example, [89]). This approach seems to work tolerably well in a hypothetical universe without weak interactions. However, given the experimental fact that the weak interactions are chiral, one must abandon the notion of non-orientable wormholes and non-orientable spacetime generally.

(The bloody-minded might take a contrary view: Since non-orientable wormholes are so "nice" it probably just means that the standard model of particle physics is wrong! Maybe the weak interactions only appear to be chiral? Maybe we should rewrite all of weak interaction physics? I expect the particle physics community to resist any such perverse notion with vigor. The experimental data take precedence over theoretical musings.)

Caveat: The mathematics just says that you cannot have both chiral fermions and non-orientable spacetime. It is then a physics judgement as to which you would rather abandon: all known weak interaction physics, or non-orientable manifolds.

Caveat: Experimentally, direct observations on the chirality of matter, and all other physics, are limited to the surface of the planet Earth and its immediate environs in the solar system. Beyond the solar system we are forced to use less direct observational techniques. In arguing for the orientability of spacetime I am of course implicitly making an assumption that the basic rules of physics are the same throughout all of spacetime—this "principle of uniformity" (the cosmological principle) has in the past proved both useful and reliable.

Caveat: This theorem does of course assume applicability of the manifold picture to describe spacetime. Thus the theorem applies classically, it also applies to semiclassical quantum gravity, *and* it applies to that class of prospective theories of quantum gravity that can be thought of as some functional integral over spacetime metrics. If one chooses to abandon the manifold structure for spacetimes the theorem of course becomes moot. Of course, abandoning the manifold picture is even more radical than merely rewriting all of weak interaction physics.

Even if one were so perverse as to insist that the rules suddenly change "somewhere out there", the fact that we *locally* observe chiral fermions places limits on acceptable *local* microphysics of the spacetime foam:

Theorem 31 *Any Lorentzian wormhole in the spacetime foam must be orientable.*

After all, if the spacetime foam contains non-orientable wormholes, then fermion-doubling problems will become apparent in the here and now.

20.4   Time-orientability?

An argument similar to the above, based on the microphysics of the standard model of particle physics, shows that spacetime must also be time-orientable.

Theorem 32 *Time-non-orientable spacetimes are incompatible with the standard model of particle physics.*

The starting point of the analysis is the observation that, if spacetime is not time-orientable, then one cannot tell the difference between past and future because by definition no such distinction can be drawn on the manifold as a whole.

The observed fact that we can distinguish past from future then forces spacetime to be time-orientable. I do not need to invoke messy semi-philosophical arguments about the "arrow of time", the increase in entropy, and the like—it suffices to observe that the microphysics of the weak interactions experimentally breaks time reversal invariance (that is, the T of the CPT theorem is experimentally violated); and that we cannot even begin to model this phenomenon in a time-non-orientable manifold because we cannot even define what time reversal means in such a spacetime.

Caveat: The mathematics just says that you cannot have both T violation and time-non-orientable spacetime. It is then a physics judgement as to which you would rather abandon: the experimental results on T violation, or time-non-orientable manifolds.

Caveat: The same warning about the implicit use of the "principle of uniformity" (cosmological principle) applies.

Caveat: Again, this theorem does of course assume applicability of the manifold picture to describe spacetime.

Provided we are willing to retain the manifold picture, the fact that we locally observe a breaking of time reversal invariance places limits on the acceptable microphysics of the spacetime foam:

Theorem 33 *Any Lorentzian wormhole in the spacetime foam must be time-orientable.*

20.5 Space-orientability?

The same argument, now invoking the observed parity-violations of the standard model of particle physics, shows that spacetime must also be space-orientable.

Theorem 34 *Space-non-orientable spacetimes are incompatible with the standard model of particle physics.*

The observed fact that we can distinguish left from right forces spacetime to be space-orientable. Let me present a very simple version of the argument: Consider a Möbius strip. As every schoolchild knows, a Möbius strip has only one side. Therefore it is a mathematical and logical impossibility to set up any physical theory on the Möbius strip that distinguishes the two sides—after all the apparent existence of two sides to the Möbius strip is an illusion: the Möbius strip only has one side. One cannot break parity invariance on a Möbius strip simply because it is impossible to tell left from right; there is no difference between left and right.

Subject to the caveats given previously, the breaking of left-right invariance places limits on the acceptable microphysics of the spacetime foam:

Theorem 35 *Any Lorentzian wormhole in the spacetime foam must be space-orientable.*

Summary: The net result of these considerations is perhaps somewhat counterintuitive:

- The experimentally observed breakdown of time reversal invariance implies that spacetime is time-orientable.

- The experimentally observed breakdown of parity invariance implies that spacetime is space-orientable.

- The experimentally observed chiral nature of the weak interactions implies that spacetime is orientable.

Comments similar to these may be found in the survey article by Geroch and Horowitz [107]. Somewhat weaker statements based on a curved-space variant of the CPT theorem are found in Hawking and Ellis [132, pp. 181–182] and in reference [89].

20.6   ANEC violations

Earlier, when discussing topological censorship (p. 195), I enunciated a theorem which can be more precisely stated as

Theorem 36 *In any Lorentzian spacetime, for any conformally-coupled quantum field, in any conformal quantum state, if the scale anomaly is nonzero, then the renormalization scale can be chosen in such a way that the ANEC is violated.*

The time has come to discuss this result more fully. (This theorem is based on an extension of the brief "note added in proof" that appears on p. 415 of a paper by Wald and Yurtsever [277]. See also [271].) To understand the proof, one should start by realizing that there is a tremendous amount of technical machinery available for analyzing the behavior of the quantum stress-energy tensor under conformal deformations of the metric.

Exercise: (Long) Read, study, and understand the books by Birrell and Davies [17] and Fulling [99]. See also the recent book by Wald [276].

A scale transformation is just the special case of a conformal transformation when the conformal rescaling factor is position independent—

$$g(x) \rightarrow \bar{g}(x) = \Omega^2 g(x). \tag{20.21}$$

Under a scale transformation in $(3+1)$ dimensions the stress-energy tensor undergoes an anomalous transformation

$$\langle T^\mu{}_\nu(\bar{g}) \rangle = \Omega^{-4} \left(\langle T^\mu{}_\nu(g) \rangle - 8a\hbar \ln \Omega \left[\nabla_\alpha \nabla^\beta + \frac{1}{2} R^\alpha{}_\beta \right] C^{\alpha\mu}{}_{\beta\nu} \right). \tag{20.22}$$

This is a standard result. For example, this is a special case of equation (66) of Page [200]. The symbol C denotes Weyl's conformal curvature tensor. Technically, for a conformally-coupled quantum field in a conformal quantum state, the renormalized vacuum expectation value of the quantum stress-energy tensor exhibits the above scale anomaly.

The coefficient a is exactly the same as that arising in the trace (conformal) anomaly,

$$\langle T^\mu{}_\mu \rangle = \hbar \left\{ a \left(C_{\mu\nu\lambda\rho} C^{\mu\nu\lambda\rho} \right) + b \left(*R_{\mu\nu\lambda\rho} * R^{\mu\nu\lambda\rho} \right) + c \,\Box R + dR^2 \right\}. \tag{20.23}$$

For future convenience define

$$Z^\mu{}_\nu \equiv \left[\nabla_\alpha \nabla^\beta + \frac{1}{2} R^\alpha{}_\beta \right] C^{\alpha\mu}{}_{\beta\nu}. \tag{20.24}$$

The anomalous scaling behavior of the stress-energy tensor under a rescaling of the metric is a consequence of the fact that in regularizing the conformal quantum field one has had to introduce a cutoff. This cutoff breaks the conformal invariance and, after proper renormalization to remove explicit cutoff dependence, results in a dimensional transmutation effect whereby the expectation value depends on a so-called renormalization scale μ. (See, for example, [142]). One may make this renormalization scale dependence explicit by writing

$$\langle T^\mu{}_\nu(\Omega g; \mu) \rangle = \langle T^\mu{}_\nu(g; \mu/\Omega) \rangle. \tag{20.25}$$

Thus, the scale anomaly may be recast as

$$\langle T^\mu{}_\nu(g; \mu/\Omega) \rangle = \Omega^{-4} \left(\langle T^\mu{}_\nu(g; \mu) \rangle - 8a\hbar \, \ln \Omega \, Z^\mu{}_\nu \right). \tag{20.26}$$

Now consider the effect of the scale anomaly on the ANEC integral. Let γ be any null curve of the metric g parameterized by a generalized affine parameter λ. Then

$$I_\gamma(\mu/\Omega) \equiv \int_\gamma \langle T^\mu{}_\nu(g;\mu/\Omega)\rangle \, k_\mu k^\nu d\lambda \qquad (20.27)$$

$$= \int_\gamma \Omega^{-4} [\, \langle T^\mu{}_\nu(g;\mu)\rangle - 8a\hbar \, \ln\Omega \, Z^\mu{}_\nu] \, k_\mu k^\nu d\lambda$$

$$= \Omega^{-4} [I_\gamma(\mu) - 8a\hbar \, \ln\Omega \, J_\gamma]. \qquad (20.28)$$

Here I have defined

$$J_\gamma \equiv \int_\gamma Z^\mu{}_\nu \, k_\mu k^\nu \, d\lambda \equiv \int_\gamma \left[\nabla_\alpha \nabla^\beta + \frac{1}{2}R^\alpha{}_\beta\right] C^{\alpha\mu}{}_{\beta\nu} \, k_\mu k^\nu \, d\lambda. \qquad (20.29)$$

Now if $J_\gamma \neq 0$ it is always possible to choose $\ln\Omega$ sufficiently large (either positive or negative) to force $I_\gamma < 0$. This means that ANEC is guaranteed to be violated for the null curve γ for a suitable choice of the renormalization scale μ. (Technical point: For simplicity I have assumed that all the integrals converge.) Note that I have explicitly retained factors of $\hbar$ to emphasize the fact that this is a *quantum* effect.

Exercise: Perform the appropriate technical incantations to deal with the case where the ANEC integral does not converge. Follow the analysis of Wald and Yurtsever [277, pp. 404–405].

This completes the proof and we can reformulate the theorem as

Theorem 37 (ANEC no-go theorem) *In $(3+1)$-dimensional Lorentzian spacetime the ANEC is guaranteed to be violated whenever*

$$Z^\mu{}_\nu \equiv \left[\nabla_\alpha \nabla^\beta + \frac{1}{2}R^\alpha{}_\beta\right] C^{\alpha\mu}{}_{\beta\nu} \neq 0. \qquad (20.30)$$

More precisely: If $Z^\mu\nu \neq 0$, then even if the ANEC is satisfied on the manifold with its original choice of metric, there exists a rescaled metric on the manifold in which the ANEC is violated.

Notes and caveats:

- If $Z^\mu{}_\nu = 0$ we cannot deduce that the ANEC *must* be satisfied. We can only deduce that it is *possible* for the ANEC to be satisfied. Cases are known where $Z^\mu{}_\nu = 0$, with the ANEC being satisfied on null geodesics, but with the ANEC violated along certain classes of non-geodesic null curves [166].

- In Minkowski space $Z^\mu{}_\nu = 0$. Thus Minkowski space evades the no-go theorem. This result is compatible with the theorems proved by Klinkhammer [166].

- In $(1+1)$ dimensions the scale anomaly vanishes. [For example, take equation (6.134) of Birrell and Davies [17] and set $\nabla\Omega = 0$.] This happy accident is due to the fact that all two-dimensional manifolds are conformally flat. Thus all $(1+1)$-dimensional spacetimes evade the no-go theorem. This result is compatible with the theorems proved by Yurtsever [294] and Wald and Yurtsever [277].

- In any Einstein spacetime $Z^\mu{}_\nu = 0$. To see this, recall that an Einstein spacetime is defined by

$$R_{\mu\nu} = \Lambda\, g_{\mu\nu}. \tag{20.31}$$

Then note

$$R^\alpha{}_\beta\, C^{\alpha\mu}{}_{\beta\nu} = \Lambda\, g^\alpha{}_\beta\, C^{\alpha\mu}{}_{\beta\nu} = 0. \tag{20.32}$$

Furthermore, starting from the Bianchi identities, a single contraction yields (see, for example, [132, equation (2.32), p. 43])

$$\nabla_\mu C^\mu{}_{\nu\sigma\rho} = -R_{\nu[\sigma;\rho]} + \frac{1}{6}g_{\nu[\sigma}R_{;\rho]}. \tag{20.33}$$

In view of the definition of an Einstein spacetime this implies

$$\nabla_\mu C^\mu{}_{\nu\sigma\rho} = 0. \tag{20.34}$$

Therefore $Z^\mu{}_\nu = 0$; all Einstein spacetimes have zero scale anomaly and evade the no-go theorem.

- In particular, since Schwarzschild spacetime has $Z^\mu{}_\nu = 0$, Schwarzschild spacetime evades the no-go theorem. This is compatible with the discussion on p. 128.

- Any Ricci flat spacetime has $Z^\mu{}_\nu = 0$ and evades the no-go theorem.

- On the other hand, generic perturbations of any spacetime with zero scale anomaly will in general lead to a nonzero scale anomaly. So generic perturbations of manifolds satisfying the ANEC lead to manifolds where the ANEC is violated by a suitable choice of renormalization scale. In this sense violations of the ANEC are generic. The case of linearized perturbations around flat Minkowski space was addressed in [277].

- This result is encouraging for wormhole physics—a potentially serious problem has been neatly avoided.

Research Problem 15 *Is the scale anomaly the only obstruction to proving some form of the ANEC from first principles?*

20.7   Fluctuations in stress-energy

An important issue of principle is the existence of fluctuations in the expectation value of the stress-energy. (See [171]. For general background information on these issues see [85, 86].) Even in situations where one can calculate $\langle T^{\mu\nu} \rangle$ exactly, there are other quantities of interest. For instance, define root-mean-square measures of the stress-energy tensor by

$$(\tau_1)^2 \equiv \langle T^{\mu\nu} \rangle \langle T_{\mu\nu} \rangle, \tag{20.35}$$

$$(\tau_2)^2 \equiv \langle T^{\mu\nu} T_{\mu\nu} \rangle, \tag{20.36}$$

$$(\Delta\tau)^2 \equiv \left\langle \left(T^{\mu\nu} - \langle T^{\mu\nu} \rangle \right) \left(T_{\mu\nu} - \langle T_{\mu\nu} \rangle \right) \right\rangle \tag{20.37}$$

$$= (\tau_2)^2 - (\tau_1)^2. \tag{20.38}$$

The quantity $\Delta\tau$ measures the size of the fluctuations around the expectation value. If the ratio $\Delta\tau/\tau_1$ or the ratio $\Delta\tau/\tau_2$ is large, say of order one or larger, one might begin to wonder about the assumed applicability of semiclassical quantum gravity.

A priori, it is quite conceivable that $\Delta\tau/\tau_1$ or $\Delta\tau/\tau_2$ might become large long before entering the Planck regime. This suggests the possibility of a failure mode for semiclassical quantum gravity that might have nothing to do with Planck scale physics.

These fluctuations in stress-energy have been investigated for the Casimir effect geometry by Kuo and Ford [171]. They calculate quantities related to but not identical with the $\Delta\tau$ discussed above, and find disturbingly large values for these fluctuations. At this stage it is difficult to evaluate how important this result is. There are several open questions:

Research Problem 16 *Is $\Delta\tau/\tau_1 > 1$ generic for spacetimes that violate the ANEC?*

If this property can be shown to be generic, it would *seem* to indicate a disaster in that one would no longer be justified in applying semiclassical quantum gravity to spacetimes containing traversable wormholes.

Research Problem 17 *Is $\Delta\tau/\tau_1 > 1$ really a disaster for semiclassical quantum gravity? Is there any way in which one might still make sense of semiclassical quantum gravity even if fluctuations in the stress-energy are relatively large?*

This is an open area of continuing research.

Chapter 21

van Vleck determinants: Formalism

In previous chapters, I have often had occasion to mention the van Vleck determinant without going into any technical details. The van Vleck determinant is a truly ubiquitous object. It arises in many physically interesting situations, such as the following:

1. Wentzel–Kramers–Brillouin (WKB) approximations to quantum time evolution operators and Green functions.

2. The theory of Green functions for curved-space wave equations.

3. Adiabatic approximations to heat kernels.

4. One-loop approximations to functional integrals.

5. The theory of caustics in geometric optics and ultrasonics.

6. The focusing and defocusing of geodesic flows in Riemannian manifolds.

7. Estimating semiclassical quantum gravity effects.

While all of these topics are interrelated, the present monograph is particularly concerned with the last case. As will soon be apparent, in this particular context the van Vleck determinant is essentially a measure of the tidal focusing and/or defocusing of geodesic flows in spacetime [269].

Introduced by van Vleck in 1928 [259], the van Vleck determinant was first utilized in elucidating the nature of the classical limit of quantum mechanics via WKB techniques. Further formal developments were due to

295

Morette [189]. A nice discussion of this original line of development can be found in Pauli's lecture notes [202].

This chapter presents a general framework that permits the estimation of the van Vleck determinant associated with geodesic flows in Lorentzian spacetimes. *A fortiori* these developments have important implications for the entire array of topics indicated. The next chapter will apply some of these techniques to traversable wormhole spacetimes.

21.1   General definition

Consider an arbitrary mechanical system, of n degrees of freedom, governed by a Lagrangian $L(\dot{q}, q)$. Solve the equations of motion to find the (possibly not unique) path γ passing through the points (q_i, t_i) and (q_f, t_f). Then calculate the action of that path

$$S_\gamma(q_i, t_i; q_f, t_f) \equiv \int_\gamma L(\dot{q}, q) dt. \qquad (21.1)$$

The van Vleck determinant is then defined as

$$\Delta_\gamma(q_i, t_i; q_f, t_f) \equiv (-1)^n \det \left\{ \frac{\partial^2 S_\gamma(q_i, t_i; q_f, t_f)}{\partial q_i \, \partial q_f} \right\}. \qquad (21.2)$$

The van Vleck determinant, in this original incarnation, occurs as a prefactor in the WKB approximation to the quantum time evolution operator [259, 189, 202]. In the Schrödinger picture

$$\langle q_f | \exp\{-iH(t_f - t_i)/\hbar\} | q_i \rangle \approx$$
$$(2\pi i \hbar)^{-n/2} \sum_\gamma \sqrt{\Delta_\gamma(q_i, t_i; q_f, t_f)} \, \exp\{+iS_\gamma(q_i, t_i; q_f, t_f)/\hbar\}.$$

$$(21.3)$$

In the particular case of a geodesic flow on a Lorentzian or Riemannian manifold the appropriate action to be inserted in this definition is the geodetic interval between a pair of points.

21.2   Geodetic interval

Consider a Lorentzian spacetime of total dimensionality $(d + 1)$, that is, d space dimensions and 1 time dimension. With appropriate changes, the following results apply also to a Riemannian manifold of total dimensionality $n = (d + 1)$. To go from the Lorentzian to the Riemannian case the only significant change is that one should forget all the subtleties associated with lightlike (null) geodesics.

Let γ be a geodesic from x to y.

- If this geodesic is timelike define the geodesic distance to be

$$s_\gamma(x, y) \equiv \int_\gamma d\tau. \qquad (21.4)$$

- If this geodesic is spacelike define the geodesic distance to be

$$s_\gamma(x, y) \equiv \int_\gamma ds. \qquad (21.5)$$

- If this geodesic is null define the geodesic distance to be

$$s_\gamma(x, y) \equiv 0. \qquad (21.6)$$

With this definition the geodesic distance $s_\gamma(x, y)$ is always positive semi-definite.

The geodetic interval is now defined by

$$\sigma_\gamma(x, y) \equiv \pm \frac{1}{2}[s_\gamma(x, y)]^2. \qquad (21.7)$$

Here we take the upper $(+)$ sign if the geodesic γ from the point x to the point y is spacelike. We take the lower $(-)$ sign if this geodesic γ is timelike. The geodetic interval is the quantity Synge refers to as the "world function" [247].

Provided the geodesic from x to y is not lightlike, one has

$$\begin{aligned} \nabla_\mu^x \sigma_\gamma(x, y) &= \pm s_\gamma(x, y) \, \nabla_\mu^x s_\gamma(x, y) \\ &= +s_\gamma(x, y) \, t_\mu(x; \gamma; x \leftarrow y) \\ &= +\sqrt{2|\sigma_\gamma(x, y)|} \; t_\mu(x \leftarrow y). \end{aligned} \qquad (21.8)$$

Here

$$t^\mu(x; \gamma; x \leftarrow y) \equiv \pm g^{\mu\nu} \, \nabla_\nu^x s_\gamma(x, y) \qquad (21.9)$$

denotes the unit tangent vector at the point x pointing along the geodesic γ away from the point y. When no confusion results we may abbreviate this by $t^\mu(x \leftarrow y)$ or even $t^\mu(x)$.

If the geodesic from x to y is lightlike things are somewhat messier. One easily sees that for lightlike geodesics $\nabla_\mu^x \sigma_\gamma(x, y)$ is a null vector. To proceed further one must introduce a canonical observer, described by a unit timelike vector V^μ at the point x. By parallel transporting this canonical observer along the geodesic (that is, $(\nabla_\mu \sigma_\gamma)\nabla^\mu V^\nu = 0$), one can set up a canonical frame that picks out a particular canonical affine parameter:

$$\nabla_\mu^x \sigma_\gamma(x, y) = +\zeta_\gamma(x, y) \, l_\mu(x; \gamma; x \leftarrow y), \qquad (21.10)$$

$$l_\mu(x; \gamma; x \leftarrow y)\nabla^\mu V^\nu(x) = 0, \qquad (21.11)$$

$$l_\mu(x; \gamma; x \leftarrow y) \, V^\mu(x) = -1, \qquad (21.12)$$

$$\zeta_\gamma(x, y) = -V^\mu(x)\nabla_\mu^x \sigma_\gamma(x, y). \qquad (21.13)$$

This affine parameter ζ can, crudely, be thought of as a distance along the null geodesic as measured by a canonical observer with four-velocity V^μ. By combining the null vector l^μ with the timelike vector V^μ, one can construct a second canonical null vector: $m^\mu \equiv 2V^\mu - l^\mu$. A trivial calculation shows

$$m_\mu(x)\, V^\mu(x) \;=\; -1; \tag{21.14}$$

$$l_\mu(x)\, m^\mu(x) \;=\; -2. \tag{21.15}$$

More interestingly

$$\nabla^z_\mu \zeta_\gamma(x,y) = -\frac{1}{2}\, l_\mu(x; \gamma; x \leftarrow y). \tag{21.16}$$

Unfortunately, while spacelike and timelike geodesics can be treated in a unified formalism as alluded to above, the subtleties involved with lightlike geodesics will require the presentation of several tedious variations on the general analysis.

Exercise 1: Prove that ζ really does provide an affine parameterization for the null geodesic γ. Show that $l^\mu \nabla_\mu l^\nu = 0$. Hint: we already know that $l^\mu \nabla_\mu l^\nu = f l^\nu$, that $l^\mu \nabla_\mu V^\nu = 0$, and that $l_\mu V^\mu = -1$. Calculate f. Finally, show that $l^\mu \nabla_\mu m^\nu = 0$.

Exercise 2: Prove equation (21.16). Hint: Show that $l^\mu \nabla_\mu \zeta = 0$. Show that $m^\mu \nabla_\mu \zeta = 1$.

Exercise 3: (Easy) Verify that all these formulas are correct in flat Minkowski spacetime.

21.3   Specific definition: Geodesic flow

Consider geodesic flow in a Lorentzian spacetime. In the present context the (scalarized) van Vleck determinant is defined by

$$\Delta_\gamma(x,y) \equiv (-1)^d \, \frac{\det\{\nabla^z_\mu \nabla^y_\nu \sigma_\gamma(x,y)\}}{\sqrt{g(x)g(y)}}. \tag{21.17}$$

This definition has a nice interpretation in terms of the Jacobian associated with the change of variables from $(x, \vec{t})$ to (x, y). One may specify a geodesic either by (1) specifying a single point on the geodesic x and the tangent vector $\vec{t}$ at that point; or by (2) specifying two separate points, (x, y), on the geodesic. (This assumes that the two points are connected by only one geodesic. This will certainly be the case if the two points are sufficiently close to each other.) The (scalarized) Jacobian associated with this change of variables is

$$J_\gamma(x,y) \equiv \frac{\det\left\{\partial(x,\vec{t}\,)/\partial(x,y)\right\}}{\sqrt{g(x)g(y)}} = \frac{\det'\left\{\partial\vec{t}/\partial y\right\}}{\sqrt{g(x)g(y)}}. \tag{21.18}$$

Here the det$'$ indicates the fact that we should ignore the known trivial zero eigenvalue(s) in determining this Jacobian.

If the points x and y are spacelike or timelike separated, then the one trivial zero arises from the fact that the tangent vector is normalized, $t^\mu t_\mu = \pm 1$, so that

$$t^\mu(x)\,\nabla_\nu^y t_\mu(x) = 0. \tag{21.19}$$

To see the connection with the van Vleck determinant, observe

$$\begin{aligned}
\nabla_\mu^x \nabla_\nu^y \sigma_\gamma(x,y) &= \nabla_\nu^y\left[s_\gamma(x,y)\,t_\mu(x;\gamma;x\leftarrow y)\right]\\
&= s_\gamma(x,y)\,\nabla_\nu^y t_\mu(x;\gamma;x\leftarrow y)\pm t_\nu(y)\,t_\mu(x). \tag{21.20}
\end{aligned}$$

By adopting suitable coordinates at x, and independently at y, one can make the tangent vectors $t_\mu(x)$ and $t_\mu(y)$ both lie in the "t" direction (if they are timelike) or both lie in the "z" direction (if they are spacelike). Then by inspection

$$\begin{aligned}
\Delta_\gamma(x,y) &= \pm[-s_\gamma(x,y)]^d\,\frac{\det'\{\nabla_\nu^y t_\mu(x;\gamma;x\leftarrow y)\}}{\sqrt{g(x)g(y)}}\\
&= \pm[-s_\gamma(x,y)]^d\,J_\gamma(x,y). \tag{21.21}
\end{aligned}$$

If the points x and y are lightlike separated, then the two trivial zeros arise from the fact that the tangent vector l^μ satisfies two constraints:

$$l^\mu l_\mu = 0 \quad \text{and} \quad l^\mu V_\mu = -1. \tag{21.22}$$

Thus both

$$l^\mu(x)\,\nabla_\nu^y t_\mu(x) = 0 \quad \text{and} \quad V^\mu(x)\,\nabla_\nu^y t_\mu(x) = 0. \tag{21.23}$$

To precisely determine the connection with the van Vleck determinant requires some subtlety. For a point x close to, but not quite on, a null geodesic emanating from y one may usefully decompose the gradient of the geodetic interval as

$$\begin{aligned}
\nabla_\mu^x \sigma_\gamma(x,y) &= +\zeta_\gamma(x,y)\,l_\mu(x;\gamma;x\leftarrow y)\\
&\quad +\xi_\gamma(x,y)\,m_\mu(x;\gamma;x\leftarrow y). \tag{21.24}
\end{aligned}$$

Here ζ and ξ are to be thought of as curvilinear null coordinates. They are the curved space generalizations of $u_\pm = (t \pm x)/2$. One is now in a position to calculate

$$
\begin{aligned}
\nabla_\mu^x \nabla_\nu^y \sigma_\gamma(x,y) &= \nabla_\nu^y [\zeta(x,y)\, l_\mu(x \leftarrow y) + \xi(x,y)\, m_\mu(x \leftarrow y)] \\
&= \zeta(x,y)\, \nabla_\nu^y l_\mu(x; \gamma; x \leftarrow y) \\
&\quad + \xi(x,y)\, \nabla_\nu^y m_\mu(x; \gamma; x \leftarrow y) \\
&\quad - \tfrac{1}{2} l_\nu(y)\, l_\mu(x) - \tfrac{1}{2} m_\nu(y)\, m_\mu(x).
\end{aligned}
\tag{21.25}
$$

Now go to the light cone, by setting $\xi = 0$. By adopting suitable coordinates at x one can arrange both:

$$
l_\mu(x) = (1,1,0,0) \quad \text{and} \quad m_\mu(x) = (1,-1,0,0). \tag{21.26}
$$

Independently, one can arrange the same to be true at y. Finally, careful inspection of the above reveals

$$
\begin{aligned}
\Delta_\gamma(x,y) &= (-1)^d [\zeta_\gamma(x,y)]^{(d-1)} \frac{\det'\{\nabla_\nu^y l_\mu(x; \gamma; x \leftarrow y)\}}{\sqrt{g(x)g(y)}} \\
&= (-1)^d [\zeta_\gamma(x,y)]^{(d-1)}\, J_\gamma(x,y).
\end{aligned}
\tag{21.27}
$$

21.4   Elementary results

In view of the equation

$$
\nabla_\mu^x \sigma_\gamma(x,y) = s_\gamma(x,y)\, t_\mu(x; \gamma; x \leftarrow y), \tag{21.28}
$$

one has as an exact result (valid also on the light cone)

$$
\sigma_\gamma(x,y) = \frac{1}{2} g^{\mu\nu}\, \nabla_\mu^x \sigma_\gamma(x,y)\, \nabla_\nu^x \sigma_\gamma(x,y). \tag{21.29}
$$

This is often called the "Hamilton–Jacobi equation" for the geodetic interval.

Repeated differentiations and contractions result in what may be termed the "master equation" governing the van Vleck determinant

$$
\nabla_x^\mu [\Delta_\gamma(x,y)\, \nabla_\mu^x \sigma_\gamma(x,y)] = (d+1)\Delta_\gamma(x,y). \tag{21.30}
$$

See, for instance [68, 59, 64, 66].

Exercise: Derive this master equation. Technical point: Remember that Δ is a scalar, but $D = \det\{\nabla\nabla\sigma\}$ is a scalar density of weight -1 [in fact, a scalar bi-density of weight $(-1,-1)$]. The covariant derivative of D is not what you might naively think it to be. Define a matrix $m = \{\nabla\nabla\sigma\}$. Show that

$$
\nabla_\mu D \;=\; \partial_\mu D - \frac{1}{\sqrt{g}}(\partial_\mu\sqrt{g})D \tag{21.31}
$$

$$
=\; \left(\mathrm{tr}[m^{-1}\partial_\mu m] - \frac{1}{\sqrt{g}}(\partial_\mu\sqrt{g}) \right) D \tag{21.32}
$$

$$
=\; \mathrm{tr}[m^{-1}\nabla_\mu m]D. \tag{21.33}
$$

Now show that

$$
\nabla_\mu \Delta = \mathrm{tr}[m^{-1}\nabla_\mu m]\Delta. \tag{21.34}
$$

This step is in fact trivial—the definition of covariant derivative for a tensor density is specifically designed to ensure that $\nabla\sqrt{g} = 0$. The derivation of the master equation is now a straightforward exercise in index gymnastics. Hint: Evaluate $(\nabla_\mu\Delta)\,(\nabla^\mu\sigma)$.

Note that the master equation is valid both on and off the light cone and may be rewritten as

$$
\nabla_x^\mu \left[\frac{\Delta_\gamma(x,y)\,\nabla_\mu^x\sigma_\gamma(x,y)}{\sigma_\gamma(x,y)^{(d+1)/2}} \right] = 0. \tag{21.35}
$$

Equivalently

$$
\nabla_x^\mu \left[\frac{\Delta_\gamma(x,y)\,t_\mu(x;\gamma;x\leftarrow y)}{s_\gamma(x,y)^d} \right] = 0, \tag{21.36}
$$

or even

$$
\nabla_x^\mu \left[\Delta_\gamma(x,y)\,\nabla_\mu^x \frac{1}{s_\gamma(x,y)^{d-1}} \right] = 0. \tag{21.37}
$$

Further manipulations based on this master equation yield additional insight into the physical significance of the van Vleck determinant. If one makes a rather strong geometrical assumption (ultrastatic spacetime) the van Vleck determinant can, in $(3+1)$ dimensions, be directly interpreted in terms of geometrically induced deviations from the inverse-square law. In more general spacetimes, the van Vleck determinant may be related to the expansion of geodesic flows. Particularly simple results hold in $(1+1)$ dimensions.

21.4.1 Ultrastatic spacetimes

Consider an ultrastatic spacetime, that is, a spacetime wherein the metric is simply given by

$$g = -dt \otimes dt + g_{ij} \, dx^i \otimes dx^j. \tag{21.38}$$

Geodesics in the spacetime can then be projected down onto geodesics on any one of the spatial slices. Conversely, geodesics on any of the spatial slices can be promoted to full spacetime geodesics. The geodetic interval simplifies to

$$\sigma_\gamma(x, y) = -\frac{1}{2} \left\{ (t - t')^2 - \tilde{s}(\vec{x}, \vec{y})^2 \right\}. \tag{21.39}$$

Here we have set $x = (t, \vec{x})$, while $y = (t', \vec{y})$. Furthermore $\tilde{s}(\vec{x}, \vec{y})$ denotes the d-dimensional distance from $\vec{x}$ to $\vec{y}$ as determined by the d-dimensional Euclidean metric g_{ij}.

The spacetime tangent vector t^μ simplifies to

$$t^\mu(x; \gamma; x \leftarrow y) = -\frac{1}{s}(t - t'; \, \tilde{s} \, \hat{t}^i) \tag{21.40}$$

Here $\hat{t}$ is the d-dimensional unit tangent vector to the spatial projection of the spacetime geodesic.

Exercise: (Trivial) Check that $t^\mu t_\mu = \pm 1$.

The master equation simplifies to

$$\nabla^x_\mu \left[\frac{\Delta_\gamma(x, y) \, (t - t'; \, \tilde{s} \, \hat{t}^i)}{([t - t']^2 - \tilde{s}^2)^{(d+1)/2}} \right] = 0. \tag{21.41}$$

Explicitly separate out the spatial and temporal derivatives

$$\vec{\nabla}_i \left[\frac{\Delta_\gamma(x, y) \, \tilde{s} \, \hat{t}^i}{([t - t']^2 - \tilde{s}^2)^{(d+1)/2}} \right] + \frac{\partial}{\partial t} \left[\frac{\Delta_\gamma(x, y) \, [t - t']}{([t - t']^2 - \tilde{s}^2)^{(d+1)/2}} \right] = 0. \tag{21.42}$$

Perform the time derivative, and evaluate the result at $t = t'$. Then

$$\vec{\nabla}_i \left[\frac{\Delta_\gamma(\vec{x}, \vec{y}) \, \hat{t}^i}{\tilde{s}^d} \right] + \frac{\Delta_\gamma(\vec{x}, \vec{y})}{\tilde{s}^{(d+1)}} = 0. \tag{21.43}$$

Finally, this gives

$$\vec{\nabla}_i \left[\frac{\Delta_\gamma(\vec{x}, \vec{y}) \, \hat{t}^i}{\tilde{s}^{d-1}} \right] = 0. \tag{21.44}$$

This is precisely the statement that the spatial vector

$$J_i = \frac{\Delta_\gamma(\vec{x}, \vec{y}) \, \hat{t}_i}{\tilde{s}^{d-1}} \propto \Delta_\gamma(\vec{x}, \vec{y}) \, \nabla_i \tilde{s}^{2-d} \tag{21.45}$$

represents a conserved spatial flux. The magnitude of the flux vector is

$$f = \|\vec{J}\| = \frac{\Delta_\gamma(\vec{x}, \vec{y})}{\tilde{s}^{d-1}}, \tag{21.46}$$

and one deduces

Theorem 38 *In a $(d+1)$-dimensional ultrastatic spacetime the equal time van Vleck determinant measures the extent to which geometry modifies the inverse-$(d-1)$-power law for flux falloff.*

Special case:

Theorem 39 *In a $(3+1)$-dimensional ultrastatic spacetime the equal time van Vleck determinant measures the extent to which geometry modifies the inverse-square law for flux falloff.*

A particularly down-to-earth and concrete physical model exhibiting this behavior is provided by geometric optics. If $n(\vec{x})$ is the refractive index at the point $\vec{x}$, then Fermat's principle assures us that the paths of light rays are given by extremizing the optical distance

$$\tilde{s}_\gamma(\vec{x}, \vec{y}) \equiv \int_{\vec{x}}^{\vec{y}} n(\vec{z}) \, \|d\vec{z}\|. \tag{21.47}$$

Fermat's principle is equivalent to the statement that light rays follow geodesics of the optical metric

$$g = n^2(\vec{x}) \, [dx \otimes dx + dy \otimes dy + dz \otimes dz]. \tag{21.48}$$

Construct the van Vleck determinant appropriate to this optical distance. That is,

$$\Delta_\gamma(\vec{x}, \vec{y}) \equiv \frac{\det\{\frac{1}{2} \frac{\partial}{\partial \vec{x}} \frac{\partial}{\partial \vec{y}} \tilde{s}_\gamma(\vec{x}, \vec{y})^2\}}{n(\vec{x})^3 \, n(\vec{y})^3}. \tag{21.49}$$

Then

Theorem 40 *In geometric optics, if a point source of luminosity L is located at the point $\vec{x}$, then (with areas expressed in terms of the optical metric) the flux of light past the point $\vec{y}$ is*

$$f = \|J\| = \frac{L \, \Delta_\gamma(\vec{x}, \vec{y})}{4\pi \, \tilde{s}_\gamma(\vec{x}, \vec{y})^2}. \tag{21.50}$$

Furthermore, poles of the van Vleck determinant correspond to foci of the optical system.

Theorem 41 *With areas expressed in terms of the physical metric, the flux of light past the point $\vec{y}$ is*

$$f_{\text{physical}} = \frac{L \, \Delta_\gamma(\vec{x}, \vec{y}) \, n(\vec{y})^2}{4\pi \, \tilde{s}_\gamma(\vec{x}, \vec{y})^2}. \tag{21.51}$$

Exercise 1: Derive these results from the general discussion on p. 303.

Exercise 2: (Easy) Check these results by considering the special case where the refractive index is independent of position. Calculate the van Vleck determinant. (It's trivial.) Now calculate the flux, both when cross-sectional areas are measured using the optical metric and when areas are measured using the physical metric. Check that these results are physically reasonable.

Exercise 3: Check these results in general by placing a small sphere around the point x and calculating the total flux through this sphere.

In more general spacetimes the particularly simple derivations and interpretations given above fail. Nevertheless, it is still possible to use more complicated analyses to relate the van Vleck determinant to deviations from the inverse square law.

21.4.2 Non-null separation

In a general spacetime, assuming that x and y are either timelike separated or spacelike separated, the master equation may be rewritten as

$$\nabla_x^\mu [\Delta_\gamma(x,y)\, s_\gamma(x,y)\, t_\mu(x)] = (d+1)\Delta_\gamma(x,y). \qquad (21.52)$$

Use of the Leibnitz rule leads to

$$s_\gamma(x,y)\, t_\mu(x)\, \nabla_x^\mu \Delta_\gamma(x,y) \pm t_\mu(x)\, t^\mu(x)\, \Delta_\gamma(x,y)$$
$$+ s_\gamma(x,y)\, \Delta_\gamma(x,y)\, \nabla_x{}^\mu t_\mu(x) = (d+1)\Delta_\gamma(x,y). \quad (21.53)$$

One notes that $t^\mu(x)$ defines a normalized spacelike or timelike vector field. Thus

$$t_\mu(x)\, \nabla_x^\mu \Delta_\gamma(x,y) = \left(\frac{d}{s_\gamma(x,y)} - [\nabla_x{}^\mu t_\mu(x)] \right) \Delta_\gamma(x,y). \qquad (21.54)$$

Now define $\Delta_\gamma(s)$ to be the van Vleck determinant calculated at a proper distance s along the geodesic γ in the direction away from y and towards x. Note the boundary condition that $\Delta_\gamma|_{(s=0)} \equiv 1$. Also, define

$$t^\mu(x)\, \nabla_\mu^x f \equiv \frac{df}{ds}. \qquad (21.55)$$

One finally obtains a first-order differential equation governing the evolution of the van Vleck determinant

$$\frac{d\Delta_\gamma(s)}{ds} = \left(\frac{d}{s} - \theta \right) \Delta_\gamma(s). \qquad (21.56)$$

Here θ is the expansion of the geodesic spray defined by the integral curves of $t^\mu(x)$. That is, $\theta \equiv \nabla_\mu t^\mu$. Direct integration yields

$$\Delta_\gamma(x,y) = s_\gamma(x,y)^d \exp\left(-\int_\gamma \theta \, ds\right). \tag{21.57}$$

The integration is to be taken along the geodesic γ from the point y to the point x. The interpretation of this result is straightforward: take a geodesic spray of trajectories emanating from the point y. If they were propagating in flat space, then after a proper distance s the (relative) transverse density of trajectories would have fallen to s^{-d}. Since they are not propagating in flat spacetime the actual (relative) transverse density of trajectories is given by the exponential of minus the integrated expansion. The van Vleck determinant is then the ratio between the actual density of trajectories and the anticipated flat space result. Note that by explicit calculation, the van Vleck determinant is completely symmetric in x and y.

21.4.3 Null separation

In general spacetimes, an analogous development holds if the points x and y are connected by a null geodesic. The details are, as one has by now grown to expect, somewhat tedious. One returns to equation (21.30) and uses equation (21.24) to derive

$$\nabla_x^\mu \left[\Delta_\gamma(x,y)\{\zeta \, l_\mu(x) + \xi \, m_\mu(x)\}\right] = (d+1)\Delta_\gamma(x,y). \tag{21.58}$$

Use of the Leibnitz rule, followed by the limit $\xi \to 0$, now leads to

$$\zeta \, l_\mu(x) \, \nabla_x^\mu \Delta_\gamma(x,y) - l_\mu(x) \, m^\mu(x) \, \Delta_\gamma(x,y)$$
$$+\zeta \, \Delta_\gamma(x,y) \, \nabla_x{}^\mu l_\mu(x) = (d+1)\Delta_\gamma(x,y). \tag{21.59}$$

One recalls that, by definition, $l^\mu(x) \, m_\mu(x) = -2$. Adopting suitable definitions, analogous to the non-null case, one derives

$$\frac{d\Delta_\gamma(\zeta)}{d\zeta} = \left(\frac{[d-1]}{\zeta} - \hat\theta\right)\Delta_\gamma(\zeta). \tag{21.60}$$

Direct integration yields

$$\Delta_\gamma(x,y) = \zeta_\gamma(x,y)^{[d-1]} \exp\left(-\int_\gamma \hat\theta \, d\zeta\right). \tag{21.61}$$

In particular, note that while the affine parameter ζ, and the expansion $\hat\theta \equiv \nabla_\mu l^\mu$ both depend on the canonical observer V^μ, the overall combination is independent of this choice. The appearance of the exponent $d-1$ for null geodesics, as contrasted to the exponent d for spacelike or timelike geodesics might at first be somewhat surprising. The ultimate reason for this is that in $(d+1)$-dimensional spacetime the set of null geodesics emanating from a point x sweeps out a d-dimensional submanifold. As one moves away from x an affine distance ζ into this submanifold the relative density of null geodesics falls as ζ^{1-d}. In $(3+1)$-dimensional spacetime, this is just the inverse-square law for luminosity.

The virtue of this tedious but elementary analysis is that one now has a unified framework applicable in all generality to cases of timelike, lightlike, or spacelike separated points x and y.

Special case: $(1+1)$ dimensions

A special result holds along null geodesics in $(1+1)$ dimensions. This simplification can be derived from the fact that all two-dimensional metrics are locally conformally flat. (A more formal derivation of this result will be possible once we further develop the formalism.) That is, in $(1+1)$ dimensions one can always locally choose coordinates in such a manner that

$$g = \Omega(t,x)^2 \left[-dt \otimes dt + dx \otimes dx\right]. \tag{21.62}$$

Since null geodesics are insensitive to overall conformal factors this implies that the null geodesics are simply $x = t$ and $x = -t$. Choose the canonical observer to be described by the vector $V = \partial/\partial t$. In coordinates, $V^\mu = (1,0)$. The affine parameter ζ is thus simply t itself. The normalized null tangent vector is

$$l^\mu = \Omega^2(1,\pm 1), \qquad \text{so that} \qquad \hat{\theta} \equiv \nabla_\mu l^\mu = 0. \tag{21.63}$$

Combining this with the general result one has

Lemma 6 *In $(1+1)$ dimensions, along any null geodesic γ,*

$$\Delta_\gamma(x,y) = 1. \tag{21.64}$$

Of course, this is nothing more than the statement that the local conformal flatness of $(1+1)$-dimensional manifolds implies that null geodesics are neither focused nor defocused .

21.5   Inequalities

Now that one has these general formulas for the van Vleck determinant, powerful inequalities can be derived by applying the Raychaudhuri equation and imposing suitable convergence conditions. Restricting attention to vorticity-free (spacelike or timelike) geodesic flows, the Raychaudhuri equation particularizes to

$$\frac{d\theta}{ds} = -\left(R_{\mu\nu}t^\mu t^\nu\right) - 2\sigma^2 - \frac{\theta^2}{d}. \tag{21.65}$$

For a concrete textbook reference, this is a special case of equation (4.26) of Hawking and Ellis [132], generalized to arbitrary dimensionality. While Hawking and Ellis are discussing timelike geodesics, the choice of notation in the present monograph guarantees that the discussion can be carried over to spacelike geodesics without alteration. The quantity σ denotes the shear of the geodesic congruence, and σ^2 is guaranteed to be positive semi-definite.

Reminder: Do not confuse the shear σ with the geodetic interval σ_γ. The notation is standard, but potentially confusing.

Definition 35 *A Lorentzian spacetime is said to satisfy the timelike, null, or spacelike convergence condition if for all timelike, null, or spacelike vectors t^μ:*

$$\left(R_{\mu\nu}t^\mu t^\nu\right) \geq 0. \tag{21.66}$$

By using the Einstein field equations, these convergence conditions can be related to the energy conditions. More details will be provided below.

Definition 36 *A Riemannian manifold is said to possess semi-positive Ricci curvature if for all vectors t^μ:*

$$\left(R_{\mu\nu}t^\mu t^\nu\right) \geq 0. \tag{21.67}$$

If the points x and y are (timelike/spacelike) separated, and the spacetime satisfies the (timelike/spacelike) convergence condition, then from the Raychaudhuri equation one may deduce the inequality

$$\frac{d\theta}{ds} \geq -\frac{\theta^2}{d}. \tag{21.68}$$

This inequality is immediately integrable:

$$\frac{1}{\theta(s)} \geq \frac{1}{\theta(0)} + \frac{s}{d}. \tag{21.69}$$

For a geodesic spray, one has $\theta(s) \to d/s$ as $s \to 0$. Therefore $[1/\theta(0)] = 0$. One has thus derived an inequality, valid for geodesic sprays, under the assumed convergence conditions

$$\theta(s) \leq +\frac{d}{s}. \tag{21.70}$$

This implies that the van Vleck determinant is a monotone function of arc length (that is, $d\Delta/ds \geq 0$). This inequality immediately integrates to an inequality on the determinant itself: $\Delta_\gamma(x,y) \geq 1$.

Now, for null geodesic flows the Raychaudhuri equation becomes

$$\frac{d\hat\theta}{d\zeta} = -(R_{\mu\nu}l^\mu l^\nu) - 2\hat\sigma^2 - \frac{\hat\theta^2}{(d-1)}. \tag{21.71}$$

(See equation (4.35) of Hawking and Ellis [132].) If the spacetime satisfies the null convergence condition one can deduce the inequality

$$\frac{d\hat\theta}{d\zeta} \geq -\frac{\hat\theta^2}{d-1}. \tag{21.72}$$

Upon integration and use of the relevant boundary condition, one derives in a straightforward manner the inequality

$$\hat\theta(\zeta) \leq +\frac{(d-1)}{\zeta}, \tag{21.73}$$

which implies that the van Vleck determinant is a monotone function of the null affine parameter, $d\Delta/d\zeta \geq 0$, and again immediately integrates to an inequality on the determinant itself: $\Delta_\gamma(x,y) \geq 1$. One is now in a position to enunciate the following results.

Theorem 42 *In any Lorentzian spacetime, if the points x and y are timelike, null, or spacelike separated, and the spacetime satisfies the timelike, null, or spacelike convergence condition, then the van Vleck determinant is a monotone function of affine parameter and is bounded from below: $\Delta_\gamma(x,y) \geq 1$.*

Theorem 43 *In any Riemannian manifold of semi-positive Ricci curvature, the van Vleck determinant is a monotone function of affine parameter and is bounded from below: $\Delta_\gamma(x,y) \geq 1$.*

To see the physical import of the (timelike/null/spacelike) convergence conditions, consider a type one stress-energy tensor (type I, see p. 115). The cosmological constant, if present, is taken to be subsumed into the definition of the stress-energy. Then

$$T_{\mu\nu} = \text{diag}[\rho;\, p_i] \equiv \begin{bmatrix} \rho & 0 & \cdots & 0 \\ 0 & p_1 & \cdots & 0 \\ \vdots & \vdots & \ddots & \vdots \\ 0 & 0 & \cdots & p_d \end{bmatrix}, \tag{21.74}$$

where the p_i are the d principal pressures. From the Einstein equations, $R_{\mu\nu} - \frac{1}{2}Rg_{\mu\nu} = 8\pi G\, T_{\mu\nu}$, one deduces

$$R_{\mu\nu} = 4\pi G \, \text{diag}\left[\rho + \sum_j p_j;\ \rho + 2p_i - \sum_j p_j\right]. \tag{21.75}$$

- The timelike convergence condition implies:

$$(1) \quad \rho + \sum_j p_j \geq 0; \qquad (2) \quad \forall i, \; \rho + p_i \geq 0. \tag{21.76}$$

This is equivalent to the strong energy condition (SEC).

- The null convergence condition implies

$$\forall i, \; \rho + p_i \geq 0. \tag{21.77}$$

This is just the null energy condition (NEC). (The WEC would require the additional constraint $\rho \geq 0$.)

- The spacelike convergence condition implies

$$(1) \quad \forall i, \; \rho + 2p_i - \sum_j p_j \geq 0; \qquad (2) \quad \forall i, \; \rho + p_i \geq 0. \tag{21.78}$$

This is a rather strong constraint that is not equivalent to any of the standard energy conditions. Particularize to $(3+1)$ dimensions, then the constraints (1) can be made more explicit as

$$\rho + p_1 - p_2 - p_3 \geq 0; \quad \rho + p_2 - p_3 - p_1 \geq 0; \quad \rho + p_3 - p_1 - p_2 \geq 0. \tag{21.79}$$

By pairwise addition of these constraints one may deduce

$$\forall i, \; \rho - p_i \geq 0. \tag{21.80}$$

Combining this with constraints (2) one deduces $\rho \geq 0$, and so

$$\forall i, \; p_i \in [-\rho, +\rho]. \tag{21.81}$$

Thus the spacelike convergence condition implies the dominant energy condition (DEC), which in turn implies the weak energy condition (WEC) and the null energy condition (NEC). The implication does not hold in the reverse direction. In particular, for an electromagnetic field of zero Poynting flux,

$$T_{\mu\nu} = \rho \cdot \mathrm{diag}[+1; \; -1; \; +1; \; +1] = \rho \cdot \begin{bmatrix} +1 & 0 & 0 & 0 \\ 0 & -1 & 0 & 0 \\ 0 & 0 & +1 & 0 \\ 0 & 0 & 0 & +1 \end{bmatrix}. \tag{21.82}$$

This stress-energy tensor satisfies the dominant energy condition but not the spacelike convergence condition. A canonical example of a stress-energy tensor that *does* satisfy the spacelike convergence condition is a perfect fluid that satisfies the dominant energy condition

$$\rho + p \geq 0; \qquad \rho - p \geq 0. \tag{21.83}$$

Note that the timelike and spacelike convergence conditions separately imply the null convergence condition.

The physical reason behind the inequality satisfied by the van Vleck determinant is now manifest: "ordinary" matter produces an attractive gravitational field. An attractive gravitational field focuses geodesics, so that they do not spread out as much as they would in flat space. Then the van Vleck determinant is bounded from below, and in fact continues to grow as one moves along any geodesic.

One should remember however, that spacetimes containing traversable wormholes violate all of these convergence conditions. For the purposes of this book this makes the inequalities derived here somewhat less useful than they might otherwise be, and suggests that it would be useful to explore the possibility of deriving more general inequalities based on weaker energy conditions. Indeed, we shall soon see that in spacetimes containing traversable wormholes the required presence of "exotic" matter leads to van Vleck determinants that are arbitrarily close to zero [164, 178, 272].

Research Problem 18 *Find inequalities constraining the van Vleck determinant based on various weakened versions of the convergence conditions. For instance, consider the averaged null energy condition.*

21.6   Reformulation

21.6.1   Tidal focusing

Direct computations using the preceding results are unfortunately rather difficult. To get a better handle on the van Vleck determinant I shall first apparently make the problem more complicated by deriving a second-order differential equation for the van Vleck determinant. By going to the second-order formalism it is possible to relate the van Vleck determinant directly to the tidal forces induced by the full Riemann tensor—more specifically to the focusing and defocusing effects induced by tidal forces.

The most direct route to this end is to pick up a standard reference such as Hawking and Ellis [132]. Equation (4.20) on p. 83 implies that

$$\int \theta \, ds = \ln \det[A^{\mu}{}_{\nu}(s)]. \qquad (21.84)$$

Here $A^{\mu}{}_{\nu}(s)$ is the $d \times d$ matrix describing the evolution of the separation of infinitesimally nearby geodesics. This formalism may also be reformulated in terms of the Jacobi fields associated with the geodesic γ. In terms of A

$$\Delta_{\gamma}(x, y) = s_{\gamma}(x, y)^d \, \det[A_d^{-1}] = \det[s_{\gamma}(x, y) A_d^{-1}]. \qquad (21.85)$$

Note that the $d \times d$ matrix $A^\mu{}_\nu(s)$ may equivalently be replaced by a $(d+1) \times (d+1)$ matrix that is trivial in the extra entries:

$$A_{(d+1)} = A_d \oplus s. \tag{21.86}$$

Then

$$\Delta_\gamma(x,y) = \det[s_\gamma(x,y)A^{-1}], \tag{21.87}$$

where the determinant can now be taken in the sense of either a $d \times d$ or a $(d+1) \times (d+1)$ matrix.

Aside: (mainly of interest to quantum field theorists). An alternative, more explicit, but also more tedious route to establishing the preceding equation is to define the object

$$\tilde{A}(s) \equiv P \exp\left\{ \int_\gamma (\vec{\nabla} \otimes \vec{t}) ds \right\}. \tag{21.88}$$

Here the symbol P denotes the path-ordering process, a generalization of the notion of time-ordering, applied now to finite matrices. By definition of path-ordering

$$\frac{d\tilde{A}}{ds} = (\vec{\nabla} \otimes \vec{t}) \cdot P \exp\{ \int_\gamma (\vec{\nabla} \otimes \vec{t}) ds \}. \tag{21.89}$$

On the other hand, $\tilde{A}(0) \equiv I$. Comparison with equation (4.10) of Hawking and Ellis now shows that $\tilde{A}(s) \equiv A(s)^\mu{}_\nu \, \partial_\mu \otimes dx^\nu$. Finally note that

$$\det \tilde{A} = \exp\left\{ \int_\gamma \text{tr}(\vec{\nabla} \otimes \vec{t}) ds \right\} = \exp\left\{ \int_\gamma \theta \, ds \right\}. \tag{21.90}$$

This reformulates the van Vleck determinant in terms of a path-ordered exponential

$$\Delta_\gamma(x,y) = s_\gamma(x,y)^d \det\left[P \exp\left\{ -\int_\gamma (\vec{\nabla} \otimes \vec{t}) ds \right\} \right]. \tag{21.91}$$

This observation serves to illustrate formal similarities (and differences) between the van Vleck determinant and the Wilson loop variables of gauge theories.

Returning to the general analysis, by taking a double derivative with respect to arc length the matrix $A(s)$ may be shown to satisfy the second-order differential equation[1]

$$\frac{d^2}{ds^2} A^\mu{}_\nu(s) = - \left(R^\mu{}_{\alpha\beta} t^\alpha t^\beta \right) A^\sigma{}_\nu. \tag{21.92}$$

[1] Technical point: The indices actually refer to components as measured using an orthonormal frame that is Fermi–Walker transported along the geodesic γ.

See Hawking and Ellis [132], equation (4.21) on p. 83, particularized to a geodesic flow. The boundary condition on $A(s)$ is that $A^\mu{}_\nu(s) \to s\delta^\mu{}_\nu$ as $s \to 0$, this being the flat-space result for a geodesic spray. This is the promised tidal formulation for the van Vleck determinant.

A particularly pleasant feature of the tidal reformulation is that the case of null geodesics can be handled without too much special case fiddling. The analog of the tidal equation is

$$\frac{d^2}{d\zeta^2} \hat{A}^\mu{}_\nu(s) = - \left(R^\mu{}_{\alpha\sigma\beta} \, l^\alpha l^\beta \right) \hat{A}^\sigma{}_\nu. \tag{21.93}$$

See Hawking and Ellis [132], equation (4.33) on p. 88. Note that two of the eigenvectors of $\hat{A}$ suffer trivial evolution. This is a consequence of the two normalization conditions on the null tangent vector. Thus $\hat{A}$ can be thought of as either a $(d+1) \times (d+1)$ matrix with two trivial entries, or as a reduced $(d-1) \times (d-1)$ matrix. The formulations are related by

$$\hat{A}_{(d+1)} = \hat{A}_{(d-1)} \oplus \zeta I_2. \tag{21.94}$$

Other results can be simply transcribed as needed.

Special case: $(1+1)$ dimensions

The special results previously discussed for $(1+1)$ dimensions can now be generalized. For null geodesics one automatically has

$$\hat{A}_{(1+1)} = \zeta \, I_2, \tag{21.95}$$

which is equivalent to

$$\Delta_\gamma(\zeta) = 1. \tag{21.96}$$

Note that the "tidal" differential equation reduces to

$$\frac{d^2}{d\zeta^2} \hat{A}^\mu{}_\nu(s) = 0. \tag{21.97}$$

For non-null geodesics one uses the special properties of the Riemann tensor in $(1+1)$ dimensions

$$R_{\alpha\beta\gamma\delta} = \frac{1}{2} R \left(g_{\alpha\gamma} \, g_{\beta\delta} - g_{\alpha\delta} \, g_{\beta\gamma} \right). \tag{21.98}$$

Thus

$$\frac{d^2 A^\mu{}_\nu(s)}{ds^2} = -\frac{1}{2} R \left(\delta^\mu{}_\sigma - t^\mu \, t_\sigma \right) A^\sigma{}_\nu. \tag{21.99}$$

But in $(1+1)$ dimensions the relationship

$$A_{(d+1)} = A_d \oplus s \tag{21.100}$$

implies that we can without further loss of generality write

$$A_{\mu\nu}(s) = A(s)\left(\delta_{\mu\nu} - t_\mu t_\nu\right) + s\, t_\mu\, t_\nu. \tag{21.101}$$

Then

$$\frac{d^2 A(s)}{ds^2} = -\frac{1}{2}R(s)A(s), \tag{21.102}$$

with the boundary condition that $A(s) \to s$ as $s \to 0$.

21.6.2 Formal solution

To solve the "tidal" differential equation (in a formal sense), introduce the one-dimensional retarded Green function

$$G_R(s_f, s_i) = \{s_f - s_i\}\Theta(s_f - s_i). \tag{21.103}$$

Here $\Theta(s)$ denotes the Heaviside step function. For notational convenience define

$$Q^\mu{}_\nu(s) = -\left(R^\mu{}_{\alpha\nu\beta}t^\alpha t^\beta\right). \tag{21.104}$$

Integration of the second-order "tidal" equation for Δ_γ, with attention to the imposed boundary condition, leads to the integral equation

$$A^\mu{}_\nu(s) = s\delta^\mu{}_\nu + \int_0^s G_R(s, s')Q^\mu{}_\sigma(s')A^\sigma{}_\nu(s')ds'. \tag{21.105}$$

More formally, one may suppress the explicit integration by regarding $G_R(s, s')$, multiplication by $Q(s)$, and multiplication by s, as functional operators. Then

$$(I - G_R Q)A = \{sI\}. \tag{21.106}$$

This has the formal solution

$$\begin{aligned} A &= (I - G_R Q)^{-1}\{sI\} \\ &= \left(I + [G_R Q] + [G_R Q]^2 + \cdots\right)\{sI\}. \end{aligned} \tag{21.107}$$

This formal solution may also be derived directly from equation (21.105) by continued iteration. To understand what these symbols mean, note that

$$[G_R Q]\{sI\} \equiv \int_0^s G_R(s, s')[Q^\mu{}_\nu(s')]s'ds' = \int_0^s (s - s')[Q^\mu{}_\nu(s')]s'ds'. \tag{21.108}$$

Similarly

$$[G_R Q]^2\{sI\} = \int_0^s ds' \int_o^{s'} ds''(s - s')[Q^\mu{}_\alpha(s')](s' - s'')[Q^\alpha{}_\nu(s'')]s''. \tag{21.109}$$

21.7 Approximations

21.7.1 Weak fields

These formal manipulations permit the derivation of a very nice weak-field approximation to the van Vleck determinant. Observe

$$
\begin{aligned}
\Delta_\gamma(s)^{-1} &= \det(s^{-1}A), \\
&= \exp \operatorname{tr} \ln\{s^{-1}(I - G_R Q)^{-1}s\} \\
&= \exp \operatorname{tr} \ln\left\{I + s^{-1}[G_R Q]s + s^{-1}[G_R Q]^2 s + \cdots\right\} \\
&= \exp \operatorname{tr}\left\{s^{-1}[G_R Q]s + O(Q^2)\right\}. \qquad (21.110)
\end{aligned}
$$

Now, recall that the determinant and trace in these formulas are to be taken in the sense of $(d+1) \times (d+1)$ matrices, so in terms of the Ricci tensor

$$
\operatorname{tr}[Q] = -\left(R_{\alpha\beta} t^\alpha t^\beta\right). \qquad (21.111)
$$

The $\operatorname{tr}[O(Q^2)]$ terms involve messy contractions depending quadratically on the Riemann tensor. In a weak field approximation one may neglect these higher-order terms and write

$$
\Delta_\gamma(s) = \exp\left\{-s^{-1}(G_R \operatorname{tr}[Q])s + O(Q^2)\right\}. \qquad (21.112)
$$

More explicitly

$$
\Delta_\gamma(x,y) = \exp\left(\frac{1}{s}\int_0^s (s - s')\left(R_{\alpha\beta} t^\alpha t^\beta\right) s'\, ds' + O([\text{Riemann}]^2)\right). \qquad (21.113)
$$

This approximation, though valid only for weak fields, has a very nice physical interpretation. The implications of constraints such as the (timelike/null/spacelike) convergence conditions can now be read off by inspection.

Another nice feature of the weak-field approximation is that the results for null geodesics can also be read off by inspection

$$
\Delta_\gamma(x,y) = \exp\left(\frac{1}{\zeta}\int_0^\zeta (\zeta - \zeta')\left(R_{\alpha\beta} l^\alpha l^\beta\right) \zeta'\, d\zeta' + O([\text{Riemann}]^2)\right). \qquad (21.114)
$$

Consider, for example, the effect of a low-density perfect fluid in an almost flat background geometry. The Ricci tensor is related to the density and pressure via the Einstein equations, with the result that for a null geodesic

$$
\Delta_\gamma(x,y) \approx \exp\left(8\pi G \int_0^\zeta [\rho + p]\frac{(\zeta - \zeta')\zeta'}{\zeta}\, d\zeta'\right). \qquad (21.115)
$$

Unsurprisingly, one sees that energy density and/or pressure located at the half-way point of the null geodesic is most effective in terms of focusing.

21.7.2 Short distances

Another result readily derivable from the formal solution (21.107) is a short distance approximation. Assume that x and y are close to each other, so that $s_\gamma(x, y)$ is small. Assume that the Riemann tensor does not fluctuate wildly along the geodesic γ. Then one may approximate the Ricci tensor by a constant, and explicitly perform the integration over arc length, keeping only the lowest-order term in $s_\gamma(x, y)$:

$$\Delta_\gamma(x, y) = 1 + \frac{1}{6} \left(R_{\alpha\beta} t^\alpha t^\beta \right) s_\gamma(x, y)^2 + O(s_\gamma(x, y)^3). \qquad (21.116)$$

This result can also be derived via a veritable orgy of index gymnastics and point-splitting techniques [68]; see equation (1.76) on p. 233.

By adopting Gaussian normal coordinates at x one may write

$$s_\gamma(x, y) \, t_\mu(x; \gamma; x \leftarrow y) = (x - y)^\mu + O(s^2) \qquad (21.117)$$

to yield

$$\Delta_\gamma(x, y) = 1 + \frac{1}{6} \left[R_{\alpha\beta}(x - y)^\alpha (x - y)^\beta \right] + O(s_\gamma(x, y)^3). \qquad (21.118)$$

In this form the result is blatantly applicable to null geodesics without further ado.

21.8 Asymptotics

To proceed beyond the weak-field approximation, consider an arbitrarily strong gravitational source in an asymptotically flat spacetime. Let the point y lie anywhere in the spacetime, possibly deep within the strongly gravitating region. Consider an otherwise arbitrary spacelike or timelike geodesic that reaches and remains in the asymptotically flat region.

Far from the source the metric is approximately flat and the Riemann tensor is of order $O(M/r^3)$. Using the tidal evolution equation, a double integration with respect to arc length provides the estimate

$$A^\mu{}_\nu(s) = (A_0)^\mu{}_\nu + s(B_0)^\mu{}_\nu + O(M/s). \qquad (21.119)$$

This estimate is valid for large values of s, where one is in the asymptotically flat region. Here (A_0) and (B_0) are constants that effectively summarize gross features of the otherwise messy strongly interacting region. For the van Vleck determinant

$$\Delta_\gamma(s)^{-1} = \det\{A/s\} = \det\{B_0 + (A_0/s) + O(1/s^2)\}. \qquad (21.120)$$

Thus the van Vleck determinant approaches a finite limit at spatial or temporal infinity, with $\Delta_\gamma(\infty) = \det\{B_0\}^{-1}$. Finally, define $J = (B_0)^{-1}(A_0)$, to obtain

$$\Delta_\gamma(s) = \frac{\Delta_\gamma(\infty)}{\det\{I + (J/s) + O(M/s^2)\}}. \tag{21.121}$$

This is our desired asymptotic estimate of the van Vleck determinant. Note that this asymptotic behavior depends on both complicated physics from the strongly interacting region, and on a term arising from the asymptotic far field. The asymptotic far field depends only on the overall mass of the system. This suggests an interesting possibility.

Research Problem 19 *Try to develop a version of the positive mass theorem that uses as input hypothesis some statement about the behavior of the van Vleck determinant near spatial infinity. That is, avoid direct use of any and all energy conditions. A modification of the Penrose–Sorkin–Woolgar [210] approach is what I have in mind.*

Chapter 22

van Vleck determinants: Wormholes

The technical machinery of the preceding chapter permits one to undertake extensive computations of the van Vleck determinant in traversable wormhole spacetimes—at least in the short-throat flat-space approximation. This chapter presents several such computations for various model spacetimes. Basic tools employed are the reformulation of the van Vleck determinant in terms of tidal focusing effects and consequent formal expansion for the van Vleck determinant in terms of the Riemann tensor. If the Riemann curvature is confined to relatively thin layers, then the extended thin-shell formalism permits reduction of this formal expansion to a finite number of terms. The number N of terms occurring in this expansion is just the number of times the closed geodesic γ intersects a layer of high curvature. Several variations on this theme are presented.

Warning: This chapter is of necessity rather turgid. On first reading one may wish to skip directly to the summary on p. 332.

22.1 First approximation

Consider a traversable wormhole: Assume that the throat of the wormhole is very short, and that curvature in the region outside the mouth of the wormhole is relatively weak. Such a wormhole can be idealized by considering Minkowski space with two regions excised, and then identifying the boundaries of those regions in some suitable manner. The Riemann tensor for such an idealized geometry is identically zero everywhere except at the wormhole mouths where the identification procedure takes place. Generi-

cally, there will be an infinitesimally thin layer of exotic matter present at
the mouth of the wormhole.

22.1.1 Preliminaries

Consider a geodesic that wraps through the mouth of the wormhole a total
of N times. From the preceding chapter we know that is is useful to consider
the quantity

$$Q^\mu{}_\nu(s) \equiv - \left(R^\mu{}_{\alpha\nu\beta} t^\alpha t^\beta \right) \qquad (22.1)$$

In the short-throat flat-space approximation the Riemann tensor is zero
except at the throat itself. Thus $Q(s)$ specializes to a sum of delta-function
contributions

$$Q^\mu{}_\nu(s) = \sum_{n=1}^{N} \delta(s - s_n)[q(n)^\mu{}_\nu]. \qquad (22.2)$$

At this stage, I am deliberately leaving the matrices $[q(n)^\mu{}_\nu]$ as general as
possible so as to obtain results that are to a large extent independent of the
particular details of the shape of the wormhole mouths and/or the identification procedure adopted at the wormhole mouths. Suitable specializations
will be introduced in due course.

This greatly simplifies the various terms in the formal expansion for the
van Vleck determinant. In particular

$$[G_R Q]\{sI\} = \sum_{n=1}^{N} (s - s_n)[q(n)^\mu{}_\nu]s_n. \qquad (22.3)$$

Furthermore

$$[G_R Q]^2\{sI\} = \sum_{n=2}^{N} \sum_{m=1}^{n-1} (s - s_n)[q(n)^\mu{}_\alpha](s_n - s_m)[q(m)^\alpha{}_\nu]s_m. \qquad (22.4)$$

At higher order in $[q]$,

$$\begin{aligned} [G_R Q]^{N-1}\{sI\} &= \sum_{n=1}^{N} (s - s_N)[q(N)^\mu{}_{\alpha_{N-2}}](s_N - s_{N-1}) \cdots \\ &\quad [q(n+1)^{\alpha_n}{}_{\alpha_{n-1}}](s_{n+1} - s_{n-1})[q(n-1)^{\alpha_{n-1}}{}_{\alpha_{n-2}}] \cdots \\ &\quad (s_2 - s_1)[q(1)^{\alpha_1}{}_\nu]s_1. \end{aligned} \qquad (22.5)$$

Ultimately

$$\begin{aligned} [G_R Q]^N\{sI\} &= (s - s_N)[q(N)^\mu{}_{\alpha_{N-1}}](s_N - s_{N-1}) \cdots \\ &\quad (s_2 - s_1)[q(1)^{\alpha_1}{}_\nu]s_1. \end{aligned} \qquad (22.6)$$

Due to the presence of a sufficient number of Heaviside functions, higher powers of $[G_R Q]$ vanish:

$$[G_R Q]^{N+n}\{sI\} = 0 \qquad \text{for} \qquad n > 0. \tag{22.7}$$

Note that the formal expansion (21.107) now terminates in a finite number of steps

$$A = \left(I + [G_R Q] + [G_R Q]^2 + \cdots + [G_R Q]^N\right)\{sI\}. \tag{22.8}$$

Here N denotes the total number of trips through the wormhole. Observe, either from the above, or from the differential equation (21.92), that in this type of geometry the matrix $A(s)$ is a piecewise linear continuous matrix function of arc length. Consequently the reciprocal of the van Vleck determinant $\Delta_\gamma(s)^{-1} = \det\{A/s\}$ is piecewise a Laurent polynomial in arc length. (Note that poles of this Laurent polynomial correspond to foci of the wormhole's gravitational lens.)

22.1.2   Single pass

For a geodesic that makes only a single pass through the wormhole ($N = 1$), one easily derives an exact closed form expression

$$A^\mu{}_\nu(s) = s\delta^\mu{}_\nu + \Theta(s - s_1)\{(s - s_1)[q(1)^\mu{}_\nu]s_1\}. \tag{22.9}$$

This result may be obtained either from the formal manipulations of the preceding section or from direct integration of the tidal equation. This leads to a closed form for the van Vleck determinant. For $s \geq s_1$:

$$\Delta_\gamma(s)^{-1} = \det\left(\delta^\mu{}_\nu + \frac{(s - s_1)s_1}{s}[q(1)^\mu{}_\nu]\right). \tag{22.10}$$

Particularize to a closed geodesic, so that $x = y$. Define $s_+ = s_1$, $s_- = s - s_1$, and let $s = s_+ + s_-$ denote the total arc length. For convenience, one may set $[q^\mu{}_\nu] \equiv [q(1)^\mu{}_\nu]$. Thus

Theorem 44 *For any closed geodesic γ that threads the throat of the wormhole once, for any arbitrary point x, the short-throat flat-space approximation yields*

$$\Delta_\gamma(x, x) = \det\left(\delta^\mu{}_\nu + \frac{s_+ s_-}{s}[q^\mu{}_\nu]\right)^{-1}. \tag{22.11}$$

Suppose that the point of interest, $x = y$, lies near the throat of the wormhole. Then either $s_+ \approx 0$ or $s_- \approx 0$.

Theorem 45 *For any closed geodesic γ that threads the throat of the wormhole once, for any point x near the throat of the wormhole, the short-throat flat-space approximation yields*

$$\Delta_\gamma(x, x) = 1 - \frac{s_+ s_-}{s}\text{tr}[q] + O(s_+^2 s_-^2 / s^2). \tag{22.12}$$

At the risk of belaboring the obvious:

Theorem 46 *For any closed geodesic γ that threads the throat of the wormhole once, for any point x on the throat of the wormhole, the short-throat flat-space approximation yields*

$$\Delta_\gamma(x, x) = 1. \qquad (22.13)$$

As we shall see in subsequent discussion, this observation, though elementary, is of some considerable interest.

On the other hand, suppose that the point $x = y$ is far away from the wormhole throat. The behavior of the van Vleck determinant is now governed by whether or not the eigenvalues of the matrix $\{s_+ s_-/s\}[q^\mu{}_\nu]$ ever become large compared to 1. This depends on the details of the wormhole's construction.

Theorem 47 *For any closed geodesic γ that threads the throat of the wormhole once, for any point x far away from throat of the wormhole, the short-throat flat-space approximation yields*

$$\Delta_\gamma(x, x) = \left[\frac{s}{s_+ s_-} \right]^\nu \det'[q]^{-1} + O\left(\left(\frac{s}{s_+ s_-} \right)^{\nu-1} \right). \qquad (22.14)$$

Here ν denotes the number of large eigenvalues of the matrix $\{s_+ s_-/s\}[q]$. Furthermore, $\det'$ denotes the determinant with small eigenvalues omitted. If all of the eigenvalues are small, then $\nu = 0$, $\det'[q] = 1$, and consequently $\Delta_\gamma(x, x) \approx 1$.

22.1.3 Double pass

For a geodesic that makes two passes through the mouth of the wormhole ($N = 2$), one has the closed form expression

$$
\begin{aligned}
A^\mu{}_\nu(s) =\ & s\delta^\mu{}_\nu \\
& + \Theta(s - s_1)\left\{ (s - s_1)[q(1)^\mu{}_\nu]s_1 \right\} \\
& + \Theta(s - s_2)\left\{ (s - s_2)[q(2)^\mu{}_\nu]s_2 \right\} \\
& + \Theta(s - s_2)\left\{ (s - s_2)[q(2)^\mu{}_\rho](s_2 - s_1)[q(1)^\rho{}_\nu]s_1 \right\}. \quad (22.15)
\end{aligned}
$$

As was the case previously, this result may be derived either from the formal expansion, or from direct integration of the tidal equation.

Now particularize to a closed geodesic, so that $x = y$. Define $s_+ = s_1$, $s_- = s - s_2$, and $s_0 = s_2 - s_1$. Then $s = s_+ + s_0 + s_-$ denotes the total arc length. For convenience, one may set $[q_+] = [q(1)]$, and $[q_-] = [q(2)]$.

Theorem 48 *For any closed geodesic γ that threads the throat of the wormhole twice, the short-throat flat-space approximation yields*

$$\Delta_\gamma(x,x)^{-1} = \det\left(\delta^\mu{}_\nu + \frac{s_+(s_- + s_0)}{s}[q^\mu_{+\nu}] + \frac{s_-(s_+ + s_0)}{s}[q^\mu_{-\nu}]\right.$$

$$\left. + \frac{s_+ s_- s_0}{s}[q^\mu_{-\rho}][q^\rho_{+\nu}]\right). \tag{22.16}$$

Now suppose the base point x of the geodesic is near the throat of the wormhole. For definiteness take $s_+ \equiv s_1 \approx 0$, though one could just as easily take $s_- \equiv s - s_2 \approx 0$. Introduce some new notation: $\epsilon \equiv s_1 \approx 0$, while $s_{1\to2} \equiv s_2 - s_1$, and $s_{2\to1} \equiv s_1 + s - s_2 \approx s - s_2$.

Theorem 49 *For any closed geodesic γ that threads the throat of the wormhole twice, for any point x near the throat of the wormhole, the short-throat flat-space approximation yields*

$$\Delta_\gamma(x,x)^{-1} = \det\left(\delta^\mu{}_\nu + \frac{s_{1\to2}\, s_{2\to1}}{s_{1\to2} + s_{2\to1}}[q^\mu_{-\nu}] + O(\epsilon)\right). \tag{22.17}$$

Taking ν to denote the number of large eigenvalues of the matrix

$$\{s_{1\to2}\, s_{2\to1}/(s_{1\to2} + s_{2\to1})\}\, [q(2)], \tag{22.18}$$

one may approximate

$$\Delta_\gamma(x,x)^{-1} \approx \left[\frac{s_{1\to2}\, s_{2\to1}}{s_{1\to2} + s_{2\to1}}\right]^\nu \det'[q_-]. \tag{22.19}$$

Far away from the wormhole throat the analysis is considerably more delicate. The critical question in this case is whether or not any of the eigenvalues of the matrix $\{(s_+ s_- s_0)/s\}[q_- q_+]$ are large.

Theorem 50 *For any closed geodesic γ that threads the throat of the wormhole twice, if any of these eigenvalues are large, the short-throat flat-space approximation yields*

$$\Delta_\gamma(x,x) = \left[\frac{s}{s_+ s_- s_0}\right]^\nu \det'[q_+ q_-]^{-1} + O\left(\left(\frac{s}{s_+ s_- s_0}\right)^{\nu-1}\right). \tag{22.20}$$

As before, let ν denote the number of large eigenvalues, this time of the matrix $\{(s_+ s_- s_0)/s\}[q_+ q_-]$. If all of the eigenvalues are small, then one must revert to

$$\Delta_\gamma(x,x)^{-1} \approx \det\left(\delta^\mu{}_\nu + \frac{s_+(s_- + s_0)}{s}[q^\mu_{+\nu}] + \frac{s_-(s_+ + s_0)}{s}[q^\mu_{-\nu}]\right). \tag{22.21}$$

For the case of a double pass through the wormhole, this is about as far as this general type of analysis can profitably be carried. To improve and extend the calculations one needs more detailed information about the matrices $[q(n)]$.

22.1.4   Multiple passes

For a multiple pass configuration, if one were to retain a completely general geodesic, the algebraic complexity involved in evaluating the van Vleck determinant would quickly rise from the merely cumbersome to the absolutely prohibitive. Even for the special case of a completely smooth closed geodesic, one that wraps smoothly around the wormhole N times, explicit calculations are tedious [272].

In view of the assumption of complete smoothness, including smoothness at the base point x, the geodesic under consideration is simply an N-fold overlay of a smooth once around the wormhole geodesic. The points where the geodesic intersects the wormhole mouth are all identical and one has

$$[q^\mu{}_\nu] \equiv [q(1)^\mu{}_\nu] = [q(2)^\mu{}_\nu] = \cdots = [q(N)^\mu{}_\nu]. \qquad (22.22)$$

Furthermore

$$s_0 \equiv s_N - s_{N-1} = \cdots = s_3 - s_2 = s_2 - s_1. \qquad (22.23)$$

Finally, define $s_+ = s_1$, and $s_- = s - s_N$. Since the total length of the geodesic is $s = N s_0$ it follows that $s_+ + s_- = s_0$. Indeed

$$s_n = s_+ + (n-1)s_0 = n s_0 - s_-. \qquad (22.24)$$

Evaluating the various terms in the expansion for the van Vleck determinant involves a turgid agony of combinatorics [272]. The closed form expression for the van Vleck determinant is

$$\begin{aligned} \Delta_\gamma^{-1} &= \det\left\{ [I] + \frac{s_+ s_-}{s_0^2}(s_0[q]) \right\} \\ &\quad \times \det\left\{ \sum_{n=0}^{N-1} \binom{N+n}{2n+1} \frac{(s_0[q])^n}{N} \right\}. \end{aligned} \qquad (22.25)$$

Provided there is no accidental zero suppressing the highest-order term, one may approximate this by

$$\Delta_\gamma^{-1} \approx \det\left\{ [I] + \frac{s_+ s_-}{s_0^2}(s_0[q]) \right\} \det'\left\{ \frac{(s_0[q])^{N-1}}{N} \right\}. \qquad (22.26)$$

Far away from the mouth of the wormhole $s_+ s_- \gg 0$, so that

$$\Delta_\gamma \approx \left(\frac{N s_0^2}{s_+ s_-} \right)^\nu \det'\{s_0[q]\}^{-N}. \qquad (22.27)$$

If the base point of the geodesic lies on the throat of the wormhole $s_+ s_- = 0$. The dominant term in the determinant is now

$$\Delta_\gamma \approx N^\nu \det' \{s_0[q]\}^{-N+1}. \tag{22.28}$$

Further simplifications would require more precise model building for the wormhole in question.

22.2 Exact results

22.2.1 Riemann tensor

It is possible to improve and make more explicit the general analysis of the van Vleck determinant by applying the extended thin-shell formalism as outlined in a previous chapter. For a wormhole in the short-throat flat-space approximation the full Riemann tensor is

$$R_{\alpha\beta\gamma\delta} = -\delta(\eta) \left[\kappa_{\alpha\gamma}\, n_\beta\, n_\delta + \kappa_{\beta\delta}\, n_\alpha\, n_\gamma - \kappa_{\alpha\delta}\, n_\beta\, n_\gamma - \kappa_{\beta\gamma}\, n_\alpha\, n_\delta \right]. \tag{22.29}$$

The delta function is presented here in terms of a Gaussian normal coordinate η. To apply this to the "tidal" formulation of the van Vleck determinant one needs to rewrite this in terms of the arc length along a geodesic penetrating the wormhole. It is easy to see that

$$\delta(\eta) = |\partial\eta/\partial s|^{-1}\, \delta(s) = |t \cdot n|^{-1}\, \delta(s). \tag{22.30}$$

The relevant source term driving the tidal evolution equation is

$$
\begin{aligned}
Q_{\mu\nu}(s) &\equiv -R_{\mu\alpha\nu\beta} t^\alpha t^\beta \\
&= \delta(\eta) \Big[\kappa_{\mu\nu}(t \cdot n)^2 + n_\mu n_\nu (\kappa_{\alpha\beta} t^\alpha t^\beta) \\
&\qquad - \{(\kappa_{\mu\alpha}t^\alpha)n_\nu + (\kappa_{\nu\alpha}t^\alpha)n_\mu\}(t \cdot n) \Big] \\
&= \delta(s) \Big[\kappa_{\mu\nu}(t \cdot n) + n_\mu n_\nu (\kappa_{\alpha\beta} t^\alpha t^\beta)(t \cdot n)^{-1} \\
&\qquad - \{(\kappa_{\mu\alpha}t^\alpha)n_\nu + (\kappa_{\nu\alpha}t^\alpha)n_\mu\} \Big].
\end{aligned}
$$

$$\text{(22.31)} \qquad \text{(22.32)}$$

Recall that the discontinuity in the second fundamental form $\kappa_{\mu\nu}$ is essentially a measure of the curvature of the wormhole mouth. We have already seen that diagonalizing κ yields

$$\kappa_{\mu\nu} = 2\, \text{diag}[A, (1/R_1), (1/R_2), 0] = 2 \begin{bmatrix} A & 0 & 0 & 0 \\ 0 & 1/R_1 & 0 & 0 \\ 0 & 0 & 1/R_2 & 0 \\ 0 & 0 & 0 & 0 \end{bmatrix}. \tag{22.33}$$

Here A is the four-acceleration of the wormhole throat—essentially the radius of curvature in the timelike direction [260]. R_1, R_2 are the usual principal radii of curvature in the three-dimensional sense. The zero eigenvalue for κ reflects the fact that $\kappa_{\mu\nu}n^\nu = 0$; the second fundamental form is by construction orthogonal to the normal.

One is now in a position to enunciate a general qualitative result:

$$\text{(small radius of curvature/large acceleration)}$$
$$\Rightarrow \text{(large } \kappa) \Rightarrow \text{(large Riemann tensor)}$$
$$\Rightarrow [\text{large } A(s)] \Rightarrow \text{(small van Vleck determinant)}. \quad (22.34)$$

To proceed, suppose that the individual wormhole mouths (though not the full spacetime) are static. This just means that the individual wormhole mouths are not undergoing any changes in internal structure, and also implies that one can safely attach the notion of a constant four-velocity V^μ to each individual wormhole mouth. In particular since each wormhole mouth is now invariant under time translations along the V^μ axis one has $\kappa_{\mu\nu}V^\nu = 0$. Further, by construction,

$$\kappa_{\mu\nu} \, n^\nu = 0 \qquad \text{and} \qquad V^\mu \, n_\mu = 0. \qquad (22.35)$$

Indeed, in this case

$$\kappa_{\mu\nu} = 2 \, \text{diag}[0, 1/R_1, 1/R_2, 0] = 2 \begin{bmatrix} 0 & 0 & 0 & 0 \\ 0 & 1/R_1 & 0 & 0 \\ 0 & 0 & 1/R_2 & 0 \\ 0 & 0 & 0 & 0 \end{bmatrix}. \qquad (22.36)$$

A brief computation shows $Q_{\mu\nu} V^\nu = 0$. The general analysis already implies $Q_{\mu\nu}t^\nu = 0$. Therefore, in this approximation two of the eigenvalues of $Q(s)$ are trivial, and the computation of the van Vleck determinant reduces to that of a (still decidedly nontrivial) 2×2 matrix.

Exercise: Check this.

22.2.2   Single pass

Consider a geodesic that starts at the point x, wraps through the wormhole once, and returns to the point x. This geodesic has two tangent vectors, $t_\pm$, one pointing toward each of the two wormhole mouths. Each of the two wormhole mouths is characterized by a four-velocity $V_\pm$ and, at the point where the geodesic impacts the mouth of the wormhole, a normal $n_\pm$. In view of the structure outlined above, one can define quantities $\gamma = (1 - \beta^2)^{-1/2}$ and θ by

$$t_\pm^\mu = \gamma \left(\beta \, V_\pm^\mu + \cos\theta \, n_\pm^\mu + \sin\theta \, z_\pm^\mu \right). \qquad (22.37)$$

Here one has used the fact that the tangent vector is (for current purposes) a spacelike unit vector, and defined another spacelike unit vector z, that lies in the plane of the wormhole throat. Because the geodesic must, by construction, pass through the wormhole mouth in a smooth manner one must have the same coefficients γ, β, and θ, occurring in each of these two equations. In particular,

$$t_+ \cdot n_+ = t_- \cdot n_- = \gamma \cos \theta. \qquad (22.38)$$

One may now define the quantity

$$\ell_\pm \equiv \gamma s_\pm = s_\pm \, (t_\pm \cdot n_\pm)/\cos\theta. \qquad (22.39)$$

This represents the physical distance between the point x and the relevant wormhole mouth, as measured in the rest frame of that wormhole mouth. By extension, one defines

$$\ell = \ell_+ + \ell_- = s \, (t_\pm \cdot n_\pm)/\cos\theta. \qquad (22.40)$$

To facilitate comparison with earlier parts of this monograph (see also [268]) one may define "time-shifts" by

$$T_\pm \equiv s_\pm \, (t_\pm \cdot V_\pm). \qquad (22.41)$$

Further, one may define $T = T_+ + T_-$. Finally, as a consistency check, observe

$$\delta(\eta_+) = (t_+ \cdot n_+)^{-1} \, \delta(s) = (t_- \cdot n_-)^{-1} \, \delta(s) = \delta(\eta_-). \qquad (22.42)$$

The van Vleck determinant is simply

$$\Delta_\gamma(x, x) = \det\left(\delta^\mu{}_\nu + \frac{s+s_-}{s}[q^\mu{}_\nu]\right)^{-1}. \qquad (22.43)$$

As a result of all the preceding definitions and calculations, one extracts the relatively compact expression

$$\begin{aligned}
[q_{\mu\nu}] &= \Big[\kappa_{\mu\nu}(t \cdot n) + n_\mu n_\nu (\kappa_{\alpha\beta} t^\alpha t^\beta)(t \cdot n)^{-1} \\
&\quad - \{(\kappa_{\mu\alpha} t^\alpha)n_\nu + (\kappa_{\nu\alpha} t^\alpha)n_\mu\}\Big] \\
&= \gamma \cos\theta \Big[\kappa_{\mu\nu} + n_\mu n_\nu \tan^2\theta(\kappa_{\alpha\beta} z^\alpha z^\beta) \\
&\quad - \tan\theta\{(\kappa_{\mu\alpha} z^\alpha)n_\nu + (\kappa_{\nu\alpha} z^\alpha)n_\mu\}\Big]. \qquad (22.44)
\end{aligned}$$

Unfortunately this expression, while exact and general, is still too algebraically messy to be tractable.

Radial impact

Utilizing the machinery defined above one can define the notion of a geodesic that radially impacts on the mouth of the wormhole. A radially impacting geodesic is simply one for which $\theta = 0$, so that one has

$$t_{\pm}^{\mu} = \gamma \left(\beta \, V_{\pm}^{\mu} + n_{\pm}^{\mu} \right). \tag{22.45}$$

For instance, for spherically symmetric wormhole mouths that directly face one another, go to any Lorentz frame where the mouths have no transverse velocity, only longitudinal velocities. In any such frame pick any point on the line joining the two wormhole mouths. The geodesic from any such point to itself will impact the wormhole mouths radially in the sense described above. This is essentially the notion of the "central geodesic" as described by Kim and Thorne [164]. Returning to the general case, the virtue of radially impacting geodesics is that for such geodesics one has the great simplification $\kappa_{\mu\nu} \, t^{\mu} = 0$, which implies

$$Q_{\mu\nu} \equiv -R_{\mu\alpha\nu\beta}t^{\alpha}t^{\beta} = \delta(\eta) \, (t \cdot n)^2 \, \kappa_{\mu\nu} = \delta(s) \, (t \cdot n) \, \kappa_{\mu\nu}. \tag{22.46}$$

Let me (temporarily) further restrict the analysis. For a spherically symmetric wormhole of radius R the discontinuity in the second fundamental forms is

$$\kappa_{\mu\nu} = \frac{2}{R} \, g_{\mu\nu}^{\perp} = \frac{2}{R} \, (g_{\mu\nu} + V_{\mu}V_{\nu} - n_{\mu}n_{\nu}). \tag{22.47}$$

Inserting all of this machinery into the previously derived closed form expression for the single pass van Vleck determinant

$$\begin{aligned}
\Delta_{\gamma}(s) &= \det \left(\delta^{\mu}{}_{\nu} + \frac{s_{+}s_{-}}{s} \frac{2(t \cdot n)}{R} [g^{\perp}]^{\mu}{}_{\nu} \right)^{-1} \\
&= \left(1 + \frac{s_{+}s_{-}}{s} \frac{2(t \cdot n)}{R} \right)^{-2} \\
&= \left(1 + \frac{\ell_{+}\ell_{-}}{\ell} \frac{2}{R} \right)^{-2}.
\end{aligned} \tag{22.48}$$

This formula nicely interpolates between the near-field and far-field results. Emphasis should be placed on the fact that the relative velocity of the wormhole mouths is completely arbitrary. Likewise, the position of the point $x = y$, though constrained by the condition of radial impact is otherwise arbitrary.

A simple generalization is to note that there is now nothing sacred about a spherical wormhole mouth. As long as the geodesic is radially impacting one may write

$$\Delta_{\gamma}(x, x) = \det \left(\delta^{\mu}{}_{\nu} + \frac{s_{+}s_{-}}{s} \gamma \kappa^{\mu}{}_{\nu} \right)^{-1}$$

$$= \det\left(\delta^\mu{}_\nu + \frac{\ell_+\ell_-}{\ell}\kappa^\mu{}_\nu\right)^{-1}$$

$$= \left(1 + \frac{\ell_+\ell_-}{\ell}\frac{2}{R_1}\right)^{-1}\left(1 + \frac{\ell_+\ell_-}{\ell}\frac{2}{R_2}\right)^{-1}. \quad (22.49)$$

Here, as previously, $R_{(1,2)}$ refer to the two principal radii of curvature, evaluated at the point where the geodesic impacts the wormhole mouth. Indeed, consider the cubical (or even polyhedral) wormholes of [261]. If the geodesic impacts radially on one of the flat faces, $R_{(1,2)} = \infty$; consequently $\Delta_\gamma(s) = 1$.

If the base point x is not on the throat of the wormhole then $\ell_+\ell_- \neq 0$. If, furthermore, the geodesic impacts radially on one of the edges, then one of the principal radii of curvature $R_{(1,2)}$ is zero (the other is infinite). Consequently $\Delta_\gamma(s) = 0$.

If the base point x is on the throat of the wormhole then $\ell_+\ell_- = 0$. In this case, if the geodesic impacts radially on one of the edges, one is faced with making sense of the indeterminate form $0/0$. Proceed as follows: The polyhedral wormholes discussed earlier in this monograph were constructed by taking smooth cut-and-paste (thin throat) wormholes and considering the limit as one of the radii of curvature tends to zero. For the issue of interest, the indeterminate $0/0$ is resolved by observing that one should let the base point x approach the wormhole mouth *before* letting the wormhole mouth acquire sharp edges by letting $R_{(1,2)} \to 0$. Consequently $\Delta_\gamma(s) = 1$.

Another way of phrasing this is as follows: Letting the wormhole mouth acquire a sharp edge by letting $R_2 \to 0$ is a somewhat dubious proposition once $R_2 < \ell_P$. Radii of curvature smaller than the Planck length are of doubtful operational significance. Accepting that the struts supporting edges of a polyhedral wormhole have thickness of order the Planck length

$$\Delta_\gamma(x,x)^{\text{polyhedral}}_{\text{edge impact}} \approx \left(1 + \frac{2\ell_+\ell_-}{\ell\,\ell_P}\right)^{-1}. \quad (22.50)$$

This "regulated" determinant is suitably well behaved as one approaches the throat. That is, as $\ell_+ \to 0$.

Generic impact

To push the analysis beyond the case of radial impact is computationally messy. It is useful, in the interests of keeping the algebra from getting too unwieldy, to return to the case of spherical symmetry. Combining spherical symmetry with a generic impact angle,

$$Q_{\mu\nu}(s) = \delta(s)\left[\kappa_{\mu\nu}(t\cdot n) + n_\mu n_\nu(\kappa_{\alpha\beta}t^\alpha t^\beta)(t\cdot n)^{-1}\right.$$

$$-\{(\kappa_{\mu\alpha}t^\alpha)n_\nu + (\kappa_{\nu\alpha}t^\alpha)n_\mu\}\Big]$$

$$= \delta(s)\frac{2\gamma\cos\theta}{R}\left[g^\perp_{\mu\nu} + n_\mu n_\nu \tan^2\theta - \tan\theta\{z_\mu n_\nu + z_\nu n_\mu\}\right].$$

$$(22.51)$$

One is faced with the task of determining the eigenvalues of the matrix

$$\begin{bmatrix} 0 & 0 & 0 & 0 \\ 0 & \tan^2\theta & -\tan\theta & 0 \\ 0 & -\tan\theta & 1 & 0 \\ 0 & 0 & 0 & 1 \end{bmatrix}.$$

$$(22.52)$$

The eigenvalues are easily determined to be

$$(0; 0; 1; 1 + \tan^2\theta) \equiv (0; 0; 1; \sec^2\theta). \qquad (22.53)$$

Consequently

$$\Delta_\gamma(s) = \left(1 + \frac{\ell_+\ell_-}{\ell}\frac{2}{R}\right)^{-1}\left(1 + \frac{\ell_+\ell_-}{\ell}\frac{2\sec^2\theta}{R}\right)^{-1}. \qquad (22.54)$$

Thus a radial impact is most effective at keeping the van Vleck determinant large. Grazing impacts lead to small values for the van Vleck determinant.

Note that as the base point of the geodesic approaches the mouth of the wormhole $\ell_+\ell_- \to 0$, so that (again) $\Delta_\gamma(x, x) \to 1$. The case $\theta = \pi/2$, which corresponds to the geodesic having a tangential impact on the mouth of the wormhole, naively leads to the indeterminate form $0/0$. To regularize this, recognize that tangential impact ($\theta = \pi/2$) occurs only if one very carefully orients the two mouths of the wormhole to be facing away from each other. But it is impossible to hold the orientation of the wormhole mouths completely fixed. If nothing else, the Heisenberg uncertainty principle provides a fundamental limitation $\Delta\theta\Delta L \approx \hbar$ relating the spread in orientation to the spread in angular momentum of the wormhole mouth. If one centers the orientation on $\theta = \pi/2$, one has

$$\langle \sec^2\theta \rangle \approx \langle \theta^2 \rangle^{-1} \approx (\Delta L/\hbar)^2 \equiv \langle J^2 \rangle. \qquad (22.55)$$

Thus a "regulated" van Vleck determinant, modified by this orientational smearing, may be estimated to be

$$\Delta_\gamma(x, x; \theta = \pi/2) \approx \left(1 + \frac{\ell_+\ell_-}{\ell}\frac{2}{R}\right)^{-1}\left(1 + \frac{\ell_+\ell_-}{\ell}\frac{2\langle J^2\rangle}{R}\right)^{-1}. \qquad (22.56)$$

This "regulated" quantity tends to 1 as x approaches the throat of the wormhole, which solves the problem of what to do with the indeterminate form.

22.2.3 Double pass

Consider a geodesic that starts at the point x, wraps through the wormhole twice, and returns to the point x. In view of the complexity of structure to be outlined below it is not particularly enlightening to even contemplate non-radially impacting geodesics. Henceforth one specializes even further by assuming collinear motion for the wormhole mouths (i.e., radial approach or recession) and restricting attention to central geodesics. The geodesic has three tangent vectors, $t_{\pm}$ and t_0. The vector t_+ points from the point x to one of the wormhole mouths, while t_- points from the point x to the other mouth. t_0 points from one wormhole mouth to another. Each of the two wormhole mouths is characterized by a four-velocity $V_{\pm}$. There are two normals to keep track of: $n_{\pm}$. Here n_+ is the normal to the wormhole mouth with velocity V_+, evaluated at the point where the geodesic from x impacts that mouth. Similarly for n_-. One can define quantities $\gamma_{\pm} = (1 - \beta_{\pm}^2)^{-1/2}$ by

$$
\begin{aligned}
t_{\pm}^{\mu} &= \gamma_{\pm} \left(\beta_{\pm} V_{\pm}^{\mu} + n_{\pm}^{\mu} \right), \\
t_0^{\mu} &= \pm \gamma_{\pm} \left(\beta_{\pm} V_{\mp}^{\mu} + n_{\mp}^{\mu} \right).
\end{aligned}
\tag{22.57}
$$

Here, as was the case for the $N = 1$ case, one has utilized the fact that the tangent vectors are (for current purposes) spacelike unit vectors. Because the geodesic must, by construction, pass through the wormhole mouth in a smooth manner one must have the same coefficients $\gamma_{\pm}$, and $\beta_{\pm}$, occurring in these equations. In particular,

$$
t_{\pm} \cdot n_{\pm} = \pm t_0 \cdot n_{\mp} = \gamma_{\pm}.
\tag{22.58}
$$

Because $V_{\pm}$ and $n_{\pm}$ are all coplanar,

$$
g_{\mu\nu}^{\perp} = (g_{\mu\nu} + V_{\mu}^{\pm} V_{\nu}^{\pm} - n_{\mu}^{\pm} n_{\nu}^{\pm}).
\tag{22.59}
$$

This obviates the otherwise messy technical requirement of keeping track of precisely which of the $g^{\perp}$ one is dealing with, and permits the radical simplification

$$
[q_{\mu\nu}^{\pm}] = \gamma_{\pm} \frac{2}{R} g_{\mu\nu}^{\perp}.
\tag{22.60}
$$

So finally

$$
\Delta_{\gamma}(x, x) = \left(1 + \frac{s_+(s_- + s_0)}{s} \frac{2\gamma_+}{R} + \frac{s_-(s_+ + s_0)}{s} \frac{2\gamma_-}{R} \right.
$$
$$
\left. + \frac{s_+ s_- s_0}{s} \frac{4\gamma_+ \gamma_-}{R^2} \right)^{-2}.
\tag{22.61}
$$

This is the promised exact result for an arbitrary central geodesic that wraps twice through the wormhole.

One may still define the quantities $\ell_\pm \equiv \gamma_\pm s_\pm = s_\pm (t_\pm \cdot n_\pm)$ and interpret these quantities as the physical distance between the point x and the relevant wormhole mouth, as measured in the rest frame of that wormhole mouth. The physical import of this result for the van Vleck determinant is unfortunately nowhere near as transparent as that derived for the case of a single pass.

One special case that is of some interest is when the point x lies on the throat of the wormhole itself. Say one takes $s_- = 0$, and furthermore approximates $s_+ \approx s_0$, then

$$\Delta_\gamma(x, x) \approx \left(1 + \frac{s_0 \gamma_+}{R}\right)^{-2}. \tag{22.62}$$

22.2.4   Multiple passes

Most of the preparatory work for this section has already been done. For a completely general geodesic the algebraic complexity would be absolutely prohibitive. Accordingly, it is useful to specialize to the case of a completely smooth closed geodesic. One that wraps around the wormhole N times.

Invoking the preceding multi-pass result, and now using the extended thin-shell formalism to probe the Riemann tensor at the wormhole throat, one has

$$s_0[q^\mu{}_\nu] = s_0 \gamma[\kappa^\mu{}_\nu] = \ell_0[\kappa^\mu{}_\nu] = \ell_0 \left[\text{diag}\{0, (2/R_1), (2/R_2), 0\}\right]. \tag{22.63}$$

The net result of the various factors of γ has been to quietly disappear, after conveniently converting the arc length along the geodesic into the physical distance between the wormhole mouths, as measured in the rest frame of the wormhole mouths. Substitution into the previous result shows

$$
\begin{aligned}
\Delta_\gamma = {} & N^2 \times \left\{1 + \frac{2\ell_+ \ell_-}{\ell_0 R_1}\right\}^{-1} \times \left\{\sum_{n=0}^{N-1} \binom{N+n}{2n+1} \left[\frac{2\ell_0}{R_1}\right]^n\right\}^{-1} \\
& \times \left\{1 + \frac{2\ell_+ \ell_-}{\ell_0 R_2}\right\}^{-1} \times \left\{\sum_{n=0}^{N-1} \binom{N+n}{2n+1} \left[\frac{2\ell_0}{R_2}\right]^n\right\}^{-1} \tag{22.64}
\end{aligned}
$$

Note that if $\ell_0 \gg R_{(1,2)}$ this expression becomes more tractable:

$$\Delta_\gamma \approx N^2 \times \left\{1 + \frac{2\ell_+ \ell_-}{\ell_0 R_1}\right\}^{-1} \times \left\{1 + \frac{2\ell_+ \ell_-}{\ell_0 R_2}\right\}^{-1} \times \left\{\frac{4\ell_0^2}{R_1 R_2}\right\}^{N-1}. \tag{22.65}$$

One could continue in this vein for a while longer but there is no pressing need for further manipulations.

22.3 Roman configurations

Evaluation of the van Vleck determinant for the "Roman configuration" spacetimes, discussed on pp. 234ff., is subtle [178, 272].

Return to the general arguments given earlier in this chapter. For a geodesic that makes a single pass through the mouths of two distinct wormholes ($N = 2$), one may still write down the closed form expression

$$
\begin{aligned}
A^\mu{}_\nu(s) = {}& s\delta^\mu{}_\nu \\
& +\Theta(s - s_1)\left\{(s - s_1)[q(1)^\mu{}_\nu]s_1\right\} \\
& +\Theta(s - s_2)\left\{(s - s_2)[q(2)^\mu{}_\nu]s_2\right\} \\
& +\Theta(s - s_2)\left\{(s - s_2)[q(2)^\mu{}_\rho](s_2 - s_1)[q(1)^\rho{}_\nu]s_1\right\}.
\end{aligned}
\tag{22.66}
$$

Particularize to a closed geodesic, so that $x = y$. Let the base point x lie on the mouth of wormhole 1. Then $s_1 = 0$, and we may define $s_{1\to2} = s_2$, $s_{2\to1} = s - s_2$, so that $s = s_{1\to2} + s_{2\to1}$. Thus

$$
\Delta_\gamma(x, x)^{-1} = \det\left(\delta^\mu{}_\nu + \frac{s_{1\to2}\, s_{2\to1}}{s}[q(2)^\mu{}_\nu]\right).
\tag{22.67}
$$

Now invoke the extended thin-shell formalism to estimate $[q(2)^\mu{}_\nu]$. For radial impact the previous arguments give

$$
[q(2)^\mu{}_\nu] = (t \cdot n)[\kappa(2)^\mu{}_\nu] = \gamma(2/R_2)[g^\perp(2)^\mu{}_\nu].
\tag{22.68}
$$

Here $\kappa(2)$ denotes the discontinuity in the second fundamental form at the throat of wormhole 2. Assuming spherical symmetry, I have taken R_2 to denote the radius of the throat of wormhole 2. The factor $\gamma = (t \cdot n)$ is easily seen to be

$$
\gamma = \frac{\ell_{1\to2}}{s_{1\to2}} = \frac{\ell_{2\to1}}{s_{2\to1}} = \frac{\ell_{1\to2} + \ell_{2\to1}}{s_{1\to2} + s_{2\to1}}.
\tag{22.69}
$$

The net result is now

$$
\gamma\frac{s_{1\to2}\, s_{2\to1}}{s} = \gamma\frac{s_{1\to2}\, s_{2\to1}}{s_{1\to2} + s_{2\to1}} = \frac{\ell_{1\to2}\, \ell_{2\to1}}{\ell_{1\to2} + \ell_{2\to1}}.
\tag{22.70}
$$

Therefore

$$
\Delta_\gamma(x, x) = \left\{1 + \frac{2\ell_{1\to2}\, \ell_{2\to1}}{(\ell_{1\to2} + \ell_{2\to1})R_2}\right\}^{-2}.
\tag{22.71}
$$

Note that this is *less* than 1. By the discussion in the previous chapter, this implies that the spacelike convergence condition must be violated.

22.4 Summary

These calculations have proved rather tedious but have served to develop a rather general formalism that is of some interest beyond the immediate concerns of this monograph.

For the purposes of this monograph however, the central results of this chapter can be neatly summarized in two observations:

1. For any short-throat wormhole residing in an otherwise approximately flat spacetime, for any point x on the wormhole throat, the geodesic γ that wraps once through the wormhole satisfies

$$\Delta_\gamma(x, x) \approx 1. \qquad (22.72)$$

2. For any Roman configuration (a pair of short-throat wormholes residing in an approximately flat spacetime), for any point x on the throat of wormhole 1, the geodesic γ that wraps once through the pair of wormholes satisfies

$$\Delta_\gamma(x, x) \approx \left\{ \frac{R_2(\ell_{1 \to 2} + \ell_{2 \to 1})}{2\ell_{1 \to 2}\, \ell_{2 \to 1}} \right\}^2. \qquad (22.73)$$

These two simple results will prove to be key issues used when obtaining estimates of the vacuum polarization.

Chapter 23

Singularity structure

In this chapter I present an explicit calculation of the dominant portions of the vacuum expectation value of the renormalized stress-energy tensor in traversable wormhole spacetimes. Point-splitting techniques are used. Particular attention is paid to the computation of the structural form of the stress-energy tensor near short self-intersecting spacelike geodesics.

Consider an arbitrary Lorentzian spacetime of nontrivial homotopy. Pick an arbitrary base point x. Since, by assumption, $\pi_1(\mathcal{M})$ is nontrivial there certainly exist self-intersecting paths not homotopic to the identity that begin and end at x. By smoothness arguments there also exist smooth self-intersecting geodesics, not homotopic to the identity, that connect the point x to itself. However, there is no guarantee that the tangent vector is continuous as the geodesic passes through the point x where it is pinned down. If any of these self-intersecting geodesics is timelike or null, then the battle against time travel is already lost, the spacetime is diseased, and it should be dropped from consideration.

To examine the types of pathology that arise as one gets "close" to building a time machine, it is instructive to construct a one-parameter family of self-intersecting geodesics that captures the essential elements of the geometry. Suppose merely that one can find a well-defined throat for one's Lorentzian wormhole. Consider the worldline swept out by a point located in the middle of the wormhole throat. At each point on this worldline there exists a self-intersecting "pinned" geodesic threading the wormhole and closing back on itself in "normal" space. This geodesic will be smooth everywhere except possibly at the place that it is "pinned" down by the throat.

If the geodetic interval from x to itself, $\sigma_\gamma(x, x)$, becomes negative, then a self-intersecting timelike curve (a fortiori—a time machine) has formed. It is this unfortunate happenstance that Hawking's chronology protection conjecture is hoped to prevent. For the purposes of this monograph it will be sufficient to consider the behavior of the vacuum expectation value of the renormalized stress-energy tensor in the limit $\sigma_\gamma(x, x) \to 0^+$.

23.1   The renormalized stress-energy tensor

To calculate the renormalized stress-energy tensor one merely inserts the Hadamard form of the Green function (propagator) into the point-split formalism. (See pp. 282–283 herein, and references [164, 167, 93, 270, 272].) Retaining only the most singular terms as $\sigma \to 0^+$ gives

$$\langle 0|T_{\mu\nu}(x)|0\rangle = \hbar \sum_{\gamma \neq \gamma_0} \frac{\Delta_\gamma(x, x)^{1/2}}{4\pi^2} \lim_{y \to x} D_{\mu\nu}(x, y; \gamma_0) \left\{ \frac{1}{\sigma_\gamma(x, y)} \right\}$$

$$+ O(\sigma_\gamma(x, x)^{-3/2}). \tag{23.1}$$

Because, for the purposes at hand, it is only the singular part of the stress-energy tensor that is of interest, it is safe to ignore any ambiguities related to finite renormalizations. Observe that

$$\nabla_\mu^x \nabla_\nu^y \left\{ \frac{1}{\sigma_\gamma(x, y)} \right\} = \frac{1}{\sigma_\gamma(x, y)^2} \{ 4 \nabla_\mu^x s_\gamma(x, y) \; \nabla_\nu^y s_\gamma(x, y)$$

$$- \nabla_\mu^x \nabla_\nu^y \sigma_\gamma(x, y) \}. \tag{23.2}$$

Similar equations hold for other combinations of derivatives. Let $y \to x$, and define the limits

$$t_\mu^1 \equiv \lim_{y \to x} t_\mu^x \equiv \lim_{y \to x} \nabla_\mu^x s_\gamma(x, y);$$

$$t_\mu^2 \equiv \lim_{y \to x} t_\mu^y \equiv \lim_{y \to x} \nabla_\mu^y s_\gamma(x, y). \tag{23.3}$$

Note that

$$g_{\alpha\beta}(x) t_\alpha^z t_\beta^z = +1, \tag{23.4}$$

since the self-connecting geodesics are all taken to be spacelike. Then, keeping only the most singular term

$$\langle 0|T_{\mu\nu}(x)|0\rangle = \hbar \sum_{\gamma \neq \gamma_0} \frac{\Delta_\gamma(x, x)^{1/2}}{4\pi^2 \sigma_\gamma(x, x)^2} [t_{\mu\nu}(x; \gamma) + s_{\mu\nu}(x; \gamma; \gamma_0)]$$

$$+ O(\sigma_\gamma(x, x)^{-3/2}). \tag{23.5}$$

Here the dimensionless tensor $t_{\mu\nu}(x; \gamma)$ is constructed solely out of the metric and the tangent vectors to the geodesic γ as follows:

$$t_{\mu\nu}(x; \gamma) = \frac{2}{3} \left(t_\mu^1 t_\nu^2 + t_\mu^2 t_\nu^1 - \frac{1}{2} g_{\mu\nu} (t^1 \cdot t^2) \right)$$

$$- \frac{1}{3} \left(t_\mu^1 t_\nu^1 + t_\mu^2 t_\nu^2 - \frac{1}{2} g_{\mu\nu} \right). \tag{23.6}$$

The dimensionless tensor $s_{\mu\nu}(x; \gamma; \gamma_0)$ is defined by

$$s_{\mu\nu}(x; \gamma; \gamma_0) \equiv \lim_{y \to x} D_{\mu\nu}(x, y; \gamma_0) \{\sigma_\gamma(x, y)\}. \qquad (23.7)$$

In many cases of physical interest the tensor $s_{\mu\nu}(x; \gamma; \gamma_0)$ either vanishes identically or is subdominant in comparison to $t_{\mu\nu}(x; \gamma)$. A general analysis has so far unfortunately proved elusive. This is an issue of some delicacy that clearly needs further clarification. Nevertheless the neglect of $s_{\mu\nu}(x; \gamma; \gamma_0)$ in comparison to $t_{\mu\nu}(x; \gamma)$ appears to be a safe approximation which shall be adopted forthwith.

Note that this most singular contribution to the stress-energy tensor is in fact traceless—there is a good physical reason for this. We remind the reader that once the length of the self-intersecting spacelike geodesic becomes smaller than the Compton wavelength of the particle under consideration, $s \ll \hbar/mc$, one expects such a physical particle to behave in an effectively massless fashion. (Aside: The trace (conformal) anomaly does not affect this observation since the trace anomaly is finite, and one is here interested only in the dominant singular contributions to the stress-energy.) Indeed, based on such general considerations, one expects the singular part of the stress-energy tensor to be largely insensitive to the type of particle under consideration. Despite the fact that the calculation has been carried out only for conformally-coupled massless scalars, one expects this leading singularity to be generic. Indeed, in terms of the geodesic distance from x to itself:

$$\langle 0|T_{\mu\nu}(x)|0\rangle = \hbar \sum_{\gamma \neq \gamma_0} \frac{\Delta_\gamma(x, x)^{1/2}}{\pi^2 s_\gamma(x, x)^4} t_{\mu\nu}(x; \gamma) + O(s_\gamma(x, x)^{-3}). \qquad (23.8)$$

A formally similar result was obtained by Frolov [93]. That result was obtained for points near the N'th polarized hypersurface of a locally-static spacetime. It is important to observe that in the present context it has not proved necessary to introduce any (global or local) static restriction on the spacetime. Neither is it necessary to introduce the notion of a polarized hypersurface. All that is needed at this stage is the existence of at least one short, nontrivial, self-intersecting, spacelike geodesic. A related calculation due to Klinkhammer [167] considers both scalar fields and spinor fields. Klinkhammer shows that the contribution due to the scalar field is opposite in sign to that due to the spinor field but that the coefficients are different and that generically there is no cancellation.

To convince oneself that the apparent s^{-4} divergence of the renormalized stress-energy is neither a coordinate artifact nor a Lorentz frame artifact consider the scalar invariant

$$T_0 = \sqrt{\langle 0|T_{\mu\nu}(x)|0\rangle \langle 0|T^{\mu\nu}(x)|0\rangle}. \qquad (23.9)$$

By noting that

$$t_{\mu\nu}(x;\gamma)t^{\mu\nu}(x;\gamma) = \frac{1}{3}\left[3 - 4(t^1 \cdot t^2) + 2(t^1 \cdot t^2)^2\right] \qquad (23.10)$$

one sees that there is no "accidental" zero in $t^{\mu\nu}$, and that T_0 does in fact diverge as s^{-4}. Thus the s^{-4} divergence encountered in the stress-energy tensor associated with the Casimir effect is generic to any multiply connected spacetime containing short self-intersecting spacelike geodesics. (See [268], and pp. 121–126.)

23.2   Other calculations

The characteristic s^{-4} divergence encountered in this analysis [268, 272] is, at first blush, somewhat difficult to reconcile with the "$(\delta t)^{-3}$" behavior described in references [164, 167, 93]. These apparent differences are due to the fact that those analyses introduce a special choice of canonical observer. When expressed in terms of quantities measured by this canonical observer, the stress-energy tensor *seems* quite different. To see how this happens, one first has to add considerably more structure to the discussion in the form of extra assumptions.

To begin the comparison, one must beg the original question by assuming that a time machine does in fact succeed in forming. Furthermore, one must assume that the resulting chronology horizon is compactly generated [129, 130, 164]. The generators of the compactly generated chronology horizon all converge in the past on a unique closed null geodesic that shall be referred to as the "fountain" and shall be denoted by $\tilde{\gamma}$. The question of interest is now the behavior of the renormalized stress-energy tensor in the neighborhood of the fountain.

To that end, pick a point x "close" to the fountain $\tilde{\gamma}$. Pick a point x_0 that is on the fountain, with x_0 being "close" to x, and with x_0 being in the future of x. Then the geodesic $\gamma_\perp$ from x to x_0 is by construction timelike. One defines $\delta t = s_{\gamma_\perp}(x, x_0)$, and $V^\mu = -\nabla^\mu_x \sigma_{\gamma_\perp}(x, x_0)$. One interprets these definitions as follows: a geodesic observer at the point x, with four-velocity V^μ, will hit the fountain $\tilde{\gamma}$ after a proper time δt has elapsed. One now seeks a computation of the stress-energy tensor at x in terms of various quantities that are Taylor series expanded around the assumed impact point x_0 with δt as the (hopefully) small parameter.

Consider, initially, the geodetic interval $\sigma_\gamma(x, y)$. Taylor series expand this as

$$\begin{aligned}
\sigma_\gamma(x, x) &= \sigma_{\tilde{\gamma}}(x_0, x_0) \\
&\quad + (\delta t\, V^\mu)\left[\nabla^x_\mu \sigma_{\tilde{\gamma}}(x, y) + \nabla^y_\mu \sigma_{\tilde{\gamma}}(x, y)\right]\big|_{(x_0, x_0)} \\
&\quad + O(\delta t^2).
\end{aligned} \qquad (23.11)$$

First, by definition of the fountain as a closed null geodesic, $\sigma_{\tilde{\gamma}}(x_0, x_0) = 0$. Second, take the vector V^μ, defined at the point x, and parallel transport it along $\gamma_\perp$ to x_0. Then parallel transport V along the fountain $\tilde{\gamma}$. This now provides us with a canonical choice of affine parameter on the fountain, defined by choosing ζ to satisfy

$$\frac{dx^\mu}{d\zeta} V_\mu = 1. \tag{23.12}$$

Naturally this canonical affine parameter is not unique, but depends on our original choice of V^μ at x, or, what amounts to the same thing, depends on our choice of x_0 as a "reference point". In terms of this canonical affine parameter, and writing

$$l^\mu = \frac{dx^\mu}{d\zeta}, \tag{23.13}$$

one sees

$$\nabla^z_\mu \sigma_{\tilde{\gamma}}(x, y)\big|_{(x_0, x_0)} = -\zeta_{-n}\, l_\mu; \tag{23.14}$$

$$\nabla^y_\mu \sigma_{\tilde{\gamma}}(x, y)\big|_{(x_0, x_0)} = -\zeta_{+n}\, l_\mu. \tag{23.15}$$

Here the notation ζ_{-n} denotes the lapse of affine parameter on going around the fountain a total of n times in the left direction, while ζ_{+n} is the lapse of affine parameter for n trips in the right direction. The fact that these total lapses are different is a reflection of the fact that the tangent vector to the fountain undergoes a boost on traveling around the fountain. Hawking has shown that the boost is simply given by [129, 130]

$$\zeta_{-n} = -e^{nh} \zeta_{+n}. \tag{23.16}$$

Here the parameter h characterizes the net gain in redshift in going once around the closed null geodesic.

Evaluating the geodetic interval:

$$\sigma_\gamma(x, x) = +\delta t\, \zeta_n \left[e^{nh} - 1\right] + O(\delta t^2). \tag{23.17}$$

In an analogous manner, one estimates the tangent vectors t^1 and t^2 in terms of the tangent vector at x_0:

$$\nabla^z_\mu \sigma_\gamma(x, y)\big|_{y \to x} = \nabla^z_\mu \sigma_{\tilde{\gamma}}(x, y)\big|_{(x_0, x_0)} + O(\delta t) \tag{23.18}$$

$$= -\zeta_n l_\mu + O(\delta t). \tag{23.19}$$

This leads to the estimate

$$t^\mu_1 \approx -e^{nh}\, t^\mu_2 \approx -(\zeta_n/s)\, l_\mu. \tag{23.20}$$

Warning: This estimate should be thought of as an approximation for the dominant *components* of the various vectors involved. If one takes the norm of these vectors one finds

$$1 \approx e^{nh} \approx 0. \tag{23.21}$$

This is true in the sense that other components are larger, but indicates forcefully the potential difficulties in this approach.

One is now ready to tackle estimation of the structure tensor $t_{\mu\nu}(x;\gamma)$. Using equations (23.6) and (23.20) one obtains

$$t_{\mu\nu}(x;\gamma) \approx -\frac{\zeta_n^2}{3s^2}\left[1 + 4e^{nh} + e^{2nh}\right]l_\mu l_\nu. \tag{23.22}$$

Pulling these various estimates together, the approximate stress-energy tensor is seen to be

$$\langle 0|T_{\mu\nu}(x)|0\rangle = -\hbar \sum_{n=1}^{\infty}\left\{\frac{\Delta_{\gamma_n}(x,x)^{1/2}}{24\pi^2\zeta_n}\frac{\left[1 + 4e^{nh} + e^{2nh}\right]}{\left[e^{nh} - 1\right]^3}\right\}\frac{l_\mu l_\nu}{(\delta t)^3} + O(\delta t^{-2}). \tag{23.23}$$

This, finally, is exactly the estimate obtained by Klinkhammer [167]—his equation (8). Furthermore this result is consistent with that of Kim and Thorne [164]—their equation (67). The somewhat detailed presentation of this derivation has served to illustrate several important points.

- First, the present result is a special case of the more general result (23.8), the present result being obtained only at the cost of many additional technical assumptions. The previous analysis has shown that the singularity structure of the stress-energy tensor may profitably be analyzed without having to restrict attention to regions near the fountain of a compactly generated chronology horizon. The existence of at least one short, self-intersecting, nontrivial, spacelike geodesic is a sufficient requirement for the extraction of useful information.

- Second, the introduction of this canonical observer allows one to extract information that is different from that obtained by the more geometric analysis of the previous section. The two approaches are complementary. Warning: the approximations required to go from (23.8) to (23.23) are subtle and potentially misleading. For instance, calculating the scalar invariant T_0 from (23.23), the leading $(\delta t)^{-3}$ term vanishes (because l^μ is a null vector). The potential presence of a subleading $(\delta t)^{-5/2}$ cross term cannot be ruled out from the present approximation, (23.23). Fortunately, we already know [from the original general analysis, (23.8)] that the dominant behavior of

T_0 is $T_0 \propto \sigma^{-2} \propto s^{-4}$. In view of the fact that, under the present restrictive assumptions, $\sigma \propto \delta t$, one sees that $T_0 \propto (\delta t)^{-2}$. The cross term, whatever it is, must vanish.

- Third, another warning—this derivation serves to expose, in excruciating detail, that the calculations encountered in this problem are sufficiently subtle that two apparently quite different results may nevertheless be closely related.

23.3 Wormhole disruption?

To get a feel for how this divergence in the vacuum polarization modifies the geometry, recall that a traversable wormhole must be threaded by some exotic stress-energy to prevent the throat from collapsing. In particular, at the throat itself (working in Schwarzschild coordinates) the total stress-energy tensor takes the form

$$T_{\mu\nu} = \frac{\hbar}{\ell_P^2 R^2} \begin{bmatrix} \xi & 0 & 0 & 0 \\ 0 & \chi & 0 & 0 \\ 0 & 0 & \chi & 0 \\ 0 & 0 & 0 & -1 \end{bmatrix}. \tag{23.24}$$

[See equation (11.47) on p. 108.] On general grounds one knows that $\xi \leq 1$; on the other hand χ is unconstrained. In particular, to prevent collapse of the wormhole throat, the scalar invariant T must satisfy

$$T = \frac{\hbar}{\ell_P^2 R^2} \sqrt{1 + \xi^2 + 2\chi^2} \approx \frac{\hbar}{\ell_P^2 R^2}. \tag{23.25}$$

On the other hand, consider the geodesic that starts at a point on the throat and circles around to itself passing through the throat of the wormhole exactly once. That geodesic, by itself, contributes to the vacuum polarization effects just considered an amount

$$\langle 0|T_{\mu\nu}(x)|0\rangle = \hbar \frac{\Delta_\gamma(x,x)^{1/2}}{\pi^2 s_\gamma(x,x)^4} t_{\mu\nu}(x;\gamma) + O(s_\gamma(x,x)^{-3}). \tag{23.26}$$

So the contribution of the single pass geodesic to the invariant T_0 is already

$$T_0 = \hbar \frac{\Delta_\gamma(x,x)^{1/2}}{\pi^2 s_\gamma(x,x)^4} \sqrt{1 - \frac{4}{3}(t^1 \cdot t^2) + \frac{2}{3}(t^1 \cdot t^2)^2} \approx \hbar \frac{\Delta_\gamma(x,x)^{1/2}}{\pi^2 s_\gamma(x,x)^4}. \tag{23.27}$$

Therefore, vacuum polarization effects dominate over the wormhole's internal structure once

$$s_\gamma(x,x)^2 \ll \Delta_\gamma(x,x)^{1/4} \ell_P R. \tag{23.28}$$

The onset of vacuum polarization induced wormhole disruption is thus seen
to be governed by the size of the van Vleck determinant. This is ultimately
the reason that such extensive computations of the van Vleck determinant
have been included in this monograph.

Consider a geometry containing a single traversable wormhole. Pick a
point x on the wormhole throat, and a geodesic γ that loops once around
the wormhole. The thin-wall approximation for the throat of the wormhole
leads to a van Vleck determinant equal to unity: $\Delta_\gamma(x,x) = 1$. Vacuum
polarization effects overwhelm the wormhole's internal structure once

$$s_\gamma(x,x)^2 \ll \ell_P R. \qquad (23.29)$$

Since by assumption $R \gg \ell_P$ (the wormhole is assumed to be macroscopic),
this disruption occurs for $s_\gamma(x,x) \gg \ell_P$, which fact I interpret as supporting
Hawking's chronology protection conjecture. (Warning: There are some
disagreements in the literature as to where exactly Planck scale physics
sets in. See p. 341 herein and references [164, 129, 130, 268, 270].)

For a "Roman configuration" [a configuration of two wormholes, where
each wormhole (when considered individually) is not near to forming a
time machine], the estimate of the van Vleck determinant provided in the
previous chapter permits the wormhole disruption criterion to be given as

$$s|_{\text{disrupt}} \approx \left[R_2 \left(\frac{1}{\ell_{1\to2}} + \frac{1}{\ell_{2\to1}} \right) \right]^{1/4} \sqrt{\pi \ell_P R_1} \qquad (23.30)$$

$$\approx \ell_P \left[\frac{R_1^2 R_2}{\ell_P^2} \left(\frac{1}{\ell_{1\to2}} + \frac{1}{\ell_{2\to1}} \right) \right]^{1/4}. \qquad (23.31)$$

Here R_1 is the radius of the mouth of wormhole 1 and R_2 that of wormhole 2. The disruption criterion just quoted characterizes the disruption
of wormhole 1 by the vacuum polarization induced via the presence of
wormhole 2. This criterion is not, and need not, be symmetric under the
interchange $1 \leftrightarrow 2$.

The best possible case for the possibility of time travel is obtained if
one makes $\Delta_\gamma(x,x)$ and hence $s|_{\text{disrupt}}$ as small as possible, and pushes
the disruption scale down into the Planck slop. This may be achieved by
making R as small as possible, and ℓ as large as possible. If one tries to
force $s|_{\text{disrupt}} < \ell_P$ one acquires a constraint on the relevant wormhole
parameters. Indeed to do so one must enforce

$$R_1^2 R_2 < \ell_P^2 \frac{\ell_{1\to2}\ell_{2\to1}}{\ell_{1\to2} + \ell_{2\to1}} < \ell_P^2 \ \max(\ell_{1\to2}, \ell_{2\to1}) < \ell_P^2 R_{\text{universe}}. \qquad (23.32)$$

Here the best possible case for time travel has been made by relaxing the
separation of the wormholes as much as possible—surely the radius of the

universe is a good upper bound on the distance between the wormholes. Take $R_{universe} \approx 3$ Gpc $\approx 6 \times 10^{60} \ell_P$. Then

$$R_1^2 R_2 < 6 \times (10^{20} \ell_P)^3 \approx (3 \times 10^{-15} \text{ m})^3. \tag{23.33}$$

So even with these ludicrously large separations between the wormhole mouths, one can only push the disruption scale down into the Planck slop by building the putative time machine out of ludicrously small "traversable" wormholes with a radius of the order of *femtometres*. Since any would-be time traveler would have to traverse the distance between wormholes 1 and 2 in normal space, he/she/it had better also be patient—a lifetime on the order of gigayears would be appropriate.

If one attempts to build a "two-wormhole time machine" on a more modest human scale one might take as a good bound

$$\max(\ell_{1 \to 2}, \ell_{2 \to 1}) < 1 \text{ AU} \approx 9 \times 10^{45} \ell_P. \tag{23.34}$$

In this case, pushing the disruption scale down into the Planck slop requires

$$R_1^2 R_2 < (2 \times 10^{15} \ell_P)^3 \approx (10^{-20} \text{ m})^3 \approx \left(\frac{\hbar}{20 \text{ TeV}/c}\right)^3. \tag{23.35}$$

So even if one lays hands on a couple of traversable wormholes, and initiates suitable solar system scale engineering projects, any decent sized wormhole will be disrupted by vacuum polarization effects before time travel is achieved. If the wormholes in question are small enough one can push the disruption scale down below the Planck regime. This, of course does not mean one has proved that time travel is actually possible—all it means is that one has "fine-tuned" the system sufficiently to be in a parameter regime where one cannot trust even rudimentary calculations. Even if one were to then succeed in building a closed timelike loop, the small size of the wormhole would preclude anything short of a 20 TeV quantum from getting through. This implies that one would need two accelerators of energy comparable to the superconducting supercollider just to get a one-bit message through the putative time machine. Even if all of these constraints are satisfied, any single trip is limited to a maximum backward time-jump of less than $(1 \text{ AU}/c) \approx 8$ minutes. This does not seem to be a useful workable recipe for studying tomorrow's *Wall Street Journal*.

23.4 Where is the Planck slop?

While all researchers in the field agree that calculations should no longer be trusted once one enters the Planck regime, there is some disagreement as to where exactly the Planck slop is first encountered. There are various manners in which one might attempt to characterize the onset of Planck scale physics. The estimates of Kim and Thorne [164] make extensive use of the "proper time to the chronology horizon" as measured by an observer who is stationary with respect to one of the wormhole mouths. Hawking [129, 130] advocates the use of an "invariant distance to the chronology horizon", which is equal to the proper time to the chronology horizon as measured by an observer who is stationary with respect to the frame of simultaneity (synchronous frame). On the other hand, in this monograph I focus attention on the invariant length of self-intersecting geodesics.

To see the inadequacy of the use of "proper time to the chronology horizon" at a conceptual level, observe that the use of this parameter intrinsically "begs the question" of the creation of a time machine by explicitly asserting the existence of a chronology horizon and then proceeding to measure distances from that presumed horizon. In particular, one may consider a wormhole in which one keeps the distance between the mouths fixed but arbitrarily close to the onset of time machine formation. (i.e., set the relative velocity of the mouths to zero, but set $\ell = T + \epsilon$). By definition, this implies that the chronology horizon never forms and that the "proper time to the chronology horizon" is infinite even though

$$s^2 = \ell^2 - T^2 \approx 2T\epsilon \qquad (23.36)$$

is arbitrarily small.

The precise definition of the "proper time to the chronology horizon" is as follows: Consider a pair of wormhole mouths moving with relative velocity $\beta_{\text{rel}}^{\text{rest}}$ (not to be confused with the totally different $\beta \equiv T/\ell$ associated with the transformation from the rest frame to the synchronous frame). Go to the rest frame of one mouth. In that rest frame the velocity of the other mouth will be $\beta_{\text{rel}}^{\text{rest}}$. For a small relative velocity, the wormhole is modeled by the identification of world-lines

$$(t, 0, 0, 0) \equiv (t + T, 0, 0, \ell - \beta_{\text{rel}}^{\text{rest}} t). \qquad (23.37)$$

Thus

$$s^2(t) = (\ell - \beta_{\text{rel}}^{\text{rest}} t)^2 - T^2. \qquad (23.38)$$

(There is nothing particularly sacred about taking the relative velocity to be small—it just simplifies life in that one only has a simple linear equation to solve instead of a quadratic.) The chronology horizon forms when $s^2(\Delta t) = 0$, that is, when

$$\Delta t = \frac{(\ell - T)}{\beta_{\text{rel}}^{\text{rest}}}. \qquad (23.39)$$

Note that this is a proper time as measured by an observer comoving with the first wormhole mouth.

To understand Hawking's "invariant distance to the chronology horizon", go to the synchronous frame, that is, make a Lorentz transformation of velocity $\beta \equiv T/\ell$, so that at time $t = 0$ the wormhole mouths connect equal times. In the synchronous frame, the individual wormhole mouths are moving with velocities β_1 and β_2. Then the wormhole may be described by the identification

$$(\gamma_1 t, 0, 0, \gamma_1 \beta_1 t) \equiv (\gamma_2 t, 0, 0, \gamma_2 \beta_2 t + s_0). \qquad (23.40)$$

Here s_0 denotes the initial separation (invariant interval) at time $t = 0$. The relative separation of the two wormhole mouths is

$$([\gamma_2 - \gamma_1]t, \ 0, \ 0, \ [\gamma_2 \beta_2 - \gamma_1 \beta_1]t + s_0), \qquad (23.41)$$

so that the invariant interval is

$$s(t)^2 = -[\gamma_2 - \gamma_1]^2 t^2 + (s_0 + [\gamma_2 \beta_2 - \gamma_1 \beta_1]t)^2. \qquad (23.42)$$

This invariant interval is zero once

$$[\gamma_2 - \gamma_1]t = s_0 + [\gamma_2 \beta_2 - \gamma_1 \beta_1]t, \qquad (23.43)$$

that is, at

$$t = \frac{s_0}{[\gamma_2 - \gamma_1] - [\gamma_2 \beta_2 - \gamma_1 \beta_1]}. \qquad (23.44)$$

To simplify things, define $\beta_{rel}^{synch} \equiv \beta_1 - \beta_2$, and assume this relative velocity is small. (Warning: $\beta_{rel}^{synch} \neq \beta_{rel}^{rest}$.) Then

$$[\gamma_2 - \gamma_1] \approx -(d\gamma/d\beta) \, \beta_{rel}^{synch};$$
$$[\gamma_2 \beta_2 - \gamma_1 \beta_1] \approx -[\beta(d\gamma/d\beta) + \gamma^{synch}] \, \beta_{rel}^{synch}. \qquad (23.45)$$

The chronology horizon forms at

$$\Delta \tau \approx \frac{s_0}{\gamma^{synch} \beta_{rel}^{synch}}. \qquad (23.46)$$

Note that $\Delta \tau$ is a proper time as measured by an observer sitting in the synchronous frame. (Technical point: observe that this is only a synchronous frame at $t = 0$.)

Hawking takes this object, $\Delta \tau = s_0/(\gamma^{synch} \beta_{rel}^{synch})$, which is indeed an invariant measure of the distance to the chronology horizon, as the parameter governing the strength of the singularities encountered when trying to build a time machine. Of course, this parameter suffers from deficiencies analogous to the Kim–Thorne parameter in that it is intrinsically incapable of properly reflecting the divergence structure that is known to occur in simple stationary (or indeed quasistationary) models.

Turning to the question of the quantum gravity cutoff, Kim and Thorne asserted that this cutoff occurs at

$$\Delta t = \frac{(\ell - T)}{\beta_{\text{rel}}^{\text{rest}}} \approx \ell_P \qquad \Rightarrow \qquad \ell - T \approx \beta_{\text{rel}}^{\text{rest}} \, \ell_P. \qquad (23.47)$$

In terms of the invariant interval, the Kim–Thorne cutoff is

$$s_0 = \sqrt{\ell^2 - T^2} \approx \sqrt{2 \beta_{\text{rel}}^{\text{rest}} \, \ell \, \ell_P}. \qquad (23.48)$$

Hawking suggested that the cutoff occurs at

$$\Delta \tau = \frac{s_0}{\gamma^{\text{synch}} \beta_{\text{rel}}^{\text{synch}}} \approx \ell_P \qquad \Rightarrow \qquad s_0 \approx \gamma^{\text{synch}} \beta_{\text{rel}}^{\text{synch}} \, \ell_P. \qquad (23.49)$$

I beg to differ. I feel that both of these proposed cutoffs exhibit unacceptable dependence on the relative motion of the wormhole mouths.

As an improved alternative cutoff, consider the following:

- Pick a point x in spacetime. Since, by hypothesis, the spacetime is assumed to have nontrivial topology there will be at least one closed geodesic of nontrivial homotopy that runs from x to itself. If the length of this geodesic is less than a Planck length, then the region surrounding the point x should no longer be treated semiclassically. (This version of the cutoff applies only to spacetimes containing traversable wormholes.)

- Presumably, one should also supplement this requirement by a bound on the curvature: If

$$R_{\alpha\beta\gamma\delta} R^{\alpha\beta\gamma\delta} \gtrsim \ell_P^{-4} \qquad (23.50)$$

then the region surrounding the point x should no longer be treated semiclassically. (This version of the cutoff applies even if there are no traversable wormholes in the spacetime.)

In the presence of wormholes, the second version of the cutoff arguably implies the first (modulo several technical assumptions). Suppose the conformal structure of the manifold is deemed to be held fixed (Weyl conformal tensor is held fixed). Then the condition $R_{\alpha\beta\gamma\delta} R^{\alpha\beta\gamma\delta} \gtrsim \ell_P^{-4}$ implies $R_{\alpha\beta} R^{\alpha\beta} \gtrsim \ell_P^{-4}$, which via the Einstein equations becomes

$$\sqrt{T_{\alpha\beta} T^{\alpha\beta}} \gtrsim \hbar \ell_P^{-4} = m_P \ell_P^{-3}, \qquad (23.51)$$

which is now the very sensible restriction that one does not want invariant measures of the size of the stress-energy tensor to exceed one Planck mass per Planck volume. By the analysis just completed, the point-split estimates for the expectation value of the quantum stress-energy tensor relate this back to a constraint on the length of the closed spacelike geodesics.

The advantage of the cutoff expressed in this manner is that this version of the cutoff continues to give meaningful answers even in the case of a stationary pair of wormhole mouths. Furthermore, this cutoff does not beg the question by initially requiring the existence of a chronology horizon to formulate the cutoff. Finally this cutoff can be concisely and clearly stated in complete generality for arbitrary spacetimes. For these reasons I believe this version of the cutoff to be eminently physically reasonable.

Warning: In case it has skipped the reader's attention, this issue still causes some mild disagreement among the various participants.

Chapter 24

Minisuperspace wormholes

24.1 Background

We have repeatedly seen, that whenever one is working with traversable wormholes, one encounters violations of the averaged null energy hypothesis. Since we know that the null energy hypothesis is experimentally violated by quantum effects, this suggests that a full-fledged quantum mechanical analysis of Minkowski signature wormholes is in order. The calculation presented in this chapter is an example of an attempt at evading the limitations of semiclassical quantum gravity.

This chapter—concentrating on the quantum mechanical aspects of Lorentzian wormholes—must immediately address the question of quantum gravity. Now, the fundamental principles of quantum gravity are as yet obscure. No satisfactory formulation of the problem exists. When confronted with interpretational and formal problems of this magnitude, one's only hope of being able to calculate is to resort to some (drastic) approximation scheme. The approximation scheme to be adopted is the minisuperspace restriction of the canonical Wheeler–DeWitt formalism.

Warning: General relativistic superspace (Wheeler's superspace) has nothing to do with particle physics superspace. General relativistic superspace is the set of all possible three-dimensional geometries (three-metrics). Modulo appropriate technical incantations general relativistic superspace can be thought of as the configuration space for classical general relativity. On the other hand, particle physics superspace is a technical formalism used in keeping track of the various technical issues encountered in models that possess supersymmetry (SUSY). These two uses of the word "superspace" are logically disjoint from each other.

The basic idea of the minisuperspace approach is to separate the three-metric into "modes" and then insist that all but a finite number of these "modes" (often one) be forced to satisfy the *classical* Einstein field equations. The remaining finite number of "modes" (*not* satisfying the Einstein field equations) are then quantized by following the standard prescription of canonical quantization. This approach is most commonly adopted in quantum cosmology calculations. Typically all the "translational" modes of the three-metric are "frozen out" by using the classical field equations, leaving only the "radius of the universe" (more precisely, the scale factor) to be quantized. In the approach adopted in this chapter, a Lorentzian wormhole is modeled by two spacetimes connected by a "hole". Everywhere except the "hole", the classical field equations are assumed to be satisfied, leaving only *the radius* of the hole as the *one* degree of freedom subject to quantization. (See, for example, [262, 263, 265, 266].)

By exactly solving the Wheeler–DeWitt equation for Einstein gravity on this minisuperspace the quantum mechanical wave function of the wormhole is obtained in closed form. These model wormholes are shown to be quantum mechanically stabilized with an average radius of order the Planck length. The analysis, based on the surgical grafting of two Reissner–Nordström spacetimes, proceeds by using a minisuperspace model to approximate the geometry of these wormholes. The thin-shell formalism is applied to this minisuperspace model to extract the effective Lagrangian appropriate to this one-degree-of-freedom system. This effective Lagrangian is then quantized and the wave function for the wormhole is explicitly exhibited. A slightly more general class of wormholes—corresponding to the addition of some "dust" to the wormhole throat—is analyzed by recourse to WKB techniques. In all cases discussed in this chapter the expectation value of the wormhole radius is calculated to be of order the Planck length. Accordingly, though these quantum wormholes are of theoretical interest they do not appear to be useful as a means for interstellar travel. The results of this chapter, if they prove to persist beyond the minisuperspace approximation, may also have a bearing on the question of topological fluctuations in quantum gravity. These calculations serve to suggest that topology-changing effects might in fact be *suppressed* by quantum gravity effects.

24.2 The model

To construct the class of wormholes of interest, use the "junction condition" formalism, also known as the "boundary layer" formalism, as discussed earlier in this monograph. The model wormhole is a minor generalization of the surgically modified Schwarzschild spacetime considered on pp. 177–186. Consider two copies of the Reissner–Nordström geometry. Both geometries are taken to be of mass M, while one geometry has charge $+Q$ and the other has charge $-Q$. One may temporarily wish to assume that $|Q| > M$ so that no event horizons exist and the Schwarzschild coordinate patch covers the complete geometry. This helps in visualizing the construction but our results are not restricted to the $|Q| > M$ case. Recall that the metric of the Reissner–Nordström geometry is

$$ds^2 = -(1 - 2M/r + Q^2/r^2)dt^2 + \frac{dr^2}{(1 - 2M/r + Q^2/r^2)}$$
$$+ r^2 \left(d\theta^2 + \sin^2 \theta \, d\varphi^2 \right). \tag{24.1}$$

The electromagnetic field is simply $E = Q/r^2$. From each copy of the Reissner–Nordström geometry one removes identical four-dimensional regions of the form

$$\Omega \equiv \{(t, r, \theta, \phi) | r < a(\tau)\}. \tag{24.2}$$

Here τ denotes proper time along the throat. One is then left with two geodesically incomplete manifolds whose boundaries are given by the time-like hypersurfaces

$$\partial\Omega \equiv \{(t, r, \theta, \phi) | r = a(\tau)\}. \tag{24.3}$$

Identify these two hypersurfaces (by "sewing" them together). The resulting spacetime $\mathcal{M}$ is geodesically complete. It possesses two asymptotically flat regions connected by a wormhole. The throat of this wormhole is at $r = a(\tau)$. Note that this spacetime is completely singularity-free because the region surrounding $r = 0$ has been explicitly excluded from the manifold. In particular, it is sometimes convenient to take $|Q| > M$; in this case the otherwise *naked* singularity has been excluded by this construction and need not further concern us. Because this manifold is piecewise Reissner–Nordström, the Ricci scalar is everywhere zero, except at the throat itself. Observe that the electric flux lines thread the throat of the wormhole and do not terminate. Thus, there is no electric charge present anywhere in the model wormhole. (This is an example of Wheeler's "charge without charge".) At the throat the Riemann curvature tensor is proportional to a delta function. The Ricci tensor at the junction can be calculated in terms of the extrinsic curvature (also known as the second fundamental form, or second metric groundform)

$$K_{ij}^\pm = \frac{1}{2} \left. \frac{\partial g_{ij}^\pm}{\partial \eta} \right|_{\eta=0}. \tag{24.4}$$

Here η denotes the normal coordinate to the throat. The Ricci tensor is almost everywhere (except at the throat) that of a Reissner–Nordström

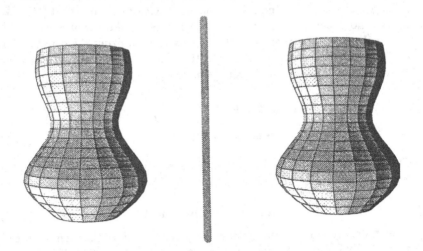

Figure 24.1: Minisuperspace wormhole. From two copies of the Reissner–Nordström geometry, delete two spherical regions of arbitrary time-dependent radius. Identify the surfaces of the excised regions. The resulting wormhole has only one degree of freedom.

geometry [260, 265]. Applying the thin-shell formalism:

$$R^\mu{}_\nu(x) = R^{\mathrm{RN}\,\mu}{}_\nu(x) - 2 \cdot \begin{bmatrix} K^i{}_j(x) & 0 \\ 0 & K(x) \end{bmatrix} \cdot \delta(\eta). \qquad (24.5)$$

The Einstein–Hilbert action reduces to

$$S_g = \frac{1}{16\pi} \int_{\mathcal{M}} \sqrt{g}\, R = -\frac{1}{4\pi} \int_{\partial\Omega} \sqrt{^3 g}\, K. \qquad (24.6)$$

By spherical symmetry, the extrinsic curvature contains only two nontrivial components: $K^\theta{}_\theta \equiv K^\phi{}_\phi$ and $K^\tau{}_\tau$. These components may conveniently be extracted from [260, 265] ($\dot a \equiv \frac{da}{d\tau}$: an overdot denotes a proper time derivative):

$$K^\theta{}_\theta \;\;\equiv\;\; K^\phi{}_\phi = \frac{1}{a} \cdot \sqrt{1 - \frac{2M}{a} + \frac{Q^2}{a^2} + \dot a^2}\,; \qquad (24.7)$$

$$K^\tau{}_\tau \;\;=\;\; \frac{\ddot a + \frac{M}{a^2} - \frac{Q^2}{a^3}}{\sqrt{1 - \frac{2M}{a} + \frac{Q^2}{a^2} + \dot a^2}}$$

$$= \frac{d}{d\tau} \sinh^{-1} \left(\frac{\dot{a}}{\sqrt{1 - \frac{2M}{a} + \frac{Q^2}{a^2}}} \right) + \frac{1}{a^2} \left(M - \frac{Q^2}{a} \right) \frac{dt}{d\tau}. \quad (24.8)$$

For completeness I point out that

$$\frac{dt}{d\tau} = \frac{\sqrt{1 - \frac{2M}{a} + \frac{Q^2}{a^2} + \dot{a}^2}}{1 - \frac{2M}{a} + \frac{Q^2}{a^2}}, \quad (24.9)$$

$$U^\mu \equiv \left(\frac{dt}{d\tau}, \frac{da}{d\tau}, 0, 0 \right)$$

$$= \left(\frac{\sqrt{1 - \frac{2M}{a} + \frac{Q^2}{a^2} + \dot{a}^2}}{1 - \frac{2M}{a} + \frac{Q^2}{a^2}}, \dot{a}, 0, 0 \right), \quad (24.10)$$

$$n^\mu = \left(\frac{\dot{a}}{1 - \frac{2M}{a} + \frac{Q^2}{a^2}}, \sqrt{1 - \frac{2M}{a} + \frac{Q^2}{a^2} + \dot{a}^2}, 0, 0 \right). \quad (24.11)$$

Exercise: Instead of looking up these formulas, derive them. That is, generalize the discussion of dynamic wormholes (pp. 176–184) to the present case by systematically replacing the neutral Schwarzschild spacetime with the charged Reissner–Nordström spacetime.

Now, $\sqrt{^3g}\, d^3x \rightarrow 4\pi\, a^2\, d\tau$, so an integration by parts leads to

$$S_g = 2 \int \left\{ a\, \dot{a}\, \sinh^{-1} \left[\dot{a} \left(1 - \frac{2M}{a} + \frac{Q^2}{a^2} \right)^{-1/2} \right] \right.$$

$$\left. - a \sqrt{1 - \frac{2M}{a} + \frac{Q^2}{a^2} + \dot{a}^2} \right\} d\tau$$

$$- \int \left(M - \frac{Q^2}{a} \right) dt. \quad (24.12)$$

It should be noted that this calculation is a direct analog of the mini-superspace techniques more commonly used in quantum cosmology. The gravitational action has been written in this way with malice aforethought. We shall see that the second integral (the $\int dt$) is effectively absent from the total action. This integral is partly canceled by the electromagnetic contribution to the action, and the remaining portion $\int M\, dt$ is merely a reflection of the fact that the mass of the system imposes an "imprint at infinity" on the metric [186]. This is merely an alternative way of describing the Gibbons–Hawking surface contribution to the gravitational action. In each asymptotically flat region the electromagnetic contribution to the action is

$$S_{em}^{(1)} = S_{em}^{(2)} = -\frac{1}{16\pi} \int F^2,$$

$$= -\frac{1}{16\pi} \int 4\pi r^2 \cdot \frac{2Q^2}{r^4} dr \, dt$$

$$= \frac{Q^2}{2} \int_a^\infty \frac{dr}{r^2} dt = -\frac{Q^2}{2} \int \frac{dt}{a(t)}. \qquad (24.13)$$

A convenient technical trick at this stage is to cut off the Reissner–Nordström geometry at some large fixed radius ρ. Then attach this truncated Reissner–Nordström geometry to a piece of Minkowski space. This procedure contributes to the total gravitational action an amount

$$S_\rho^{(1)} = S_\rho^{(2)} = \frac{1}{2} \int [K] \rho^2 dt = \frac{1}{2} \int M \, dt + O(1/\rho). \qquad (24.14)$$

Here $[K]$ denotes the discontinuity in the second fundamental form at $r = \rho$. Now let $\rho \to \infty$ and combine terms to obtain

$$S_{\text{eff}} = S_{\text{g}} + S_{\text{em}}^{(1)} + S_{\text{em}}^{(2)} + S_\infty^{(1)} + S_\infty^{(2)}$$

$$= 2 \int \left\{ a \, \dot{a} \, \sinh^{-1} \left(\frac{\dot{a}}{\sqrt{1 - \frac{2M}{a} + \frac{Q^2}{a^2}}} \right) \right.$$

$$\left. - a \sqrt{1 - \frac{2M}{a} + \frac{Q^2}{a^2} + \dot{a}^2} \right\} d\tau. \qquad (24.15)$$

The effective Lagrangian is

$$L_{\text{eff}} = 2a \left\{ \dot{a} \, \sinh^{-1} \left(\frac{\dot{a}}{\sqrt{1 - \frac{2M}{a} + \frac{Q^2}{a^2}}} \right) - \sqrt{1 - \frac{2M}{a} + \frac{Q^2}{a^2} + \dot{a}^2} \right\}.$$

$$(24.16)$$

This effective Lagrangian includes contributions from the Einstein–Hilbert gravitational Lagrangian, the Maxwell electromagnetic Lagrangian, and the "imprint at infinity". Note that I have not used the Einstein field equations at any stage in this analysis: All that has been done is that the curvature tensor has been calculated as a function of the one degree of freedom $a(\tau)$, and that this curvature tensor has been substituted into the action to get a reduced minisuperspace action.

Exercise: Verify the steps used to calculate this effective Lagrangian.

24.3   Dynamics

24.3.1   Classical behavior

In order to get a better understanding of the classical behavior of this wormhole, I shall now add some matter to the model. This matter should be thought of as a "regulator", used to improve the classical dynamics of the wormhole (since, classically, pure vacuum is incompatible with wormhole behavior). I choose pressureless dust that is confined to lie on the throat of the wormhole. Then $m = 4\pi\sigma a^2$, the mass of the dust shell, is a constant of the motion. To deduce this I have used the conservation of stress-energy $T^{\mu\nu}{}_{;\nu} = 0$. Note that for any arbitrary geometry one automatically has $G^{\mu\nu}{}_{;\nu} = 0$. What I have not yet used, at this stage, are the classical Einstein field equations $G^{\mu\nu} = 8\pi G\, T^{\mu\nu}$. The matter Lagrangian simply reduces to $L_m = -m$, and the total Lagrangian describing the matter plus gravity system is

$$
L_{\text{tot}} = 2\left\{a\,\dot{a}\,\sinh^{-1}\left(\frac{\dot{a}}{\sqrt{1 - \frac{2M}{a} + \frac{Q^2}{a^2}}}\right) - a\sqrt{1 - \frac{2M}{a} + \frac{Q^2}{a^2} + \dot{a}^2}\right\}
$$
$$
-m. \tag{24.17}
$$

Aside: This way of treating the matter action is equivalent to that of DeWitt [61]. Another way of dealing with the matter action, adopted by Redmount and Suen [222, 223] is to use the classical Einstein equations plus some assumed equation of state for the matter to reduce the entire classical system to one equation of motion. One then searches for some suitable Lagrangian that reproduces the classical equation of motion and quantizes that Lagrangian. This procedure is typically not unique in that many different Lagrangians can lead to the same classical equation of motion. The precise relationship between these two techniques is still somewhat obscure.

Aside: A perhaps more legitimate way of introducing matter is by adding a new quantum degree of freedom, such as a conformally-coupled scalar field. It is not clear how to do this.

The total "reduced" action is, in my approach, still a function of the sole remaining, completely unconstrained, degree of freedom $a(\tau)$. The classical Wheeler–DeWitt Hamiltonian is now easily extracted; the momentum conjugate to a is

$$
p \equiv \frac{\partial L_{\text{tot}}}{\partial\dot{a}} = 2a\,\sinh^{-1}\left(\frac{\dot{a}}{\sqrt{1 - \frac{2M}{a} + \frac{Q^2}{a^2}}}\right). \tag{24.18}
$$

This relation may be easily inverted to yield

$$
\dot{a} = \sqrt{1 - \frac{2M}{a} + \frac{Q^2}{a^2}} \cdot \sinh(p/2a), \tag{24.19}
$$

so that the Wheeler–DeWitt Hamiltonian is

$$H_{\text{tot}}(p,a) \equiv p\dot{a} - L_{\text{tot}} \tag{24.20}$$

$$= 2a \cdot \sqrt{1 - \frac{2M}{a} + \frac{Q^2}{a^2}} \cdot \cosh\left(p/2a\right) + m \tag{24.21}$$

$$= 2a \cdot \sqrt{1 - \frac{2M}{a} + \frac{Q^2}{a^2} + \dot{a}^2} + m. \tag{24.22}$$

Exercise: Verify this derivation. These results are surprisingly simple given the rather messy form of the total Lagrangian .

The classical dynamics of the wormhole is now obtained by setting $H_{\text{tot}} = 0$. The fact that the Hamiltonian is zero is a standard consequence of the reparameterization invariance of the theory. To check the correctness of this calculation, observe that the constraint equation $H_{\text{tot}} = 0$ reproduces the classical Einstein field equations for the motion of the wormhole throat [260, 265]

$$\sqrt{1 - \frac{2M}{a} + \frac{Q^2}{a^2} + \dot{a}^2} = -\frac{m}{2a}. \tag{24.23}$$

Classically, it is easy to see from the Einstein equations of motion that m must be negative. We shall soon see that quantum effects permit well-behaved wave functions for $m = 0$ negative, zero, and even positive. This calculation of H_{tot} is a nontrivial check of correctness in that it shows that classically we can either:

1. consider arbitrary metric variations in the action to obtain the full Einstein field equations, and then use symmetry to simplify these field equations [260, 265], or

2. use symmetry to simplify the action *ab initio*, and then use restricted variations of the metric to obtain the same physics. This procedure does not always work. That it does work in this case is essential to the quantization procedure given below.

In fact, if a classical analysis is all that is required, then proceeding from the Einstein field equations is both more direct and less subject to subtle interpretational disputes. However, when it comes to quantizing the system, the Hamiltonian approach just exhibited will be much more useful.

Finally, I add some extra technical comments concerning the truncation procedure. The truncation procedure forces the geometry at large radius to be exactly Minkowski (rather than asymptotically Minkowski), thus allowing use of the simple version of the Hamiltonian constraint $H_{\text{eff}} = 0$. An alternative procedure is available. If one writes the Lagrangian and

Hamiltonian in terms of the Schwarzschild time coordinate and does not truncate the geometry at large radius, then it is a standard result of the ADM formalism that $H_{\text{Sch.}} = 2M$, where each asymptotically flat region contributes an amount M to the ADM mass (this is the "imprint at infinity"). By defining a new "effective" Lagrangian

$$S_{\text{eff}} = \int (L_{\text{Sch.}} - 2M)dt \qquad (24.24)$$

one finds that the "effective" Hamiltonian corresponding to this "effective" action satisfies $H_{\text{eff}} = 0$. It is this S_{eff} which, after it is rewritten in terms of the proper time coordinate, is equivalent to the truncation procedure previously outlined. This makes explicit our previous comment that the truncation procedure is related to the "imprint at infinity" of the wormhole's mass. With the Hamiltonian of the wormhole model in hand, we now turn to the quantum dynamics of the model wormhole.

24.3.2 Canonical quantization

From the classical equations of motion we see that

$$\dot{a} = \pm\sqrt{-1 + \frac{2M}{a} - \frac{Q^2 - m^2/4}{a^2}}. \qquad (24.25)$$

Thus large values of a are classically forbidden ($\dot{a}$ is imaginary), while for small a the behavior depends on the relative magnitudes of $Q^2 - \frac{m^2}{4}$ and M^2. The classical turning points occur at

$$a_{\pm} = M \pm \sqrt{M^2 - Q^2 + m^2/4}. \qquad (24.26)$$

Case $(Q^2 - m^2/4) < M^2$: There are two real classical turning points and the system oscillates between these turning points. For large enough values of $|m|$ only one of these turning points is physical, at $a_{\max} \approx 2M + |m|/2$. The second "turning point" is then $a_{\min} = 0$, so that our picture of the motion is simple: the wormhole "emerges" from $a = 0$ with infinite velocity, expands to a maximum radius of order $|m|$, and recollapses to $a = 0$ in finite proper time (also of order $|m|$). Note that even if $Q > M$, we can with large enough $|m|$ ensure that $(Q^2 - m^2/4) < M^2$, so that the comments of this paragraph apply.

Case $Q^2 - m^2/4 = M^2$: The two turning points coalesce at $a = M$.

Case $Q^2 - m^2/4 > M^2$: Both turning points are unphysical (complex). The entire range $a \in (0, \infty)$ is classically forbidden, and careful attention to suitable limiting procedures indicates that the classical solution is $a \equiv 0$.

Case $Q = M = 0$: This case is interesting in its own right because of its
simplicity. See [262]. The turning points are

$$a_{\min} = 0 \quad \text{and} \quad a_{\max} = |m|/2, \tag{24.27}$$

while at small times and small distances

$$a(\tau) \approx \sqrt{|m|\tau/2}. \tag{24.28}$$

Then $a_{\max} \to 0$ as $|m| \to 0$. It follows that for $m = 0$ the classical wormhole
always remains at $a = 0$. Adding quantum effects serves to "smear out"
this classically pointlike object.

With the classical dynamics of this model now understood, and the
relevant Wheeler–DeWitt Hamiltonian in hand, quantization is straightforward. The only remarkable aspect of the analysis is that in some cases
closed form exact expressions are obtained. Canonical quantization proceeds via the usual replacement

$$p \mapsto -i\hbar \frac{\partial}{\partial a}. \tag{24.29}$$

Naturally, the resulting quantum Hamiltonian has a factor ordering ambiguity. This factor ordering ambiguity may be removed in a natural (though
not unique!) way by demanding that the quantum Hamiltonian be Hermitian:

$$\hat{H}_{\text{tot}} = \sqrt[4]{1 - \frac{2M}{a} + \frac{Q^2}{a^2}} \cdot 2a \cdot \cos\left(\frac{\ell_P^2}{2}\frac{1}{a}\frac{\partial}{\partial a}\right) \cdot \sqrt[4]{1 - \frac{2M}{a} + \frac{Q^2}{a^2}}$$

$$+ m. \tag{24.30}$$

That this Hamiltonian is Hermitian may formally be seen by Taylor series
expansion of the cosine. A more careful statement, taking into account appropriate boundary conditions, is that this Hamiltonian acts on $L^2[0, \infty)$,
the space of square integrable functions on $[0, \infty)$. However, when we discuss the $m = 0$ case we shall see that in this instance the boundary conditions may be deduced from the Wheeler–DeWitt equation rather than
being put in by hand. The wave function of the wormhole is determined in
the usual fashion by the Wheeler–DeWitt equation

$$\hat{H}_{\text{tot}}\,\psi(a) = 0. \tag{24.31}$$

The Wheeler–DeWitt equation is the generalization of the Schrödinger
equation appropriate to canonically quantized gravity. This equation may
be rewritten as

$$\hat{H}_{\text{eff}}\,\psi = -m\,\psi. \tag{24.32}$$

Thus m may be interpreted as an eigenvalue of the effective Hamiltonian
associated with L_{eff}. The mass of the dust shell is therefore quantized in

this formalism. This behavior is similar to that seen by DeWitt [61], where a minisuperspace quantization of a Friedmann–Robertson–Walker (FRW) universe led to a quantization condition on the mass of the dust which that universe contained.

24.4 Wave functions

24.4.1 $m = 0$: Exact solution

For nonzero values of m exact solutions of the Wheeler–DeWitt equation have proved elusive, and one must resort to WKB techniques. For the special eigenvalue of $m = 0$ exact solutions of the Wheeler–DeWitt equation may be written down by inspection:

$$\psi_n(a) = \frac{\exp\left[-(n + \frac{1}{2})\pi(a/\ell_P)^2\right]}{\sqrt[4]{1 - \frac{2M}{a} + \frac{Q^2}{a^2}}}. \tag{24.33}$$

Here n is integer-valued quantum number describing the internal state of the wormhole. Wave functions with negative values of n, not being normalizable, are discarded in the usual fashion. Note that the dynamics thus implies that $\psi(0) = 0$.

Exercise: (Easy) Verify that this wave function satisfies the Wheeler–DeWitt equation.

The expectation value of the wormhole radius is

$$\langle a \rangle \equiv \frac{\langle \psi_n|a|\psi_n \rangle}{\langle \psi_n|\psi_n \rangle} \approx \ell_P. \tag{24.34}$$

The apparent occurrence of singularities at the classical horizons is not at all a problem in that the wave function is square integrable over these "poles". Physical quantities do not pick up infinities from the horizons and the presence of these horizons is not a matter of concern.

While not a cause for concern, the presence of horizons does complicate the global geometry of the wormhole. Recall, that if $|Q| > M$, then horizons do not occur and the global geometry is correspondingly simple. On the other hand, if $|Q| \leq M$, the Schwarzschild coordinate patch does not completely cover the Reissner–Nordström geometry, and the global geometry of the wormhole is more complex. For instance if $M > 0$, $Q = 0$, the global geometry may be described as follows:

1. Take two Kruskal diagrams (appropriate to the description of the Kruskal–Szekeres maximally extended Schwarzschild solution).

2. Trim a small fringe off the future singularity of one diagram (the black hole).

3. Trim a small fringe off the past singularity of the other diagram (the white hole).

4. Then sew the two diagrams together along the trimmed fringes.

The result is a model whereby one can make physical sense of the oft repeated hope that matter which falls down a black hole will reappear in a white hole somewhere else in the "multiverse". Unfortunately, were one to undertake such a trip one would reappear in a future incarnation of the universe, rather than in a distant part of our own universe, in the meantime having been squeezed down to sizes of order the Planck length. This is not a useful method of interstellar travel.

While it is clear from the exponential decay of the wave function that the average radius of the wormhole will be of order the Planck length, it is possible is some cases to make more precise statements. The general $M \neq 0$, $Q \neq 0$ case is intractable. However, for $M = 0$, $Q \neq 0$ exact results may be obtained:

$$
\begin{aligned}
\langle a \rangle &= \frac{\int a \, |\psi_n|^2 \, da}{\int |\psi_n|^2 \, da} \\[2mm]
&= \frac{\int \frac{a^2}{\sqrt{a^2+Q^2}} e^{-\{(2n+1)\pi a^2/\ell_P^2\}} \, da}{\int \frac{a}{\sqrt{a^2+Q^2}} e^{-\{(2n+1)\pi a^2/\ell_P^2\}} \, da} \\[2mm]
&= Z^2 \alpha \frac{\sqrt{2n+1}}{2} e^{-\{Z^2\alpha(n+\frac{1}{2})\}} \\[2mm]
&\quad \times \frac{K_1(Z^2\alpha\pi(n+\frac{1}{2})) - K_0(Z^2\alpha\pi(n+\frac{1}{2}))}{\operatorname{erfc}(\sqrt{Z^2\alpha\pi(2n+1)})}
\end{aligned}
\tag{24.35}
$$

Here K_0 and K_1 are modified Bessel functions, erfc is the complementary error function, and $Q \equiv Ze \equiv Z\sqrt{\alpha}\ell_P$. And yes, α really is the fine structure constant. This result, while exact, is in its present form unenlightening. The situation may be somewhat improved by using asymptotic expansions to show that for large n:

$$
\langle a \rangle \approx \frac{1}{4} \frac{\ell_P}{\sqrt{2n+1}}.
\tag{24.36}
$$

Alternatively, for the $n = 0$ mode, one may use the smallness of the fine structure constant ($\alpha \approx 1/137$) to obtain

$$
\langle a \rangle = \frac{\ell_P}{\pi} + O(Z^2\alpha).
\tag{24.37}
$$

To see the physical regime in which these calculations may be of interest, recall that for elementary particles such as the electron $Q/M \approx 10^{+22}$. It is only in the realm of charged elementary particles that the $M = 0$, $Q \neq 0$ case is likely to be a good approximation to physics.

Aside: If one tries to take this too seriously—as a real model for real elementary particles—one will at a minimum have to add spin angular momentum to the discussion. And no, I do not think that one can build a realistic model for elementary particles this way.

One may, on the other hand, consider the "astrophysical case". Naturally occurring black holes are expected to have $Q \ll M$ and $M \gg m_P$. Let us approximate by setting $Q = 0$, and estimate $\langle a \rangle$ by Taylor series expanding the square root occurring in the integral for $\langle a \rangle$. It is this "astrophysical case" whose global geometry was considered previously. A brief calculation yields

$$\langle a \rangle = \frac{\Gamma(5/4)}{\Gamma(3/4)\sqrt{\pi}} \cdot \frac{\ell_P}{\sqrt{2n+1}} + O(\ell_P^2/[(2n+1)M]). \qquad (24.38)$$

Again, though the detailed calculations are tedious, they support the assertion that $\langle a \rangle \approx \ell_P$. It might be argued that with hindsight this result is not surprising on grounds of dimensional analysis. To see that this is not quite true, observe that the model wormhole possesses *three* independent length scales:

1. the Planck length ℓ_P,

2. the Schwarzschild radius $2M$, and

3. the "charge radius" $Q \equiv Z \ell_Q \equiv Z\sqrt{\alpha}\,\ell_P$.

It is worth pointing out the exact sense in which I am claiming the wormhole to be stable—it is *a priori* quite possible that the minisuperspace calculation could have led to a Wheeler–DeWitt equation whose solution was a non-normalizable wave function that blew up as $a \to 0$. With such a wave function one could at best define $\langle a \rangle = 0$, indicating that the wormhole would be overwhelmingly likely to have collapsed to a point. In fact, of course, the situation is very much better than that unpleasant possibility, the calculated wave function behaving as $\psi(a) \to \sqrt{a/Q}$ as $a \to 0$. This boundary condition, coming directly from solving the Wheeler–DeWitt equation, does not have to be put in "by hand".

24.4.2 $m \neq 0$: Degeneracy

Once one adds dust to the wormhole throat relatively few exact statements can be made. It is, however, possible to show that the mass eigenvalues possess an infinite degeneracy—this can be traced back to the fact that the Hamiltonian is essentially a trigonometric function. Let us proceed by noting the identity:

$$
\begin{aligned}
\cos\left(\partial_x\right)\left[f(x)g(x)\right] &\equiv \cos\left(\partial_y + \partial_z\right)\left[f(y)g(z)\right]\Big|_{x=y=z} \\
&= \Big\{ \cos\left(\partial_y\right)\cos\left(\partial_z\right) \\
&\quad - \sin\left(\partial_y\right)\sin\left(\partial_z\right)\Big\}\left[f(y)g(z)\right]\Big|_{x=y=z} \\
&= \left[\cos\left(\partial_x\right)f(x)\right]\left[\cos\left(\partial_x\right)g(x)\right] \\
&\quad - \left[\sin\left(\partial_x\right)f(x)\right]\left[\sin\left(\partial_x\right)g(x)\right].
\end{aligned}
\tag{24.39}
$$

With a little bit of work one can use this identity to evaluate

$$
\begin{aligned}
\hat{H}_{\text{eff}}\left[e^{-\beta(a/\ell_P)^2}\psi\right] &= \\
\cos(\beta)e^{-\beta(a/\ell_P)^2}\hat{H}_{\text{eff}}[\psi] &+ \sin(\beta)\cdot e^{-\beta(a/\ell_P)^2} \\
\times \sqrt[4]{1 - \tfrac{2M}{a} + \tfrac{Q^2}{a^2}}\,2a\,\sin&\left(\frac{\ell_P^2}{2a}\frac{\partial}{\partial a}\right)\left[\sqrt[4]{1 - \tfrac{2M}{a} + \tfrac{Q^2}{a^2}}\,\psi(a)\right].
\end{aligned}
\tag{24.40}
$$

Thus, suppose we take ψ to be an eigenfunction of $\hat{H}_{\text{eff}}$ with $\hat{H}_{\text{eff}}\psi = m\psi$, then

$$
\hat{H}_{\text{eff}}\left[e^{-2\pi n(a/\ell_P)^2}\psi\right] = +m\left[e^{-2\pi n(a/\ell_P)^2}\psi\right],
\tag{24.41}
$$

$$
\hat{H}_{\text{eff}}\left[e^{-2\pi(n+\frac{1}{2})(a/\ell_P)^2}\psi\right] = -m\left[e^{-2\pi(n+\frac{1}{2})(a/\ell_P)^2}\psi\right].
\tag{24.42}
$$

So one sees that the infinite degeneracy occurring in the $m = 0$ case is no accident. Moreover, this informs us that quantum mechanically positive values of m are as well-behaved as negative values.

24.4.3 $m \neq 0$: WKB techniques.

For the case $m \neq 0$ exact solutions have proved elusive, and recourse has been made to WKB techniques (see, for instance, [15]). Since the Hamiltonian is not quadratic in momenta, a slight variant of the usual WKB technology is appropriate. Consider an *arbitrary* classical Hamiltonian $H(p, q)$, we wish to find the WKB approximations to the true energy eigenvalues and eigenfunctions

$$\hat{H}(\hat{p}, \hat{q})\psi = E\,\psi. \tag{24.43}$$

Proceed as follows:

1. Set $H(p, q) = E$ and invert to obtain $p(E, q)$, the momentum at the point q of a classical trajectory of energy E.

2. Quantize the energy eigenvalues by setting

$$\oint p(E, a)da = (\ell + \delta)\hbar. \tag{24.44}$$

Here δ is a number that depends on both boundary conditions and the Hamiltonian $H(p, q)$. In the old Bohr–Sommerfeld quantization δ is just taken to be zero. For a Hamiltonian quadratic in momenta, the usual WKB method shows that δ is typically a simple fraction (e.g., 1/2, 3/4, etc.). For a Hamiltonian non-quadratic in momenta, δ must be evaluated on a case by case basis (δ is often transcendental). Since, for the purposes of this chapter, a precise calculation of δ would add little to our understanding, δ will not be evaluated but shall merely be carried along as an arbitrary constant.

3. In the classically allowed region

$$\psi_{\text{WKB}}(q) = \frac{1}{\sqrt{\left|\frac{\partial H}{\partial p}(p(E, q), q)\right|}} \exp\left\{\pm i \int^q p(E, x)dx/\hbar\right\}, \tag{24.45}$$

while in the classically forbidden region

$$\psi_{\text{WKB}}(q) = \frac{1}{\sqrt{\left|\frac{\partial H}{\partial p}(p(E, q), q)\right|}} \exp\left\{\pm \int^q |p(E, x)|dx/\hbar\right\}. \tag{24.46}$$

Observe that in the allowed region

$$|\psi_{\text{WKB}}(q)|^2 = \frac{1}{\left|\frac{\partial H}{\partial p}(p(E, q), q)\right|} = \frac{1}{|\dot{q}(E, q)|}. \tag{24.47}$$

As is usual, the particle is most likely to be in those regions where classically it travels the slowest. It is easy to see that when $H = \frac{p^2}{2m} + V(q)$ that this generalized prescription reduces to the usual WKB approximation. This generalized WKB approximation may be systematically derived in the usual manner from the first two terms of a formal power series expansion in $\hbar$.

Applying this formalism to the problem at hand, in place of E we write m, the mass of the dust shell which is to be quantized. We note that $p(m, a)$ is a *multivalued* function

$$
\begin{aligned}
p(m, a) &= 2a \, \text{arccosh} \left(-\frac{m}{2a} \cdot \frac{1}{1 - \frac{2M}{a} + \frac{Q^2}{a^2}} \right) \\
&= 2a \left[\pm \cosh^{-1} \left(-\frac{m}{2a} \cdot \frac{1}{1 - \frac{2M}{a} + \frac{Q^2}{a^2}} \right) + 2\pi i \cdot n \right] . \text{(24.48)}
\end{aligned}
$$

Here the function $\cosh^{-1}$ is taken to map the domain $[1, \infty)$ to the range $[0, \infty)$, and the $+/-$ denotes outgoing/ingoing directions. The quantization condition on m reads

$$
\oint p(m, a) da = 2 \int_{a_{\min}}^{a_{\max}} 2a \, \cosh^{-1} \left(-\frac{m}{2a} \cdot \frac{1}{1 - \frac{2M}{a} + \frac{Q^2}{a^2}} \right) = (\ell + \delta)\hbar.
$$

$$(24.49)$$

Note that the imaginary contribution to $p(m, a)$, being a total derivative, does not contribute to the quantization condition. The WKB estimate for the eigenvalue m is thus implicitly given as a function of ℓ: $m = m(\ell)$. Note that each m eigenvalue has an infinite degeneracy with respect to n. The quantum number n does, however, contribute when estimating the WKB wave function,

$$
\psi_{\text{WKB}}(a) = \frac{\exp\{i \int^a p \, dx / \hbar\}}{\sqrt{\dot{a}(m, a)}} = \frac{\exp\{-2\pi n (a/\ell_P)^2\}}{\sqrt[4]{1 - \frac{2M}{a} + \frac{Q^2 - m^2/4}{a^2}}} \cdot e^{i\Theta(a)}, \quad (24.50)
$$

so that in the classically allowed region the wave function is indeed oscillatory, but with an envelope that for $n > 0$ is exponentially damped. Note that we have in this manner recovered the degenerate modes discussed previously. If we now flip $m \to -m$ and use

$$
\cosh^{-1}(-x) = \cosh^{-1}(x) + i\pi, \quad (24.51)
$$

we find that

$$
\psi_{\text{WKB}}(a) = \frac{\exp\{i \int^a p \, dx / \hbar\}}{\sqrt{\dot{a}(m, a)}} = \frac{\exp\{-(2n + 1)\pi (a/\ell_P)^2\}}{\sqrt[4]{1 - \frac{2M}{a} + \frac{Q^2 - m^2/4}{a^2}}} \cdot e^{i\Theta(a)} \quad (24.52)
$$

are WKB eigenmodes corresponding to an eigenmass $-m$. Suppose we go outside the classically allowed region, to large values of a, then

$$
\cosh^{-1} \left(-\frac{m}{2a} \cdot \frac{1}{1 - \frac{2M}{a} + \frac{Q^2}{a^2}} \right) \approx \cosh^{-1}(0) = i\pi/2 \quad (24.53)
$$

so that as $a \to \infty$ one has

$$p(m, a) \to 2a \cdot 2\pi i(n + \tfrac{1}{4}); \tag{24.54}$$

$$\psi_{\text{WKB}}(a) \to \exp\{-2\pi(a/\ell_P)^2(n + \tfrac{1}{4})\}, \tag{24.55}$$

independent of ℓ and $|m|$.

For the "flipped" eigenvalues $(-m)$, one sees

$$p(m, a) \to 2a \cdot 2\pi i(n + \tfrac{3}{4}); \tag{24.56}$$

$$\psi_{\text{WKB}}(a) \to \exp\{-2\pi(a/\ell_P)^2(n + \tfrac{3}{4})\}. \tag{24.57}$$

Thus the large radius asymptotic behavior of the WKB solution is identical to that of the exact $m = 0$ solution.

Turning to the other extreme, as $a \to 0$ the "velocity factor" $1/\sqrt{a}$ dominates and

$$\psi_{\text{WKB}}(a) \to \frac{\sqrt{a}}{\sqrt[4]{Q^2 - m^2/4}}. \tag{24.58}$$

Again, this is similar to the exact $m = 0$ solution but with a "shifted" value of the charge, $Q \mapsto \sqrt{Q^2 - m^2/4}$.

In fact, if one just blithely sets $m = 0$ one can recover the exact solutions from the WKB estimates. For $m = 0$,

$$p(a) = 2a \cdot \text{arccosh}(0) = 2a \cdot i\pi \left(n + \frac{1}{2}\right), \tag{24.59}$$

while

$$|\dot{a}(a)| = \sqrt{1 - \frac{2M}{a} + \frac{Q^2}{a^2}}, \tag{24.60}$$

the entire real line being classically forbidden. The WKB wave function is then

$$\psi_{\text{WKB}}(a) = \frac{\exp\{-\pi(a/\ell_P)^2(n + \tfrac{1}{2})\}}{\sqrt[4]{1 - \frac{2M}{a} + \frac{Q^2}{a^2}}}, \tag{24.61}$$

in agreement with the exact calculation.

Returning to estimates of the eigenmass of the dust shell, the quantization integral in the general $Q \neq 0$, $M \neq 0$ case is intractable. Though the integral may be computed in closed form for the $Q \neq 0$, $M = 0$ case, the result is not enlightening. A feel for the physics may best be obtained by considering the $Q = 0$, $M = 0$ case:

$$2 \int_0^{m/2} 2a \, \cosh^{-1}\left(-\frac{m}{2a}\right) da = (\ell + \delta)\hbar. \tag{24.62}$$

Rescaling to dimensionless variables, the integral is trivial with the result that

$$m(\ell) = -m_P\sqrt{2(\ell + \delta)}. \tag{24.63}$$

Thus the mass of the dust shell is quantized in terms of the Planck mass, as is only to be expected.

In summary, the WKB analysis indicates that the results obtained for the exact $m = 0$ eigenfunctions are generic. The wave function is well behaved at the origin and exponentially damped at large radius. The average radius is of order the Planck length.

24.5 Comments

I encourage the reader to make a very careful "reality check" as to how much of these calculations to actually believe. Perhaps the most damaging technical criticism that can be made concerning this calculation is that it is performed in minisuperspace instead of using Wheeler's full superspace. It is quite possible (maybe even likely) that the brutal truncation from an infinite number of degrees of freedom $g_{ij}(\vec{x}, t)$ down to one degree of freedom $a(\tau)$ has also brutally truncated the real physics. Unfortunately, given our current lack of calculational abilities, we simply have no choice. In mitigation of this point, observe that although the application is unique, the minisuperspace technology employed is a standard quantum gravity technique.

I should mention some other potentially serious problems:

- Though the factor ordering choice made in $\hat{H}_{\text{tot}}$ is in some sense "natural", it is by no means unique. Fortunately, this criticism does not apply to the WKB analysis—and the WKB analysis indicates that the qualitative features of the exact solutions continue to hold for $m \neq 0$.

- Since the expected wormhole radius is of order the Planck length, it is far from clear that the Einstein–Hilbert action is an appropriate description for gravity. If R^2 terms are present (and this is expected on rather general grounds) the analysis of this chapter is incomplete [137]. In particular a naive application of the thin-shell formalism is no longer appropriate since this would now involve squares of delta functions.

The "bottom line" is this: This calculation seems to indicate that minisuperspace models of Wheeler wormholes are quantum mechanically stable with a natural radius of order the Planck length. It is this qualitative feature of the analysis that should be taken as the main thrust of this chapter, rather than any of the particular details. Unfortunately it is rather difficult to judge to what extent if any these results might survive if one attempts to go beyond the minisuperspace approximation.

The implication that Wheeler wormholes are stable and amenable to some limited calculational techniques bears upon a secondary question—the quantum mechanical process of topology fluctuation [265, 266]. The occurrence of a fluctuation in topology may be viewed as equivalent to the collapse (and subsequent detachment) of a Wheeler wormhole. However, the minisuperspace calculation presented in this chapter can be interpreted as indicating that the required collapse does not occur. Thus, if we are willing to believe this result beyond the minisuperspace approximation (and this is a big if), it is possible to argue that the putative quantum mechanical stability of the Wheeler wormhole might in fact *prevent* fluctuations in topology [265, 266]. Actually, Wheeler himself seems to have had such a *dénouement* in mind when he advocated the possibility that elementary charges could be viewed as electric flux linkages *trapped* in the topology of space. If such a Wheeler wormhole were to undergo collapse, its collapse would necessarily produce a pair of true point charges, this possibility being completely antithetical to Wheeler's announced viewpoint that "geometry is everything". Naturally, this particular line of argument does not extend to electrically neutral wormholes.

Recall that Wheeler's conjectures regarding topology change were based on the fact that quantized linearized gravity predicts that the components of the linearized metric undergo fluctuations of order $\Delta h \sim (\ell_P/L)$, where L is the scale at which measurements of the fluctuation are taken. Thus for length scales less than that of the Planck length, fluctuations in the linearized metric are *large*. This indicates the breakdown of the linearized theory at the Planck scale. While it is certainly true that the metric (i.e., the geometry) becomes strongly interacting at this scale, this does not necessarily imply that the topology changes.

Furthermore, I wish to remind the reader, that even if the topology of space is fixed it is still possible to have very interesting *geometrical*, rather than topological, effects that "mimic" what is usually thought of as a topology change. While keeping the topology fixed it is possible for spacetime to develop narrow necks with diameters (say) of order the Planck length that connect relatively large relatively flat regions. Since a neck of diameter R will be impervious to physical probes of energies less than $E_{\text{crit}} = \hbar c/R$, a coarse-grained "physicist's topology" might well see a case of apparent topology change, this apparent change in topology being due to the finite resolution of his/her probes.

Note that insofar as any calculations have been carried out [5, 6, 67], the behavior of the geometry at the "crotch singularity" is the cornerstone upon which the physics of topology change is based. For example, Anderson and DeWitt, in a very nice exposition that complements the point of view taken in this chapter, have shown that in an idealized trousers geometry the "crotch singularity" is the source of an infinite-strength "flash" of radiation

which they interpret as forbidding topology change by "aborting" the birth
of the baby universe.

Though the idea that quantum gravity engenders topological fluctuations has been current in the community for a rather long time, the number
of calculations that can actually be carried out is distressingly small. Those
few calculations that are tractable in fact argue *against* topology change.
Adopting the viewpoint that topology change should not be included in
quantum gravity greatly simplifies our mental picture of the problem and
may lead to an improvement in our ability to calculate quantum gravitational effects. There is considerable disagreement on these issues and I warn
the reader that my own views are in the minority.

Part VI

Reprise

Chapter 25

Where we stand

25.1   General observations

In drawing this monograph to a close I feel that I should provide a few words to assess the overall situation. How far should one believe and trust the models and ideas sketched in this monograph?

On the one hand, Lorentzian wormholes are definitely to be regarded as speculative physics—there is *zero* direct experimental evidence to support the existence of Lorentzian wormholes.

On the other hand, we have seen that the theoretical analysis of Lorentzian wormholes is "merely" an extension of *known physics*—no new physical principles or fundamentally new physical theories are involved. Much of the analysis can be carried out using ordinary classical Einstein gravity. Portions of the analysis require semiclassical quantum gravity. On less firm footing, the more exotic minisuperspace analysis of the wormhole system steps outside the constraints of semiclassical quantum gravity but is still well within the mainstream of minor extensions of known physics.

In a certain sense the status of Lorentzian wormholes with respect to quantum gravity is similar to the status of the grand unified theories (GUTs) with respect to particle physics. GUTs are "merely" extensions of the *known physics* of the SU(3) × SU(2) × U(1) standard model of particle physics.

The chief virtue of GUTs is that they are generic: given the assumption that at energies significantly below the Planck scale particle physics is adequately described by a quantum field theory (present-day experimental data, after all, only goes up to $E \approx 1\,\text{TeV} = 10^3\,\text{GeV}$), the further *observation* that the running coupling constants converge at about $m_{\text{GUT}} \approx 10^{15}\,\text{GeV}$, almost guarantees some GUT-like behavior. This is also the major weakness of GUT models—one has to assume no major surprises between 10^3 and 10^{15} GeV.

Similarly with regard to Lorentzian wormholes, one makes the working hypothesis that at energy scales significantly below the Planck scale quantum gravity is adequately described by semiclassical quantum gravity (quantum field theory on a fixed background manifold). By imposing the Einstein field equations on the expectation value of the stress-energy tensor the existence of a classical limit that reproduces Einstein gravity is assured. With the observation that semiclassical quantum gravity often violates the classical energy conditions; a scenario involving Lorentzian wormholes is, while not guaranteed, at least highly plausible. Everything else is just details—unfortunately the devil is in the details.

There are many questions and open research questions still to be addressed. For instance,

- **Research Problem 20** *Find, from first principles, the precise conditions under which the averaged null energy condition holds. (At least there is widespread agreement on what this question means, and what the relevant mathematical framework is.)*

- **Research Problem 21** *Do macroscopic Lorentzian wormholes exist in nature? (A nice experimental question.)*

- **Research Problem 22** *Failing that, do microscopic Lorentzian wormholes exist in nature? (Another nice experimental question.)*

- **Research Problem 23** *Can one prove or disprove, from first principles, the existence or otherwise of quantum fluctuations in topology? (The answer to this question will depend on one's choice of microphysics for describing quantum gravity.)*

- **Research Problem 24** *Can one prove or disprove, from first principles, the truth or falsity of the Novikov consistency conjecture? (Again, the answer to this question will depend on one's choice of microphysics for describing quantum gravity.)*

- **Research Problem 25** *Can one prove or disprove, from first principles, the truth or falsity of Hawking's chronology protection conjecture? (Surprise, the answer to this question will again depend on one's choice of microphysics for describing quantum gravity.)*

Even though the topic of Lorentzian wormholes seems to be tightly circumscribed, and though one might not have expected such a narrowly defined topic to be capable of supporting a four hundred page monograph, I hope that I have convinced the reader that this monograph serves a useful function. A topic as seemingly limited in scope as Lorentzian wormholes, because it forces us to come to grips with fundamental issues regarding quantum gravity, can still teach us a lot. I hope that I have succeeded in convincing the reader that the successful marriage of quantum physics with gravity (that is, Planck scale physics) is a major theoretical issue—an issue that confronts both particle physics and gravity physics.

Experimental data are sparse. The construction of a Planck scale accelerator does not appear feasible in the immediate future. Recall that in the mid 1950s typical atom-smasher energies were of order 1 GeV. By the mid-1990s accelerator energies have risen to about 1 TeV. That's three orders of magnitude in forty years. Since we have about sixteen orders of magnitude to go to reach the Planck scale, a linear extrapolation *in the exponent* of the available energy suggests that a Planck scale accelerator might be on-line in about 210 years, circa 2200 A.D. This estimate is very crude and should in no way be considered definitive. [Aside: One hopeful sign—we have already achieved available energy increases of twelve orders of magnitude since the days of rubbing amber on silk ($E \approx 1$ eV).] Given this time scale, interested researchers should plan either on an exceedingly long lifetime, or on being satisfied by indirect probes of Planck scale physics.

Indirect probes of GUT scale physics are well-known: the running of the coupling constants and the non-decay of the proton. Now the GUT scale is tolerably close to the Planck scale (a factor of about 10^4). What the community really needs is some bright ideas on indirect experimental probes of the Planck scale—presumably via high-precision but low-energy experiments. The question of the orientability of the spacetime foam is an example of such an indirect probe. At this stage I have nothing further to offer along these lines.

Exercise: Estimate the minimum conceivable length for a Planck scale linear accelerator. Hints: The only known way of accelerating particles is via electromagnetic effects. For a particle of charge $Q = Ze$, accelerated in a linear accelerator with electric field E, this implies

$$L \gtrsim \frac{m_P c^2}{QE}. \tag{25.1}$$

But the maximum conceivable macroscopic electric field is limited by vacuum decay into positron–electron pairs. (Positrons and electrons are relevant since they are the lightest charged particles.) This vacuum decay process implies

$$eE \lesssim \frac{m_e c^2}{(\hbar/m_e c)} = \frac{m_e^2 c^3}{\hbar}. \tag{25.2}$$

Combining these results, one sees that

$$L \gtrsim \frac{m_P \hbar}{Z m_e^2 c} = \left(\frac{m_P}{m_e}\right)^2 \frac{\ell_P}{Z}. \tag{25.3}$$

Evaluate $E_{max} \equiv (m_e^2 c^3)/(\hbar e)$ numerically in volts/metre. Evaluate $L_{min} \equiv \ell_P(m_P/m_e)^2$ numerically in metres. How reasonable is this figure? Compare this to, say, the Earth–Moon separation.

Research Problem 26 *Develop and analyze in detail a few indirect probes of Planck scale physics. Subtlety and careful thought are called for.*

25.2   Exotic theories of gravity

In this monograph I have steered clear of certain exotica. There is a school of thought that holds that quantum gravity can only be attacked by changing the classical theory of gravity and thereby adopting a radically new microphysics—a new microphysics that is not guided by experimental evidence. This school in essence advocates doing an end run on the physics of the Planck scale. Some of the more exotic models investigated over the last few decades include:

- Kaluza–Klein theories.

- Supergravity theories (SUGRA).

- Superstring theories.

- Twistors.

- Loop quantization.

- Lattice gravity.

These exotic attempts at a theory of quantum gravity should be considered long-shot gambles. Note that for energies below the Planck scale all these exotica reduce to point-like quantum field theories in $(3 + 1)$-dimensional spacetime. Thus all of these exotica are compatible with the scenario involving Lorentzian wormholes that is outlined in this monograph.

(Aside: Establishing this result for superstring theory is a little tricky. There are two ways of proceeding. (1) If one studies superstrings propagating in flat spacetime it can be shown that the particle spectrum contains a massless spin-2 particle. This massless spin-2 particle has the correct self-interactions to be interpreted as the graviton arising from Einstein gravity linearized around a flat background. (2) If one studies superstrings propagating in a curved spacetime, and one demands that the world-sheet conformal anomaly vanish, then the resulting sigma-model approach imposes constraints on the spacetime manifold. These constraints can be shown to mimic Einstein gravity with certain higher-curvature modifications.)

This observation is one of the primary reasons I have been willing to put so much work into what might otherwise be considered a highly abstruse field. For the purposes of this monograph it does not really matter what the detailed microphysics of quantum gravity is. Any acceptable quantization of gravity must, in the low-energy limit (energies less than the Planck scale), reduce to some version of quantum field theory on curved spacetime, and in that limit the various techniques discussed and questions raised in this monograph (topology change, chronology protection, consistency constraints, multiple timelines) must be confronted.

If any putative theory of quantum gravity is not capable of addressing the issues raised in this monograph, then I submit that such a model for quantum gravity is not physically acceptable. I do not wish to prejudge the particular answers to be demanded of any candidate theory of quantum gravity—I only wish to demand that putative theories of quantum gravity be sufficiently well defined to permit the questions to be well-posed and in principle answerable.

25.3 Functional integration?

In earlier chapters I invoked the working hypothesis that the quantization of gravity might be achieved by some suitable functional integration over Lorentzian metrics. At this level, the debate is over exactly what the appropriate class of metrics to be integrated over is. At a deeper level, it is reasonable to question whether or not the metric is even the appropriate quantity to be quantized.

For instance, it may be more likely that the fundamental quantity to be integrated over is the vierbein (tetrad) rather than the metric. While such a choice makes no difference at the classical level, the quantum physics could be quite different. (See, for example, [124]). Since one cannot construct general relativistic fermions without using vierbeins one has rather strong reasons for suspecting the vierbein to be more fundamental than the metric. (For the purposes of this monograph exclusive use of the vierbein formalism would have been an unnecessary complication.)

Even more complicated choices are possible: in classical Einstein gravity the use of the Palatini formalism allows the metric and connexion to be treated as independent variables. One could conceive of quantizing gravity by functionally integrating over metric and connexion separately.

Proceeding further along these lines, it is conceivable that the right thing to do is to functionally integrate over vierbein and spin-connexion separately. (I have not explained what a spin-connexion is: experts will already know, and non-experts will have a lot of work to do.)

Functional integration over metrics or vierbeins is in some sense a minimalist approach to quantizing gravity—minimalist in that it keeps as close as possible to the classical formulation of Einstein gravity. I feel that the minimalist approach is still well worth pursuing.

25.4   Choices

One of the messages that I have tried to get across in this monograph
is that when attempting to quantize gravity the theorist is faced with a
wide variety of choices. Because the experimental data are rather sparse
and limited in purview, one quickly enters a realm where one loses direct
experimental guidance. In addition, calculations are often horrendously
difficult, and their interpretations may sometimes be rather subtle. In such
a situation, the theorist must, for purely practical reasons, make a few
choices to cut down the clutter and circumscribe the area of discourse.

These choices, while sometimes made explicitly, are often implicitly imposed at various stages of the calculation, with the result that there is no
clear-cut widely accepted definition of what exactly quantum gravity is,
or is supposed to be. There is even room for disagreement over the basic
phenomena that a successful theory should encompass. Because the choices
are often implicit, rather than explicit, there is sometimes an unfortunate
tendency for researchers to talk past each other rather than to each other.

Two particular choices have been explored in some detail in this monograph: topology change and causal structure.

- It is commonly believed that quantum gravity should permit the
  topology of space to fluctuate as a function of time. This is not
  a logical necessity derived from an underlying microphysical theory,
  but is instead a *choice* made based on one's best guess of what sub-
  Planckian metric fluctuations *might* entail.

- It is commonly believed that the universe is *causal*. The disputes
  arise when one attempts to decipher what the word causal means in
  the context of quantum gravity. It would be lovely to have a proof of
  causality based on some underlying microphysics—unfortunately to
  do so one would first have to decide upon and precisely specify one's
  choice of microphysical assumptions. Depending on the choices one
  adopts one is likely to be lead to one of the four standard responses to
  the time travel paradoxes that I sketched earlier in this monograph.

What are my personal choices? I think that all the possibilities discussed
in this monograph should be investigated to some extent. It is only by
testing one's own choice of microphysics against the alternatives that any
real progress can be made.

Tentatively, my own views are

- Topology change is bad.

- Traversable wormholes are good.

- Time travel is bad.

These views could easily change as better arguments and calculations become available. A lack of calculational ability and direct experimental evidence currently precludes any more definite statement.

I strongly encourage all researchers in the field to think carefully about their own personal choices and urge them to be as explicit as possible about the choices they make. I strongly encourage the community to eschew phrases such as "quantum gravity predicts ... " in favor of more modest statements such as "adopting choices A, B, and C, the XYZ model of quantum gravity predicts ... "

25.5 Coda

Finally, I wish to point out that while this monograph might seem rather speculative in content, the point of view adopted is downright restrained compared to what is sometimes encountered in the literature. In my own way, I have tried to keep the presentation as conservative as possible. If the reader feels at this stage that the situation is still rather murky and confusing, well, congratulations, welcome to research!

For your final task —

Research Problem 27 *Quantize gravity!*

Bibliography

[1] E. G. Adelberger, B. R. Heckel, G. Smith, Y. Su, and H. E. Swanson. Eötvös experiments, lunar ranging, and the strong equivalence principle. *Nature*, 347:261–263, 1990.

[2] E. G. Adelberger, B. R. Heckel, C. W. Stubbs, and Y. Su. Does antimatter fall with the same gravitational acceleration as ordinary matter? *Phys. Rev. Lett.*, 66:850–853, 1991.

[3] E. G. Adelberger, C. W. Stubbs, B. R. Heckel, Y. Su, H. E. Swanson, G. Smith, J. H. Grundlach, and W. F. Rogers. Testing the equivalence principle in the field of the earth: particle physics at masses below 1 μeV? *Phys. Rev. D*, 42:3267–3292, 1990.

[4] D. Z. Albert. Bohm's alternative to quantum mechanics. *Sci. Am.*, pages 58–67, May 1994.

[5] A. Anderson and B. S. DeWitt. Does the topology of space fluctuate? *Found. Phys.*, 16:91–105, 1986. Reprinted in [296].

[6] A. Anderson and B. S. DeWitt. Does the topology of space fluctuate? In Zurek et al. [296], pages 74–89. Reprint of [5].

[7] P. R. Anderson, W. A. Hiscock, and D. A. Samuel. Stress–energy tensor of quantized scalar fields in static black hole spacetimes. *Phys. Rev. Lett.*, 70:1739–1742, 1993.

[8] P. R. Anderson, W. A. Hiscock, J. Whitesell, and J. W. York, Jr. Semiclassical black hole in thermal equilibrium with a nonconformal scalar field. *Phys. Rev. D*, 50:6427–6434, 1994.

[9] T. Applequist, A. Chodos, and P. G. O. Freund. *Modern Kaluza–Klein Theories*. Frontiers in Physics No. 65. Addison–Wesley, Reading, Massachusetts, 1987.

[10] C. Barrabès. Singular hypersurfaces in general relativity: a unified description. *Class. Quantum Grav.*, 6:581–588, 1989.

[11] C. Barrabès and W. Israel. Thin shells in general relativity and cosmology: The lightlike limit. *Phys. Rev. D*, 43:1129–1142, 1991.

[12] J. D. Barrow and F. J. Tipler. *The Anthropic Cosmological Principle*. Oxford University Press, Oxford, England, 1986.

[13] R. Bartnik. Existence of maximal surfaces in asymptotically flat spacetimes. *Commun. Math. Phys.*, 94:155–175, 1984.

[14] J. S. Bell. *Speakable and Unspeakable in Quantum Mechanics.* Cambridge University Press, Cambridge, England, 1987.

[15] C. M. Bender and A. A. Orszag. *Advanced Mathematical Methods for Scientists and Engineers.* McGraw–Hill, New York, 1978.

[16] B. Bhawal and S. Kar. Lorentzian wormholes in Einstein–Gauss–Bonnet theory. *Phys. Rev. D*, 46:2464–2468, 1992.

[17] N. D. Birrell and P. C. W. Davies. *Quantum Fields in Curved Space.* Cambridge University Press, Cambridge, England, 1982.

[18] J. D. Bjorken and S. D. Drell. *Relativistic Quantum Fields.* McGraw–Hill, New York, 1965.

[19] S. K. Blau, M. Visser, and A. Wipf. Zeta functions and the Casimir energy. *Nucl. Phys. B*, 310:163–180, 1988.

[20] D. Bohm. A suggested interpretation of the quantum theory in terms of "hidden" variables. I. *Phys. Rev.*, 85:166–179, 1952.

[21] D. Bohm. A suggested interpretation of the quantum theory in terms of "hidden" variables. II. *Phys. Rev.*, 85:180–193, 1952.

[22] W. B. Bonnor. The rigidly rotating relativistic dust cylinder. *J. Phys. A*, 13:2121–2132, 1980.

[23] A. Borde. Geodesic focusing, energy conditions and singularities. *Class. Quantum Grav.*, 4:343–356, 1987.

[24] A. Borde. Topology change in classical general relativity. 1994. gr-qc/9406053.

[25] D. G. Boulware. Quantum field theory in spaces with closed timelike curves. *Phys. Rev. D*, 46:4421–4441, 1992.

[26] P. R. Brady, J. Louko, and E. Poisson. Stability of a shell around a black hole. *Phys. Rev. D*, 44:1891–1894, 1991.

[27] V. B. Braginskii and V. I. Panov. Verification of the equivalence of inertial and gravitational mass. *Sov. Phys. JETP*, 34:463–466, 1972.

[28] S. Braunstein. Unpublished. Discussed in [190], 1987.

[29] D. Brill and T. Dray. Spell it Nordström. *Gen. Rel. Gravit*, 25:435–436, 1993.

[30] M. R. Brown, A. C. Ottewill, and D. N. Page. Conformally invariant quantum field theory in static Einstein space-times. *Phys. Rev. D*, 33:2840–2850, 1986.

[31] M. Calvani, F. de Felice, B. Muchotrzeb, and F. Salmistraro. Time machine and geodesic motion in Kerr metric. *Gen. Rel. Gravit.*, 9:155–163, 1978.

[32] P. Candelas. Vacuum polarization in Schwarzschild spacetime. *Phys. Rev. D*, 21:2185–2202, 1980.

[33] P. Candelas, P. Chrzanowski, and K. W. Howard. Quantization of electromagnetic and gravitational perturbations of a Kerr black hole. *Phys. Rev. D*, 24:297–304, 1981.

[34] P. Candelas and K. W. Howard. Vacuum $\langle \phi^2 \rangle$ in Schwarzschild spacetime. *Phys. Rev. D*, 29:1618–1625, 1984.

[35] S. M. Carroll, E. Farhi, and A. H. Guth. An obstacle to building a time machine. *Phys. Rev. Lett.*, 68:263–266, 1992. Erratum—*ibid.*, p. 3368.

[36] S. M. Carroll, E. Farhi, A. H. Guth, and K. D. Olum. Energy momentum restrictions on the creation of Gott time machines. *Phys. Rev. D*, 50:6190–6206, 1994.

[37] H. B. G. Casimir. On the attraction between two perfectly conducting plates. *Proc. Kon. Nederl. Akad. Wetenschap*, 51:793–795, 1948.

[38] A. Chamblin, G. W. Gibbons, and A. R. Steif. Kinks and time machines. *Phys. Rev. D*, 50:2353–2355, 1994.

[39] S. M. Christensen and S. A. Fulling. Trace anomalies and the Hawking effect. *Phys. Rev. D*, 15:2088–2104, 1977.

[40] C. J. S. Clarke and F. de Felice. Globally non–causal space–times. *J. Phys. A*, 15:2415–2417, 1978.

[41] C. J. S. Clarke and F. de Felice. Globally non–causal space–times II: Naked singularities and curvature conditions. *Gen. Rel. Gravit.*, 16:139–148, 1984.

[42] G. Clement. A class of wormhole solutions to higher dimensional gravity. *Gen. Rel. Gravit.*, 16:131–138, 1984.

[43] M. Cline. Does the wormhole mechanism for vanishing cosmological constant work in Lorentzian gravity? *Phys. Lett. B*, 224:53–57, 1989.

[44] S. Coleman. Why there is nothing rather than something: A theory of the cosmological constant. *Nucl. Phys. B*, 310:643–668, 1988. Reprinted in [133].

[45] S. Coleman and S. Hughes. Black holes, wormholes, and the disappearance of global charge. *Phys. Lett. B*, 309:246–251, 1993.

[46] S. Coleman and K. Lee. Escape from the menace of the giant wormholes. *Phys. Lett. B*, 221:242–249, 1989.

[47] S. Coleman and K. Lee. Wormholes made without massless matter fields. *Nucl. Phys. B*, 329:387–409, 1990.

[48] J. G. Cramer. The transactional interpretation of quantum mechanics. *Rev. Mod. Phys.*, 58:647–687, 1986.

[49] J. G. Cramer, R. L. Forward, M. S. Morris, M. Visser, G. Benford, and G. A. Landis. Natural wormholes as gravitational lenses. *Phys. Rev. D*, 51:3117–3120, 1995.

[50] C. Cutler. Global structure of Gott's two–string spacetime. *Phys. Rev. D*, 45:487–494, 1992.

[51] P. C. W. Davies and S. A. Fulling. Radiation from moving mirrors and from black holes. *Proc. R. Soc. London Ser. A*, 356:237–257, 1977.

[52] P. C. W. Davies and S. A. Fulling. "Strange" expectation values of $T_{\mu\nu}$ due to interference between amplitudes for different particle numbers. *Proc. R. Soc. London Ser. A*, 356:255–256, 1977. (Appendix to [51]).

[53] F. de Felice. Timelike nongeodesic trajectories which violate causality: A rigorous derivation. *Nuovo Cimento*, 65B:224–232, 1981.

[54] F. de Felice and M. Calvani. Causality violation in the Kerr metric. *Gen. Rel. Gravit.*, 9:155–163, 1978.

[55] S. Deser, R. Jackiw, and G. 't Hooft. Three dimensional Einstein gravity: Dynamics of flat space. *Ann. Phys. (NY)*, 152:220–235, 1984.

[56] S. Deser, R. Jackiw, and G. 't Hooft. Physical cosmic strings do not generate closed timelike curves. *Phys. Rev. Lett.*, 68:267–269, 1992.

[57] D. Deutsch. Quantum mechanics near closed timelike lines. *Phys. Rev. D*, 44:3197–3127, 1991.

[58] D. Deutsch and M. Lockwood. The quantum physics of time travel. *Sci. Am.*, pages 68–74, March 1994.

[59] B. S. DeWitt. Dynamical theory of groups and fields. In DeWitt and DeWitt [72]. See especially pp. 734–745. See also [60].

[60] B. S. DeWitt. *Dynamical Theory of Groups and Fields*. Documents on modern physics. Gordon and Breach, New York, 1965. (Expanded version of [59]).

[61] B. S. DeWitt. Quantum theory of gravity. I. The canonical theory. *Phys. Rev.*, 160:1113–1148, 1967. (First appearance of the Wheeler-DeWitt equation.) Errata—*ibid.*, 171:1834–1834, 1968.

[62] B. S. DeWitt. Quantum theory of gravity. II. The manifestly covariant theory. *Phys. Rev.*, 162:1195–1239, 1967. Errata—*ibid.*, 171:1834–1834, 1968.

[63] B. S. DeWitt. Quantum theory of gravity. III. Applications of the covariant theory. *Phys. Rev.*, 162:1239–1256, 1967. Errata—*ibid.*, 171:1834–1834, 1968.

[64] B. S. DeWitt. Quantum field theory in curved spacetime. *Phys. Rep.*, 19:295–357, 1975. See especially pp. 342–345.

[65] B. S. DeWitt. Quantum gravity: The new synthesis. In Hawking and Israel [134], pages 680–745.

[66] B. S. DeWitt. Spacetime approach to quantum field theory. In DeWitt and Stora [71], pages 381–738. See especially pp. 531–536.

[67] B. S. DeWitt. Changing topology. In T. Goldman and M. M. Nieto, editors, *Proceedings of the Santa Fe Meeting*, pages 432–436, Singapore, 1985. World Scientific.

[68] B. S. DeWitt and R. W. Brehme. Radiation damping in a gravitational field. *Ann. Phys. (NY)*, 9:220–259, 1960.

[69] B. S. DeWitt and N. Graham, editors. *The Many-Worlds Interpretation of Quantum Mechanics*. Princeton University Press, Princeton, 1973.

[70] B. S. DeWitt, C. F. Hart, and C. J. Isham. Topology and quantum field theory. *Physica*, 96A:197–211, 1979.

[71] B. S. DeWitt and R. Stora, editors. *Relativity, Groups, and Topology II [Les Houches 1983]*. North Holland, Amsterdam, 1984.

[72] C. M. DeWitt and B. S. DeWitt, editors. *Relativity, Groups, and Topology [Les Houches 1963]*, New York, 1964. Gordon and Breach.

[73] C. M. DeWitt and J. A. Wheeler, editors. *Battelle Recontres: 1967 Lectures in Mathematics and Physics*. Benjamin, New York, 1968.

[74] J. F. Donoghue, E. Golowich, and B. R. Holstein. *Dynamics of the Standard Model*. Cambridge University Press, Cambridge, England, 1992.

[75] F. Echeverria, G. Klinkhammer, and K. S. Thorne. Billiard balls in wormhole spacetimes with closed timelike curves: Classical theory. *Phys. Rev. D*, 44:1077–1099, 1991.

[76] A. Einstein, B. Hoffmann, and L. Infeld. The gravitational equations and the problem of motion. *Ann. Math.*, 39:65–100, 1938.

[77] A. Einstein and N. Rosen. The particle problem in the general theory of relativity. *Phys. Rev.*, 48:73–77, 1935. (The first paper on wormholes).

[78] R. Eötvös, D. Pektar, and E. Fekete. Bietrage zum gesetze der proportionalitat von Tragheit und Gravitat. *Ann. Phys. (Leipzig)*, 68:11–66, 1922.

[79] H. Epstein, V. Glaser, and A. Jaffe. Nonpositivity of the energy density in quantized field theories. *Nuovo Cimento*, 36:1016–1022, 1965.

[80] H. Everett. Relative state formulation of quantum mechanics. *Rev. Mod. Phys.*, 29:454–462, 1957. Reprinted in [69].

[81] M. S. Fawcett. The energy–momentum tensor near a black hole. *Commun. Math. Phys.*, 89:103–115, 1983.

[82] W. Fischler, I. Klebanov, J. Polchinski, and L. Susskind. Quantum mechanics of the googolplexus. *Nucl. Phys. B*, 327:157–177, 1989.

[83] W. Fischler and L. Susskind. A wormhole catastrophe. *Phys. Lett. B*, 217:48–54, 1989.

[84] L. Flamm. Beiträge zur Einsteinschen Gravitationstheorie. *Phys. Z.*, 17:448–454, 1916.

[85] L. H. Ford and T. A. Roman. Moving mirrors, black holes, and cosmic censorship. *Phys. Rev. D*, 41:3662–3670, 1990.

[86] L. H. Ford and T. A. Roman. Motion of inertial observers through negative energy. *Phys. Rev. D*, 48:776–782, 1993.

[87] J. Friedman and M. S. Morris. The Cauchy problem for the scalar wave equation is well defined in a class of spacetimes with closed time like curves. *Phys. Rev. Lett.*, 66:401–404, 1991.

[88] J. Friedman, M. S. Morris, I. D. Novikov, F. Echeverria, G. Klinkhammer, K. S. Thorne, and U. Yurtsever. Cauchy problem in spacetimes with closed timelike curves. *Phys. Rev. D*, 42:1915–1930, 1990.

[89] J. L. Friedman, N. Papastamatiou, L. Parker, and H. Zhang. Non-orientable foam and an effective Planck mass for point–like fermions. *Nucl. Phys. B*, 309:533–551, 1988.

[90] J. L. Friedman, N. J. Papastamatiou, and J. Z. Simon. Failure of unitarity for interacting fields on spacetimes with closed timelike curves. *Phys. Rev. D*, 46:4456–4469, 1992.

[91] J. L. Friedman, N. J. Papastamatiou, and J. Z. Simon. Unitarity of interacting fields in curved spacetime. *Phys. Rev. D*, 46:4442–4455, 1992.

[92] J. L. Friedman, K. Schleich, and D. M. Witt. Topological censorship. *Phys. Rev. Lett.*, 71:1486–1489, 1993.

[93] V. P. Frolov. Vacuum polarization in a locally static multiply connected spacetime and a time machine problem. *Phys. Rev. D*, 43:3878–3894, 1991.

[94] V. P. Frolov, W. Israel, and W. G. Unruh. Gravitational fields of straight and circular cosmic strings: Relation between gravitational mass, angular deficit, and internal structure. *Phys. Rev. D*, 39:1084–1096, 1989.

[95] V. P. Frolov and I. D. Novikov. Physical effects in wormholes and time machines. *Phys. Rev. D*, 42:1057–1065, 1990.

[96] V. P. Frolov and I. D. Novikov. Wormhole as a device for studying a black hole's interior. *Phys. Rev. D*, 48:1607–1615, 1993.

[97] V. P. Frolov and K. S. Thorne. Renormalized stress–energy tensor near the horizon of a slowly evolving, rotating black hole. *Phys. Rev. D*, 39:2125–2154, 1989.

[98] V. P. Frolov and A. I. Zel'nikov. Killing approximation for vacuum and thermal stress–energy tensor in static space-times. *Phys. Rev. D*, 35:3031–3044, 1987.

[99] S. A. Fulling. *Aspects of Quantum Field Theory in Curved Space-Time.* Cambridge University Press, Cambridge, England, 1989.

[100] S. A. Fulling and P. C. W. Davies. Radiation from a moving mirror in two dimensional space–time: Conformal anomaly. *Proc. R. Soc. London Ser. A*, 348:393–414, 1976.

[101] D. Garfinkle and A. Strominger. Semiclassical Wheeler wormhole production. *Phys. Lett. B*, 256:146–149, 1991.

[102] S. J. Gates, Jr., M. T. Grisaru, M. Roček, and W. Seigel. *Superspace or One Thousand and One Lessons in Supersymmetry.* Frontiers in Physics No. 58. Addison–Wesley, Reading, Massachusetts, 1983.

[103] M. Gell-Mann and J. B. Hartle. Classical equations for quantum systems. *Phys. Rev. D*, 47:3345–3382, 1993.

[104] C. Gerhardt. H-surfaces in Lorentzian manifolds. *Commun. Math. Phys.*, 89:523–553, 1983.

[105] R. P. Geroch. *Singularities in the spacetime of general relativity: Their definition, existence, and local characterization.* PhD thesis, Princeton University, 1967. See also [106].

[106] R. P. Geroch. Topology in general relativity. *J. Math. Phys.*, 8:782–786, 1967.

[107] R. P. Geroch and G. T. Horowitz. Global structure of spacetimes. In Hawking and Israel [134], pages 212–293.

[108] K. Ghoroku and T. Soma. Lorentzian wormholes in higher derivative gravity and the weak energy condition. *Phys. Rev. D*, 46:1507–1516, 1992.

[109] G. W. Gibbons. Quantum field theory in curved spacetime. In Hawking and Israel [134], pages 639–679.

[110] G. W. Gibbons and S. W. Hawking. Selection rules for topology change. *Commun. Math. Phys.*, 148:345–352, 1992.

[111] S. Giddings, J. Abbot, and K. Kuchař. Einstein's theory in a three-dimensional space. *Gen. Rel. Gravit.*, 16:751–775, 1984.

[112] S. Giddings and A. Strominger. Baby universes, third quantization, and the cosmological constant. *Nucl. Phys. B*, 321:481–508, 1989.

[113] D. Giulini. Selection rules for spin–Lorentz cobordisms. *Commun. Math. Phys.*, 148:353–357, 1992.

[114] R. J. Gleiser, C. N. Kozameh, and O. M. Moreschi, editors. *GR13: General Relativity and Gravitation 1992—Proceedings of the 13th International Conference on General Relativity and Gravitation, Cordoba, Argentina, 1992*, Bristol, 1993. Insitite of Physics.

[115] J. Glimm and A. Jaffe. *Quantum Physics: A Functional Integral Point of View*. Springer-Verlag, New York, 1987.

[116] N. N. Gnedin. Unpublished. Discussed in [114, 93], 1991.

[117] K. Gödel. An example of a new type of cosmological solution of Einstein's field equation of gravitation. *Rev. Mod. Phys.*, 21:447–450, 1949.

[118] D. S. Goldwirth, M. J. Perry, and T. Piran. The breakdown of quantum mechanics in the presence of time machines. *Gen. Rel. Gravit.*, 25:7–13, 1994. Errata—see [119].

[119] D. S. Goldwirth, M. J. Perry, T. Piran, and K. S. Thorne. Quantum nonrelativistic particle in the vicinity of a time machine. *Phys. Rev. D*, 49:3951–3997, 1994.

[120] J. R. Gott, III. Gravitational lensing effects of vacuum strings: Exact solutions. *Astrophys. J.*, 288:422–427, 1985.

[121] J. R. Gott, III. Closed timelike curves produced by pairs of moving cosmic strings: Exact solutions. *Phys. Rev. Lett.*, 66:1126–1129, 1991.

[122] J. D. E. Grant. Cosmic strings and chronology protection. *Phys. Rev. D*, 47:2388–2394, 1993.

[123] M. B. Green, J. H. Schwarz, and E. Witten. *Superstring Theory*. Cambridge Universitry Press, Cambridge, England, 1987. Two volumes.

[124] J. Greensite. Dynamical origin of the Lorentzian signature of spacetime. *Phys. Lett. B*, 300:34–37, 1993.

[125] S. Harris. What is the shape of space in a spacetime? *Proc. Symp. Pure Math.*, 54, Part 2:287–296, 1993.

[126] J. B. Hartle. Unruly topologies in two-dimensional quantum gravity. *Class. Quantum Grav.*, 2:707–720, 1985.

[127] B. Hatfield. *Quantum Field Theories of Point Particles and Strings*. Frontiers in Physics No. 75. Addison–Wesley, Reading, Massachusetts, 1990.

[128] S. W. Hawking. The path integral approach to quantum gravity. In Hawking and Israel [134], pages 746–789. Reprinted in [133] and [131].

[129] S. W. Hawking. The chronology protection conjecture. In Sato [234], pages 3–13.

[130] S. W. Hawking. Chronology protection conjecture. *Phys. Rev. D*, 46:603–611, 1992. Reprinted in [131].

[131] S. W. Hawking, editor. *Hawking on the Big Bang and Black Holes*. World Scientific, Singapore, 1993.

[132] S. W. Hawking and G. F. R. Ellis. *The Large Scale Structure of Space-Time*. Cambridge University Press, Cambridge, England, 1973.

[133] S. W. Hawking and G. W. Gibbons, editors. *Euclidean Quantum Gravity*. World Scientific, Singapore, 1993.

[134] S. W. Hawking and W. Israel, editors. *General Relativity: An Einstein Centenary Survey*. Cambridge University Press, Cambridge, England, 1979.

[135] S. W. Hawking and W. Israel, editors. *300 Years of Gravitation*. Cambridge University Press, Cambridge, England, 1987.

[136] W. A. Hiscock. Exact gravitational field of a string. *Phys. Rev. D*, 31:3288–3290, 1985.

[137] D. Hochberg. Lorentzian wormholes in higher order gravity theories. *Phys. Lett. B*, 251:349–354, 1990.

[138] D. Hochberg and T. W. Kephart. Lorentzian wormholes from the gravitationally squeezed vacuum. *Phys. Lett. B*, 268:377–383, 1991.

[139] D. Hochberg and T. W. Kephart. Gauge field back reaction on a black hole. *Phys. Rev. D*, 47:1465–1470, 1993.

[140] D. Hochberg and T. W. Kephart. Wormhole cosmology and the horizon problem. *Phys. Rev. Lett.*, 70:2665–2668, 1993.

[141] D. Hochberg, T. W. Kephart, and J. W. York, Jr. Positivity of entropy in the semiclassical theory of black holes and radiation. *Phys. Rev. D*, 48:479–484, 1993.

[142] G. T. Horowitz. Semiclassical relativity: The weak–field limit. *Phys. Rev. D*, 21:1455–1461, 1980.

[143] G. T. Horowitz. Topology change in general relativity. *Class. Quantum Grav.*, 8:587–601, 1991.

[144] G. T. Horowitz and M. J. Perry. Gravitational energy cannot become negative. *Phys. Rev. Lett.*, 48:371–374, 1982.

[145] G. T. Horowitz and R. M. Wald. Quantum stress tensor in nearly conformally flat spacetimes. *Phys. Rev. D*, 21:1462–1465, 1980.

[146] K. W. Howard. Vacuum $\langle T_\mu{}^\nu \rangle$ in Schwarzschild spacetime. *Phys. Rev. D*, 30:2532–2547, 1984.

[147] K. W. Howard and P. Candelas. Quantum stress tensor in Schwarzschild space-time. *Phys. Rev. Lett.*, 53:403–406, 1984. Reprinted in [133].

[148] C. Isenberg. *The Science of Soap Films and Soap Bubbles*. Dover, New York, 1992.

[149] J. N. Islam. *Rotating fields in general relativity*. Cambridge University Press, Cambridge, England, 1985.

[150] W. Israel. Singular hypersurfaces and thin shells in general relativity. *Nuovo Cimento*, 44B [Ser. 10]:1–14, 1966. Errata—*ibid.*,48B [Ser. 10]:463–463, 1967.

[151] W. Israel. The formation of black holes in nonspherical collapse and cosmic censorship. *Can. J. Phys.*, 64:120–127, 1986.

[152] W. Israel. Must nonspherical collapse produce black holes? A gravitational confinement theorem. *Phys. Rev. Lett.*, 56:789–791, 1986.

[153] W. Israel. Effect of radiative tails on black hole interiors. *Int. J. Mod. Phys. D*, 3:71–79, 1994.

[154] C. Itzykson and J.-B. Zuber. *Quantum Field Theory*. McGraw-Hill, New York, 1980.

[155] R. Jackiw, N. N. Khuri, S. Weinberg, and E. Witten, editors. *Shelter Island II*, Cambridge, Massachusetts, 1983. MIT Press.

[156] J. D. Jackson. *Classical Electrodynamics*. Wiley, New York, second edition, 1975.

[157] M. Jammer. *The Conceptual Foundations of Quantum Mechanics*. McGraw-Hill, New York, 1966.

[158] B. P. Jensen and A. C. Ottewill. Renormalized electromagnetic stress tensor in Schwarzschild spacetime. *Phys. Rev. D*, 39:1130–1138, 1989.

[159] B. P. Jensen and H. H. Soleng. General relativistic model of a spinning cosmic string. *Phys. Rev. D*, 45:3528–3533, 1992.

[160] L. H. Kauffman. *On Knots*. Annals of Mathematics: Study No. 115. Princeton University Press, Princeton, 1987.

[161] L. H. Kauffman. *Knots and Physics*. World Scientific, Singapore, 1991.

[162] R. P. Kerr. Gravitational field of a spinning mass as an example of algebraically special metrics. *Phys. Rev. Lett.*, 11:237–238, 1963.

[163] S. W. Kim. Particle creation for time travel through a wormhole. *Phys. Rev. D*, 46:2428–2434, 1992.

[164] S. W. Kim and K. S. Thorne. Do vacuum fluctuations prevent the creation of closed time like curves? *Phys. Rev. D*, 43:3929–3947, 1991.

[165] I. Klebanov, L. Susskind, and T. Banks. Wormholes and the cosmological constant. *Nucl. Phys. B*, 317:665–692, 1989. Reprinted in [133].

[166] G. Klinkhammer. Averaged energy conditions for free scalar fields in flat spacetime. *Phys. Rev. D*, 43:2542–2548, 1991.

[167] G. Klinkhammer. Vacuum polarization of scalar and spinor fields near closed null geodesics. *Phys. Rev. D*, 46:3388–3394, 1992.

[168] E. W. Kolb and M. S. Turner. *The Early Universe.* Frontiers in Physics No. 69. Addison–Wesley, Reading, Massachusetts, 1990.

[169] D. A. Kompaneets. Unpublished. Discussed in [114, 93], 1991.

[170] D. Kramer, H. Stephani, E. Herlt, and M. MacCallum. *Exact Solutions of Einstein's Field Equations.* Cambridge University Press, Cambridge, England, 1980.

[171] C.-I. Kuo and L. H. Ford. Semiclassical gravity theory and quantum fluctuations. *Phys. Rev. D*, 47:4510–4519, 1993.

[172] K. Lanczos. Untersuching über flächenhafte verteilung der materie in der Einsteinschen gravitationstheorie. Unpublished, 1922.

[173] K. Lanczos. Flächenhafte verteilung der materie in der Einsteinschen gravitationstheorie. *Ann. Phys. (Leipzig)*, 74:518–540, 1924.

[174] L. D. Landau and E. M. Lifshitz. *The Classical Theory of Fields.* Pergamon, Oxford, fourth revised english edition, 1975.

[175] D. Laurence. Isometries between Gott's two–string spacetime and Grant's generalization of Misner space. *Phys. Rev. D*, 450:4957–4965, 1994.

[176] P. S. Letelier and A. Wang. Spherically symmetric thin shells in Brans–Dicke theory of gravity. *Phys. Rev. D*, 48:631–646, 1952.

[177] B. Linet. The static metrics with cylindrical symmetry describing a model of cosmic strings. *Gen. Rel. Gravit.*, 17:1109–1115, 1993.

[178] M. Lyutikov. Vacuum polarization at the chronology horizon of the Roman spacetime. *Phys. Rev. D*, 49:4041–4048, 1994.

[179] K. I. MacRae and R. J. Riegert. Long range antigravity. *Nucl. Phys. B*, 244:513–522, 1984.

[180] D. Marković and E. Poisson. Classical stability and quantum instability of black hole Cauchy horizons. *Phys. Rev. Lett.*, 70:1280–1283, 1995.

[181] J. E. Marsden and F. J. Tipler. Maximal hypersurfaces and foliations of constant mean curvature in general relativity. *Phys. Rep.*, 66:109–139, 1980.

[182] A. J. McConnell. *Applications of Tensor Analysis.* Dover, New York, 1957.

[183] R. A. Millikan. The isolation of an ion, a precision measurement of its charge, and the correction to Stoke's law. *Science*, 32:436–448, 1910. (Abridged version of [184]).

[184] R. A. Millikan. The isolation of an ion, a precision measurement of its charge, and the correction to Stoke's law. *Phys. Rev.*, 32:352–397, 1911. (Extended version of [183]).

[185] P. W. Milonni. *The Quantum Vacuum: An Introduction to Quantum Electrodynamics.* Academic, New York, 1994.

[186] C. W. Misner, K. S. Thorne, and J. A. Wheeler. *Gravitation.* W. H. Freeman and Company, San Francisco, 1973.

[187] C. W. Misner and J. A. Wheeler. Classical physics as geometry: gravitation, electromagnetism, unquantized charge, and mass as properties of curved empty space. *Ann. Phys. (NY)*, 2:525–603, 1957. Reprinted in [286].

[188] J. W. Moffat and T. Svoboda. Traversable wormholes and the negative-stress-energy problem in the nonsymmetric gravitational theory. *Phys. Rev. D*, 44:429–432, 1991.

[189] C. Morette. On the definition and approximation of Feynman's path integrals. *Phys. Rev.*, 81:848–852, 1951.

[190] M. S. Morris and K. S. Thorne. Wormholes in spacetime and their use for interstellar travel: A tool for teaching general relativity. *Am. J. Phys.*, 56:395–412, 1988.

[191] M. S. Morris, K. S. Thorne, and U. Yurtsever. Wormholes, time machines, and the weak energy condition. *Phys. Rev. Lett.*, 61:1446–1449, 1988.

[192] G. L. Naber. *Spacetime and Singularities: An Introduction.* London Mathematical Society Student Texts No. 11. Cambridge University Press, Cambridge, England, 1988.

[193] Y. Nambu and M. Siino. Wormhole formation in numerical cosmology. *Phys. Rev. D*, 46:5367–5377, 1992.

[194] M. M. Nieto and T. Goldman. The arguments against antigravity and the gravitational acceleration of antimatter. *Phys. Rep.*, 205:221–281, 1991.

[195] I. D. Novikov. An analysis of the operation of a time machine. *Sov. Phys. JETP*, 68:439–443, 1989. Translation of Zh. Eksp. Teor. Fiz. 95 (March) 769–776, 1989.

[196] I. D. Novikov. Time machine and self-consistent evolution in problems with self-interaction. *Phys. Rev. D*, 45:1989–1994, 1992.

[197] A. Ori. Rapidly moving cosmic strings and chronology protection. *Phys. Rev. D*, 44:R2214–R2215, 1991.

[198] A. Ori. Must time machine construction violate the weak energy condition? *Phys. Rev. Lett.*, 71:2517–2520, 1993.

[199] A. Ori and Y. Soen. Causality violation and the weak energy condition. *Phys. Rev. D*, 49:3990–3997, 1994.

[200] D. N. Page. Thermal stress tensor in Einstein spaces. *Phys. Rev. D*, 25:1499–1509, 1982. Reprinted in [133].

[201] Particle Data Group. Review of particle properties. *Phys. Rev. D*, 45:S1 [I.1–XI.8], 1992. See especially pp. III.1–III.2.

[202] W. Pauli. *Pauli Lectures on Physics: Volume 6. Selected Topics in Field Quantization.* MIT Press, Cambridge, 1973. See especially pp. 161–174.

[203] W. Pauli. *Theory of Relativity*. Dover, New York, 1981. Reprint of the 1958 English language translation of the 1921 German language original.

[204] J. P. Paz and W. H. Zurek. Environment induced decoherence, classicality and consistency of quantum histories. *Phys. Rev. D*, 48:2728–2738, 1993.

[205] P. J. E. Peebles. *Principles of Physical Cosmology*. Princeton University Press, Princeton, 1993.

[206] R. Penrose. Singularities and time–asymmetry. In Hawking and Israel [134], pages 581–638.

[207] R. Penrose. On Schwarzschild causality — a problem for "Lorentz covariant" general relativity. In Tipler [255], pages 1–12.

[208] R. Penrose. *The Emperor's New Mind*. Oxford University Press, Oxford, England, 1989.

[209] R. Penrose. *Shadows of the Mind: A Search for the Missing Science of Consciousness*. Oxford University Press, Oxford, England, 1994.

[210] R. Penrose, R. D. Sorkin, and E. Woolgar. A positive mass theorem based on the focusing and retardation of null geodesics. 1993. gr-qc/9301015.

[211] G. Plunien, B. Müller, and W. Greiner. The Casimir effect. *Phys. Rep.*, 134:97–193, 1986.

[212] E. Poisson and W. Israel. Internal structure of black holes. *Phys. Rev. D*, 41:1796–1809, 1990.

[213] J. Polchinski. Decoupling versus excluded volume or return of the giant wormholes. *Nucl. Phys. B*, 325:619–630, 1989.

[214] H. D. Politzer. Simple quantum systems in spacetimes with closed timelike curves. *Phys. Rev. D*, 46:4470–4476, 1992.

[215] H. D. Politzer. Path integrals, density matrices, and information flow with closed timelike curves. *Phys. Rev. D*, 49:3981–3989, 1994.

[216] A. M. Polyakov. Quantum geometry of bosonic strings. *Phys. Lett. B*, 103:207–210, 1981. Reprinted in [238].

[217] A. M. Polyakov. Quantum geometry of fermionic strings. *Phys. Lett. B*, 103:211–213, 1981. Reprinted in [238].

[218] A. M. Polyakov. Fine structure of strings. *Nucl. Phys. B*, 268:406–412, 1986.

[219] P. Ramond. *Field Theory: A Modern Primer*. Frontiers in Physics No. 51. Addison–Wesley, Reading, Massachusetts, 1981.

[220] I. H. Redmount. Blue–sheet instability of Schwarzschild wormholes. *Prog. Theor. Phys.*, 73:1401–1426, 1985.

[221] I. H. Redmount. Dynamics of a void–dominated universe: Cell–lattice models. *Mon. Not. R. Astron. Soc.*, 235:1301–1312, 1988.

[222] I. H. Redmount and W. M. Suen. Is quantum spacetime foam unstable? *Phys. Rev. D*, 47:R2163–2167, 1993.

[223] I. H. Redmount and W. M. Suen. Quantum dynamics of Lorentzian space-time foam. *Phys. Rev. D*, 49:5199–5210, 1994.

[224] P. G. Roll, R. Krotlov, and R. H. Dicke. The equivalence of inertial and passive gravitational mass. *Ann. Phys. (NY)*, 26:442–517, 1964.

[225] T. A. Roman. Quantum stress-energy tensors and the weak energy condition. *Phys. Rev. D*, 33:3526–3533, 1986.

[226] T. A. Roman. On the "averaged weak energy condition" and Penrose's singularity theorem. *Phys. Rev. D*, 37:546–548, 1988.

[227] T. A. Roman. Inflating Lorentzian wormholes. *Phys. Rev. D*, 47:1370–1379, 1993.

[228] T. A. Roman. The inflating wormhole: a Mathematica animation. *Comput. Phys.*, 8:480–487, 1994.

[229] B. Rose. Construction of matter models which violate the strong energy condition and may avoid the initial singularity. *Class. Quantum Grav.*, 3:975–995, 1986.

[230] B. Rose. A matter model violating the strong energy condition — the influence of temperature. *Class. Quantum Grav.*, 4:1019–1030, 1987.

[231] G. G. Ross. *Grand Unified Theories*. Frontiers in Physics No. 60. Addison-Wesley, Reading, Massachusetts, 1984.

[232] L. H. Ryder. *Quantum Field Theory*. Cambridge University Press, Cambridge, England, 1985.

[233] R. K. Sachs and H. Wu. *General Relativity for Mathematicians*. Graduate texts in mathematics. Springer–Verlag, New York, 1977.

[234] H. Sato, editor. *Proceedings of the 6th Marcel Grossmann Meeting, Kyoto, Japan*, Singapore, 1992. World Scientific.

[235] R. Schoen and S. T. Yau. On the proof of the positive mass conjecture in general relativity. *Commun. Math. Phys.*, 65:45–76, 1979.

[236] R. Schoen and S. T. Yau. Positivity of the total mass of a general spacetime. *Phys. Rev. Lett.*, 43:1457–1459, 1979.

[237] R. Schoen and S. T. Yau. Proof that the Bondi mass is positive. *Phys. Rev. Lett.*, 48:369–371, 1982.

[238] J. H. Schwarz. *Superstrings—The First Fifteen Years of Superstring Theory*. World Scientific, Singapore, 1985. Two volumes.

[239] N. Sen. Über die grenzbedingungen des schwerefeldes an unstetig keitsflächen. *Ann. Phys. (Leipzig)*, 73:365–396, 1924.

[240] A. Y. Shiekh. Does nature place a fundamental limit on strength? *Can. J. Phys.*, 70:458–462, 1992.

[241] J. Z. Simon. No Starobinsky inflation from self–consistent semiclassical gravity. *Phys. Rev. D*, 45:1953–1960, 1992.

[242] G. Smith, E. G. Adelberger, B. R. Heckel, and Y. Su. Test of the equivalence principle for ordinary matter falling toward dark matter. *Phys. Rev. Lett.*, 70:123–126, 1993.

[243] R. D. Sorkin. Topology change and monopole creation. *Phys. Rev. D*, 33:978–982, 1986.

[244] R. D. Sorkin and E. Woolgar. Causal structure and the positivity of mass. In Sato [234], pages 754–756.

[245] A. A. Starobinsky. A new type of isotropic cosmological model without singularity. *Phys. Lett. B*, 91:99–102, 1980.

[246] J. Stewart. *Advanced General Relativity*. Cambridge University Press, Cambridge, England, 1990.

[247] J. L. Synge. *Relativity: the General Theory*. North-Holland, Amsterdam, 1964.

[248] A. H. Taub. Space times with distribution valued curvature tensors. *J. Math. Phys.*, 21:1423–1431, 1970.

[249] J. C. Taylor. *Gauge Theories of Weak Interactions*. Cambridge University Press, Cambridge, England, 1976.

[250] K. S. Thorne. Closed timelike curves. In Gleiser et al. [114], pages 295–315.

[251] F. J. Tipler. Rotating cylinders and the possibility of global causality violation. *Phys. Rev. D*, 9:2203–2206, 1974.

[252] F. J. Tipler. Causality violation in asymptotically flat spacetime. *Phys. Rev. Lett.*, 37:879–882, 1976. See also [253].

[253] F. J. Tipler. Singularities and causality violation. *Ann. Phys. (NY)*, 108:1–36, 1977. See also [252].

[254] F. J. Tipler. Energy conditions and spacetime singularities. *Phys. Rev. D*, 17:2521–2528, 1978.

[255] F. J. Tipler, editor. *Essays in General Relativity*, New York, 1980. Academic.

[256] W. D. Unruh. Quantum coherence, wormholes, and the cosmological constant. *Phys. Rev. D*, 40:1053–1063, 1989.

[257] W. G. Unruh. Notes on black hole evaporation. *Phys. Rev. D*, 14:870–892, 1976.

[258] W. J. van Stockum. Gravitational field of a distribution of particles rotating about an axis of symmetry. *Proc. R. Soc. Edin.*, 57:135–154, 1937.

[259] J. H. van Vleck. The correspondence principle in the statistical interpretation of quantum mechanics. *Proc. Nat. Acad. Sci. US*, 14:178–188, 1928.

[260] M. Visser. Traversable wormholes from surgically modified Schwarzschild spacetimes. *Nucl. Phys. B*, 328:203–212, 1989.

[261] M. Visser. Traversable wormholes: Some simple examples. *Phys. Rev. D*, 39:3182–3184, 1989.

[262] M. Visser. Quantum mechanical stabilization of Minkowski signature wormholes. *Phys. Lett. B*, 242:24–28, 1990.

[263] M. Visser. Quantum wormholes in Lorentzian signature. In B. Bonner and H. Miettinen, editors, *Proceedings of the Rice Meeting, 1990 Meeting of the Division of Particles and Fields of the American Physical Society*, volume 2, pages 858–860. World Scientific, Singapore, 1990.

[264] M. Visser. Wormholes, baby universes, and causality. *Phys. Rev. D*, 41:1116–1124, 1990.

[265] M. Visser. Quantum wormholes. *Phys. Rev. D*, 43:402–409, 1991.

[266] M. Visser. Wheeler wormholes and topology change: A minisuperspace analysis. *Mod. Phys. Lett. A*, 6:2663–2667, 1991.

[267] M. Visser. Dirty black holes: Thermodynamics and horizon structure. *Phys. Rev. D*, 46:2445–2451, 1992.

[268] M. Visser. From wormhole to time machine: Remarks on Hawking's chronology protection conjecture. *Phys. Rev. D*, 47:554–565, 1993.

[269] M. Visser. van Vleck determinants: Geodesic focussing and defocussing in Lorentzian spacetimes. *Phys. Rev. D*, 47:2395–2402, 1993.

[270] M. Visser. Hawking's chronology protection conjecture: Singularity structure of the quantum stress–energy tensor. *Nucl. Phys. B*, 416:895–906, 1994.

[271] M. Visser. Scale anomalies imply violation of the averaged null energy condition. 1994. gr-qc/9409043. (To appear in *Phys. Lett. B*.).

[272] M. Visser. van Vleck determinants: Traversable wormhole spacetimes. *Phys. Rev. D*, 49:3963–3980, 1994.

[273] H. Waelbroeck. Do universes with parallel cosmic strings or two-dimensional wormholes have closed timelike curves? *Gen. Rel. Gravit.*, 23:219–233, 1991.

[274] S. M. Wagh and N. Dadhich. Absence of super–radiance of the Dirac particles in the Kerr–Newman geometry and the weak positive–energy condition. *Phys. Rev. D*, 32:1863–1865, 1985.

[275] R. M. Wald. *General Relativity*. University of Chicago Press, Chicago, 1984.

[276] R. M. Wald. *Quantum Field Theory in Curved Spacetime and Black Hole Thermodynamics*. University of Chicago Press, Chicago, 1994.

[277] R. M. Wald and U. Yurtsever. General proof of the averaged null energy condition for a massless scalar field in two dimensional spacetime. *Phys. Rev. D*, 44:403–416, 1991.

[278] D. F. Walls. Squeezed states of light. *Nature*, 306:141–146, 1983.

[279] S. Weinberg. *Gravitation and Cosmology*. Wiley, New York, 1972.

[280] J. Wess and J. Bagger. *Supersymmetry and Supergravity*. Princeton series in physics. Princeton University Press, Princeton, 1983.

[281] P. West. *Introduction to Supersymmetry and Supergravity*. World Scientific, Singapore, 1986.

[282] H. Weyl. *Philosophie der Mathematik und Naturwissenschaft*. Handbuch der philosophie. Leibniz Verlag, Munich, 1928. See p. 65 (republished in english, revised and augmented, as [283]).

[283] H. Weyl. *Philosophy of Mathematics and Natural Science*. Princeton University Press, Princeton, 1949. See p. 91 (this book is a revised and augmented English language version of [282]).

[284] J. A. Wheeler. Geons. *Phys. Rev.*, 97:511–536, 1955. (First diagram of a wormhole on p. 535).

[285] J. A. Wheeler. On the nature of quantum geometrodynamics. *Ann. Phys. (NY)*, 2:604–614, 1957.

[286] J. A. Wheeler. *Geometrodynamics*. Academic, New York, 1962.

[287] J. A. Wheeler. Superspace and the nature of quantum geometrodynamics. In DeWitt and Wheeler [73], pages 242–307.

[288] G. S. Whiston. Lorentzian characteristic classes. *Gen. Rel. Gravit.*, 6:463–475, 1975.

[289] C. M. Will. *Was Einstein Right?* Basic Books, New York, 1986.

[290] C. M. Will. *Theory and Experiment in Gravitational Physics*. Cambridge University Press, Cambridge, England, second edition, 1993.

[291] E. Witten. Fermion quantum numbers in Kaluza–Klein theories. In Jackiw et al. [155], pages 227–277.

[292] J. A. Wyler. Rasputin, science, and the transmogrification of destiny. *Gen. Rel. Gravit.*, 5:175–182, 1974.

[293] J. W. York, Jr. Black hole in thermal equilibrium with a scalar field: The back reaction. *Phys. Rev. D*, 31:775–784, 1985.

[294] U. Yurtsever. Does quantum field theory enforce the averaged weak energy condition? *Class. Quantum Grav.*, 7:L251–L258, 1990.

[295] W. H. Zurek, S. Habib, and J. P. Paz. Coherent states via decoherence. *Phys. Rev. Lett.*, 70:1187–1190, 1993.

[296] W. H. Zurek, A. van der Merwe, and W. A. Miller, editors. *Between Quantum and Cosmos*, Princeton, 1988. Princeton University Press.

Index

absolute horizon, *see* event horizon

achronal, 198

action

    classical unit of, 40

    Einstein–Hilbert, 12

        linearized, 56

    free massless scalar, 32

    Nambu–Goto, 170, 180, 190, 191

    Planck's quantum, 39

adiabatic approximation, 267, 268, 270

adiabatic techniques, 282–283

ADM formalism, 355

ADM mass, 111–113, 119, 253, 355

ADM split, 15, 17, 68, 160

    lapse function, 15, 18, 20

    shift function, 15

affine connection, *see* connexion

affine parameter, 117, 133, 298, 337

    canonical, 297, 337

    generalized, 117

algebraic numerology, 40

ANEC, *see also* energy conditions, 117, 119, 131–135, 195–199, 275, 281, 290–293, 310, 370

    definition, 117

    integral, 119, 128, 134–135, 146–148, 291

    no-go theorem, 292

    violations, 124, 195, 197–199, 255, 347

angular momentum, 75, 79, 328

    linear density, 221

    orbital, 223

anomaly

    conformal, 122, 124, 280, 291, 335, 372

    scale, 198, 290, 291, 293

    trace, 122, 124, 280, 291, 335

anthropic principle, 93

anti-wormhole legislation, 275

apparent horizon, *see also* horizon, 16–17, 220

arbitrarily advanced civilization, 144, 227

Arnowitt–Deser–Misner, *see* ADM

ASEC, *see also* energy conditions, 118–119, 136, 197

    definition, 118

asymptotically flat (definition), 17, 196

atomic mass, 124

autonomy, 212, 256

averaged null energy condition, *see* ANEC

averaged strong energy condition, *see* ASEC

averaged weak energy condition, *see* AWEC

AWEC, *see also* energy conditions, 117–118, 135–136, 197

    definition, 117–118

baby universe, 74, 90, 91, 93, 145, 366

back-reaction, 198, 246, 264, 270, 272, 273

    bootstrap, 281

    ignored, 128, 163, 256, 281

backward time-jump, 215, 221, 236, 238, 256, 341

balloon, 163

Bessel function, 37, 358

Bianchi identities, 157, 293

big bang, 95, 129, 144

393

Printed in the United States
By Bookmasters